IET CONTROL ENGINEERING SERIES 81

Optimal Adaptive Control and Differential Games by Reinforcement Learning Principles

Other volumes in this series:

Volume 8	**A history of control engineering, 1800–1930** S. Bennett
Volume 18	**Applied control theory, 2nd edition** J.R. Leigh
Volume 20	**Design of modern control systems** D.J. Bell, P.A. Cook and N. Munro (Editors)
Volume 28	**Robots and automated manufacture** J. Billingsley (Editor)
Volume 33	**Temperature measurement and control** J.R. Leigh
Volume 34	**Singular perturbation methodology in control systems** D.S. Naidu
Volume 35	**Implementation of self-tuning controllers** K. Warwick (Editor)
Volume 37	**Industrial digital control systems, 2nd edition** K. Warwick and D. Rees (Editors)
Volume 39	**Continuous time controller design** R. Balasubramanian
Volume 40	**Deterministic control of uncertain systems** A.S.I. Zinober (Editor)
Volume 41	**Computer control of real-time processes** S. Bennett and G.S. Virk (Editors)
Volume 42	**Digital signal processing: principles, devices and applications** N.B. Jones and J.D.McK. Watson (Editors)
Volume 44	**Knowledge-based systems for industrial control** J. McGhee, M.J. Grimble and A. Mowforth (Editors)
Volume 47	**A history of control engineering, 1930–1956** S. Bennett
Volume 49	**Polynomial methods in optimal control and filtering** K.J. Hunt (Editor)
Volume 50	**Programming industrial control systems using IEC 1131-3** R.W. Lewis
Volume 51	**Advanced robotics and intelligent machines** J.O. Gray and D.G. Caldwell (Editors)
Volume 52	**Adaptive prediction and predictive control** P.P. Kanjilal
Volume 53	**Neural network applications in control** G.W. Irwin, K. Warwick and K.J. Hunt (Editors)
Volume 54	**Control engineering solutions: a practical approach** P. Albertos, R. Strietzel and N. Mort (Editors)
Volume 55	**Genetic algorithms in engineering systems** A.M.S. Zalzala and P.J. Fleming (Editors)
Volume 56	**Symbolic methods in control system analysis and design** N. Munro (Editor)
Volume 57	**Flight control systems** R.W. Pratt (Editor)
Volume 58	**Power-plant control and instrumentation** D. Lindsley
Volume 59	**Modelling control systems using IEC 61499** R. Lewis
Volume 60	**People in control: human factors in control room design** J. Noyes and M. Bransby (Editors)
Volume 61	**Nonlinear predictive control: theory and practice** B. Kouvaritakis and M. Cannon (Editors)
Volume 62	**Active sound and vibration control** M.O. Tokhi and S.M. Veres
Volume 63	**Stepping motors: a guide to theory and practice, 4th edition** P.P. Acarnley
Volume 64	**Control theory, 2nd edition** J.R. Leigh
Volume 65	**Modelling and parameter estimation of dynamic systems** J.R. Raol, G. Girija and J. Singh
Volume 66	**Variable structure systems: from principles to implementation** A. Sabanovic, L. Fridman and S. Spurgeon (Editors)
Volume 67	**Motion vision: design of compact motion sensing solution for autonomous systems** J. Kolodko and L. Vlacic
Volume 68	**Flexible robot manipulators: modelling, simulation and control** M.O. Tokhi and A.K.M. Azad (Editors)
Volume 69	**Advances in unmanned marine vehicles** G. Roberts and R. Sutton (Editors)
Volume 70	**Intelligent control systems using computational intelligence techniques** A. Ruano (Editor)
Volume 71	**Advances in cognitive systems** S. Nefti and J. Gray (Editors)
Volume 72	**Control theory: a guided tour, 3rd edition** James Ron Leigh
Volume 73	**Adaptive sampling with mobile WSN** K. Sreenath, M.F. Mysorewala, D.O. Popa and F.L. Lewis
Volume 74	**Eigenstructure control algorithms: applications to aircraft/rotorcraft handling qualities design** S. Srinathkumar
Volume 75	**Advanced control for constrained processes and systems** F. Garelli, R.J. Mantz and H. De Battista
Volume 76	**Developments in control theory towards glocal control** L. Qiu, J. Chen, T. Iwasaki and H. Fujioka (Editors)
Volume 77	**Further advances in unmanned marine vehicles** G.N. Roberts and R. Sutton (Editors)
Volume 78	**Frequency-domain control design for high-performance systems** J. O'Brien

Optimal Adaptive Control and Differential Games by Reinforcement Learning Principles

Draguna Vrabie, Kyriakos G. Vamvoudakis and Frank L. Lewis

The Institution of Engineering and Technology

Published by The Institution of Engineering and Technology, London, United Kingdom

The Institution of Engineering and Technology is registered as a Charity in England & Wales (no. 211014) and Scotland (no. SC038698).

First published 2013

The Institution of Engineering and Technology
Michael Faraday House
Six Hills Way, Stevenage
Herts, SG1 2AY, United Kingdom

www.theiet.org

British Library Cataloguing in Publication Data
A catalogue record for this product is available from the British Library

ISBN 978-1-84919-489-1 (hardback)
ISBN 978-1-84919-490-7 (PDF)

Typeset in India by MPS Limited
Printed in the UK by CPI Group (UK) Ltd, Croydon, CR0 4YY

To Adrian Miron
Draguna Vrabie

To my parents, George and Evgenia. Your love and support was and will continue to be my biggest inspiration and source of strength
Kyriakos Vamvoudakis

To Galina, Roma, Chris – who have made every day exciting
Frank Lewis

Contents

Preface

This book studies dynamic feedback control systems of the sort that regulate human-engineered systems including aerospace systems, aircraft autopilots, vehicle engine controllers, ship motion and engine control, industrial processes and elsewhere. The book shows how to use reinforcement learning techniques to design new structures of adaptive feedback control systems that learn the solutions to optimal control problems online in real time by measuring data along the system trajectories.

Feedback control works on the principle of observing the actual outputs of a system, comparing them to desired trajectories, and computing a control signal based on the error used to modify the performance of the system to make the actual output follow the desired trajectory. James Watt used feedback controllers in the 1760s to make the steam engine useful as a prime mover. This provided a substantial impetus to the Industrial Revolution.

Adaptive control and *optimal control* represent two different philosophies for designing feedback control systems. These methods have been developed by the Control Systems Community of engineers. Optimal controllers minimize user-prescribed performance functions and are normally designed offline by solving Hamilton–Jacobi–Bellman (HJB) design equations. This requires knowledge of the full system dynamics model. However, it is often difficult to determine an accurate dynamical model of practical systems. Moreover, determining optimal control policies for non-linear systems requires the offline solution of non-linear HJB equations, which are often difficult or impossible to solve. By contrast, adaptive controllers learn online to control systems with unknown dynamics using data measured in real time along the system trajectories. Adaptive controllers are not usually designed to be optimal in the sense of minimizing user-prescribed performance functions.

Reinforcement learning (RL) describes a family of machine learning systems that operate based on principles used in animals, social groups and naturally occurring systems. RL methods were used by Ivan Pavlov in the 1860s to train his dogs. Methods of RL have been developed by the Computational Intelligence Community in computer science engineering. RL has close connections to both optimal control and adaptive control. It refers to a class of methods that allow the design of adaptive controllers that learn online, in real time, the solutions to user-prescribed optimal control problems. RL techniques were first developed for Markov Decision Processes having finite state spaces. RL techniques have been applied for years in the control of discrete-time dynamical systems with continuous state spaces. A family of RL methods known as *approximate dynamic programming* (ADP) was proposed by Paul Werbos and developed by many researchers. RL methods have not been extensively used in the Control Systems Community until recently. The application of RL to continuous-time dynamical systems has lagged due to the inconvenient form of the Hamiltonian function. In discrete-time systems, the Hamiltonian does not depend on the system dynamics, whereas in continuous-time systems it does involve the system dynamics.

This book shows that techniques based on reinforcement learning can be used to unify optimal control and adaptive control. Specifically, RL techniques are used to design adaptive control systems with novel structures that learn the solutions to optimal control problems in real time by observing data along the system trajectories. We call these *optimal adaptive controllers*. The methods studied here depend on RL techniques known as *policy iteration* and *value iteration*, which evaluate the performance of current control policies and provide methods for improving those policies.

Chapter 1 gives an overview of optimal control, adaptive control and reinforcement learning. Chapter 2 provides a background on reinforcement learning. After that, the book has three parts. In Part I, we develop novel RL methods for the control of continuous-time dynamical systems. Chapter 3 introduces a technique known as *integral reinforcement learning* (*IRL*) that allows the development of policy iteration methods for the optimal adaptive control of linear continuous-time systems. In Chapter 4, IRL is extended to develop optimal adaptive controllers for non-linear continuous-time systems. Chapter 5 designs a class of controllers for non-linear systems based on RL techniques known as generalized policy iteration. To round out Part I, Chapter 6 provides simplified optimal adaptive control algorithms for linear continuous-time systems based on RL methods known as value iteration.

In Part I, the controller structures developed are those familiar in the RL community. Part I essentially extends known results in RL for discrete-time systems to the case of continuous-time systems. This results in a class of adaptive controllers that learn in real time the solutions to optimal control problems. This is accomplished by learning mechanisms based on tuning the parameters of the controller to improve the performance. The adaptive learning systems in Part I are of the actor–critic structure, wherein there are two networks in two control loops – a critic network that evaluates the performance of current control policies, and an actor network that computes those current policies. The evaluation results of the critic network are used to update the parameters of the actor network so as to obtain an improved control policy. In the actor–critic topologies of Part I, the critic and actor networks are updated sequentially, that is, as one network learns and its parameters are tuned, the other is not tuned and so does not learn.

In the Control Systems Community, by contrast, continuous-time adaptive controllers operate by tuning all parameters in all control loops simultaneously in real time. Thus, the optimal adaptive controllers in Part I are not of the standard sort encountered in adaptive control. Therefore, in Part II, we use RL techniques to develop control structures that are more familiar from the feedback control systems perspective. These adaptive controllers learn online and converge to optimal control solutions by tuning all parameters in all loops simultaneously. We call this *synchronous online learning*. Chapter 7 develops the basic form of synchronous optimal adaptive controller for the basic non-linear optimal control problem. First, a policy iteration algorithm is derived using RL methods using techniques like those from Part I. Then, however, that PI structure is used to derive a two-loop adaptive control topology wherein all the parameters of the critic network and the control loop are tuned or updated simultaneously. This is accomplished by developing two learning networks that interact with each other as they learn, and so mutually tune their parameters together simultaneously. In Chapter 8, notions of integral reinforcement learning and synchronous tuning are combined to yield a synchronous adaptive control structure that converges to optimal control solutions without knowing the full system dynamics. This provides a powerful class of optimal adaptive controllers that learn in real time the solutions to the HJB design equations without knowing the full system dynamics. Specifically, the system drift dynamics need not be known. The drift term is often difficult to identify for practical modern systems.

Part III applies RL methods to design adaptive controllers for multiplayer games that converge online to optimal game theoretic solutions. As the players interact, they use the high-level information from observing each other's actions to tune the parameters of their own control policies. This is a true class of interactive learning controllers that bring the interactions of the players in the game to a second level of inter-communications through online learning. Chapter 9 presents synchronous adaptive controllers that learn in real time the Nash equilibrium solution of zero-sum two-player differential games. In Chapter 10, adaptive controllers are developed that learn online the Nash solution to multiplayer non-linear differential games. In Chapter 11 IRL methods are used to learn the solution to the two-player zero-sum games online without knowing the system drift dynamics. These controllers solve the generalized game Riccati equations online in real time without knowing the system drift dynamics.

Draguna Vrabie
Kyriakos Vamvoudakis
Frank Lewis

Acknowledgements

This work resulted from the support over several years of National Science Foundation grant ECCS-1128050, Army Research Office grant W91NF-05-1-0314, and Air Force Office of Scientific Research grant FA9550-09-1-0278.

Chapter 1

Introduction to optimal control, adaptive control and reinforcement learning

This book studies dynamic feedback control systems of the sort that regulate human-engineered systems, including aerospace systems, aircraft autopilots, vehicle engine controllers, ship motion and engine control, industrial processes and elsewhere. The book shows how to use reinforcement learning (RL) techniques to design new structures of adaptive feedback control systems that learn the solutions to optimal control problems online in real time by measuring data along the system trajectories.

Feedback control works on the principle of observing the actual outputs of a system, comparing them to desired trajectories and computing a control signal based on the error used to modify the performance of the system to make the actual output follow the desired trajectory. James Watt used feedback controllers in the 1760s to make the steam engine useful as a prime mover. This provided a substantial impetus to the Industrial Revolution. Vito Volterra showed in 1920 that feedback is responsible for the balance of predator–prey fish populations in a closed ecosystem. Charles Darwin showed in 1860 that feedback over long time periods is responsible for natural selection (Darwin, 1859). Adam Smith showed in 1776 that feedback mechanisms play a major role in the interactions of international economic entities and the wealth of nations.

Adaptive control and *optimal control* represent different philosophies for designing feedback control systems. These methods have been developed by the Control Systems Community of engineers. Optimal controllers minimize user-prescribed performance functions and are normally designed offline by solving Hamilton–Jacobi–Bellman (HJB) design equations, for example, the Riccati equation, using complete knowledge of the system dynamical model. However, it is often difficult to determine an accurate dynamical model of practical systems. Moreover, determining optimal control policies for non-linear systems requires the offline solution of non-linear HJB equations, which are often difficult or impossible to solve. By contrast, adaptive controllers learn online to control systems with unknown dynamics using data measured in real time along the system trajectories. Adaptive controllers are not usually designed to be optimal in the sense of minimizing user-prescribed performance functions. Indirect adaptive controllers use system identification techniques to first identify the system parameters, then use the obtained model to solve optimal design equations (Ioannou and Fidan, 2006). Adaptive controllers may satisfy certain inverse optimality conditions, as shown in Li and Krstic (1997).

Reinforcement learning (RL) describes a family of learning systems that operates based on principles used in animals, social groups and naturally occurring systems. RL was used by Ivan Pavlov in the 1860s to train his dogs. Methods of RL have been developed by the Computational Intelligence Community in computer science engineering. RL allows the learning of optimal actions without knowing a dynamical model of the system or the environment. RL methods have not been extensively used in the feedback control community until recently.

In this book, we show that techniques based on RL allow the design of adaptive control systems with novel structures that learn the solutions to optimal control problems in real time by observing data along the system trajectories. We call these *optimal adaptive controllers*. Several of these techniques can be implemented without knowing the complete system dynamics. Since the optimal design is performed in real time using adaptive control techniques, unknown and time-varying dynamics and changing performance requirements are accommodated. The methods studied here depend on RL techniques known as *policy iteration* and *value iteration*, which evaluate the performance of current control policies and provide methods for improving those policies.

RL techniques have been applied for years in the control of discrete-time dynamical systems. A family of RL methods known as *approximate dynamic programming* (ADP) was proposed by Paul Werbos (Werbos, 1989, 1991, 1992, 2009) and developed by many researchers (Prokhorov and Wunsch, 1997; Barto *et al.*, 2004; Wang *et al.*, 2009). Offline solution methods for discrete-time dynamical systems and Markov Processes were developed by Bertsekas and Tsitsiklis (1996). The application of RL to continuous-time systems lagged due to the inconvenient form of the Hamiltonian function. In discrete-time systems, the Hamiltonian does not depend on the system dynamics, whereas in continuous-time systems it does involve the system dynamics.

This book applies RL methods to design optimal adaptive controllers for continuous-time systems. In this chapter we provide brief discussions of optimal control, adaptive control, RL and the novel adaptive learning structures that appear in optimal adaptive control.

1.1 Optimal control

This section presents the basic ideas and solution procedures of optimal control (Lewis *et al.*, 2012). In naturally occurring systems such as the cell, animal organisms and species, resources are limited and all actions must be exerted in such a fashion as to preserve them. Optimal control formalizes this principle in the design of feedback controllers for human-engineered systems.

1.1.1 Linear quadratic regulator

The most basic sort of optimal controller for dynamical systems is the linear quadratic regulator (LQR) (Lewis *et al.*, 2012). The LQR considers the linear time-invariant dynamical system described by

$$\dot{x}(t) = Ax(t) + Bu(t) \tag{1.1}$$

with state $x(t) \in \mathbb{R}^n$ and control input $u(t) \in \mathbb{R}^m$. To this system is associated the infinite-horizon quadratic cost function or performance index

$$V(x(t_0), t_0) = \int_{t_0}^{\infty} (x^T(\tau)Qx(\tau) + u^T(\tau)Ru(\tau))\, d\tau \tag{1.2}$$

with weighting matrices $Q \geq 0$, $R > 0$. It is assumed that (A, B) is stabilizable, that is there exists a control input that makes the system stable, and that $(A, \sqrt{Q})$ is detectable, that is the unstable modes of the system are observable through the output $(y = \sqrt{Q}x)$.

The LQR optimal control problem requires finding the control policy that minimizes the cost

$$u^*(t) = \arg \min_{\substack{u(t) \\ t_0 \leq t \leq \infty}} V(t_0, x(t_0), u(t)) \tag{1.3}$$

The solution of this optimal control problem is given by the state-feedback $u(t) = -Kx(t)$, where the gain matrix is

$$K = R^{-1}B^T P \tag{1.4}$$

and matrix P is a positive definite solution of the algebraic Riccati equation (ARE)

$$A^T P + PA + Q - PBR^{-1}B^T P = 0 \tag{1.5}$$

Under the stabilizability and detectability conditions, there is a unique positive semidefinite solution of the ARE that yields a stabilizing closed-loop controller given by (1.4). That is, the closed-loop system $A - BK$ is asymptotically stable.

To find the optimal control that minimizes the cost, one solves the ARE for the intermediate matrix P, then the optimal state feedback is given by (1.4). This is an offline solution procedure that requires complete knowledge of the system dynamics matrices (A, B) to solve the ARE. Moreover, if the system dynamics change or the performance index varies during operation, a new optimal control solution must be computed.

1.1.2 Linear quadratic zero-sum games

In the linear quadratic (LQ) zero-sum (ZS) game one has linear dynamics

$$\dot{x} = Ax + Bu + Dd \tag{1.6}$$

with state $x(t) \in \mathbb{R}^n$, control input $u(t) \in \mathbb{R}^m$ and disturbance $u(t) \in \mathbb{R}^m$. To this system is associated the infinite-horizon quadratic cost function or performance index

$$V(x(t), u, d) = \frac{1}{2}\int_{t}^{\infty} (x^T Qx + u^T Ru - \gamma^2 \|d\|^2)\, d\tau \equiv \int_{t}^{\infty} r(x, u, d)\, d\tau \tag{1.7}$$

with the control weighting matrix $R = R^T > 0$, and a scalar $\gamma > 0$.

The LQ ZS game requires finding the control policy that minimizes the cost with respect to the control and maximizes the cost with respect to the disturbance

$$\begin{aligned} V^*(x(0)) &= \min_u \max_d J(x(0), u, d) \\ &= \min_u \max_d \int_0^\infty \left(Q(x) + u^T R u - \gamma^2 \|d\|^2\right) dt \end{aligned} \tag{1.8}$$

This game captures the intent that the control seeks to drive the states to zero while minimizing the energy it uses, whereas the disturbance seeks to drive the states away from zero while minimizing its own energy used.

The solution of this optimal control problem is given by the state-feedback policies

$$u(x) = -R^{-1}B^T Px = -Kx \tag{1.9}$$

$$d(x) = \frac{1}{\gamma^2} D^T Px = Lx \tag{1.10}$$

where the intermediate matrix P is the solution to the game (or generalized) algebraic Riccati equation (GARE)

$$0 = A^T P + PA + Q - PBR^{-1}B^T P + \frac{1}{\gamma^2} PDD^T P \tag{1.11}$$

There exists a solution $P > 0$ if (A, B) is stabilizable, $(A, \sqrt{Q})$ is observable and $\gamma > \gamma^*$, the H-infinity gain (Başar and Olsder, 1999; Van Der Schaft, 1992).

To solve the ZS game problem, one solves the GARE equation for the non-negative definite optimal value kernel $P \geq 0$, then the optimal control is given as a state variable feedback in terms of the ARE solution by (1.9) and the worst case disturbance by (1.10). This is an offline solution procedure that requires complete knowledge of the system dynamics matrices (A, B, D) to solve the GARE. Moreover, if the system dynamics (A, B, D) change or the performance index (Q, R, γ) varies during operation, a new optimal control solution must be computed.

1.2 Adaptive control

Adaptive control describes a variety of techniques that learn feedback controllers in real time that stabilize a system and satisfy various design criteria (Ioannou and Fidan, 2006; Astrom and Wittenmark, 1995). Control systems are learned online without knowing the system dynamics by measuring data along the system trajectories. Adaptive controllers do not generally learn the solutions to optimal control problems or game theory problems such as those just described. Adaptive controllers may provide solutions that are optimal in a least-squares sense. In this section, we present some basic structures of adaptive control systems.

In adaptive control, unknown systems are parameterized in terms of known basic structures or functions, but unknown parameters. Then, based on a suitable problem formulation, the unknown parameters are learned or tuned online to achieve various design criteria. This is accomplished in what is known as a data-based manner, that is by measuring data along the system trajectories and without knowing the system dynamics. Two broad classes of adaptive controllers are direct adaptive control and indirect adaptive control.

Direct adaptive controller. Standard adaptive control systems can have many structures. In the direct adaptive tracking controller in Figure 1.1, it is desired to make the output of the system, or plant, follow a desired reference signal. The controller is parameterized in terms of unknown parameters, and adaptive tuning laws are given for updating the controller parameters. These tuning laws depend on the tracking error, which it is desired to make small.

Indirect adaptive controller. In the indirect adaptive tracking controller in Figure 1.2, two feedback networks are used, one of which learns dynamically online. One network is a plant identifier that has the function of identifying or learning the plant dynamics model $\dot{x} = f(x) + g(x)u$ online in real time. The tuning law for the plant identifier depends on the identification error, which it is desired to

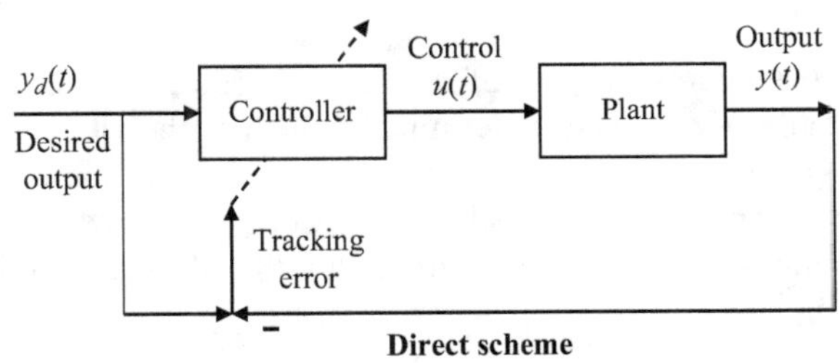

Figure 1.1 Standard form of direct adaptive controller where the controller parameters are updated in real time

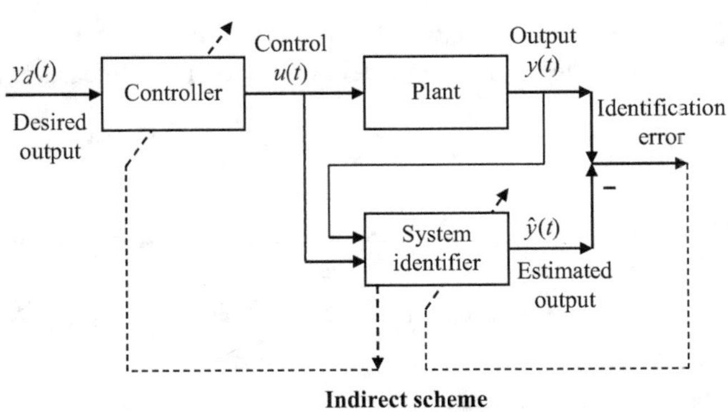

Figure 1.2 Standard form of indirect adaptive controller where the parameters of a system identifier are updated in real time

make small. After the plant has been identified, the controller parameters in the controller network can be computed using a variety of methods, including for instance solution of the Diophantine equation (Astrom and Wittenmark, 1995).

Direct model reference adaptive controller (MRAC). Adaptive controllers can be designed to make the plant output follow the output of a reference model. In the indirect MRAC, the parameters of a plant identifier are tuned so that it mimics the behavior of the plant. Then, the controller parameters are computed.

In the direct MRAC scheme in Figure 1.3, the controller parameters are tuned directly so that the plant output follows the model output. Consider the simple scalar case (Ioannou and Fidan, 2006) where the plant is

$$\dot{x} = ax + bu \tag{1.12}$$

with state $x(t) \in \mathbb{R}$, control input $u(t) \in \mathbb{R}$ and input gain $b > 0$. It is desired for the plant state to follow the state of a reference model given by

$$\dot{x}_m = -a_m x_m + b_m r \tag{1.13}$$

with $r(t) \in \mathbb{R}$ a reference input signal.

To accomplish this, take the controller structure as

$$u = -kx + dr \tag{1.14}$$

which has a feedback term and a feedforward term. The controller parameters k, d are unknown, and are to be determined so that the state tracking error $e(t) = x(t) - x_m(t)$ goes to zero or becomes small. This can be accomplished by tuning the controller parameters in real time. It is direct to show, using for instance Lyapunov techniques (Ioannou and Fidan, 2006), that if the controller parameters are tuned according to

$$\dot{k} = \alpha ex, \quad \dot{d} = -\beta er \tag{1.15}$$

where $\alpha, \beta > 0$ are tuning parameters, then the tracking error $e(t)$ goes to zero with time.

These are dynamical tuning laws for the unknown controller parameters. The direct MRAC is said to learn the parameters online by using measurements of the

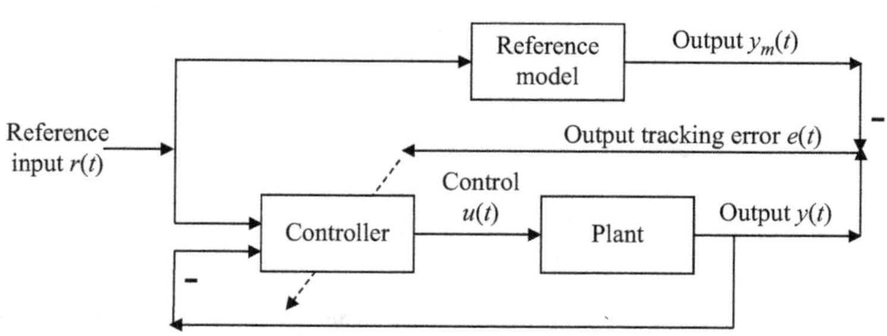

Figure 1.3 Direct MRAC where the parameters of the controller are updated in real time

signals $e(t), x(t), r(t)$ measured in real time along the system trajectories. This is called an adaptive or learning controller. Note that the tuning laws (1.15) are quadratic functions of time signals. The feedback gain k is tuned by a product of its input $x(t)$ in (1.14) and the tracking error $e(t)$, whereas the feedforward gain d is tuned by a product of its input $r(t)$ and the tracking error $e(t)$. The plant dynamics (a, b) are not needed in the tuning laws. That is, the tuning laws (1.15) and the control structure (1.14) work for unknown plants, guaranteeing that the tracking error $e(t)$ goes to zero for any scalar plant that has control gain $b > 0$.

1.3 Reinforcement learning

Reinforcement learning (RL) is a type of machine learning developed in the Computational Intelligence Community. It has close connections to both optimal control and adaptive control. RL refers to a class of methods that allow designing adaptive controllers that learn online, in real time, the solutions to user-prescribed optimal control problems. In machine learning, RL (Mendel and MacLaren, 1970; Powell, 2007; Sutton and Barto, 1998) is a method for solving optimization problems that involves an actor or agent that interacts with its environment and modifies its actions, or control policies, based on stimuli received in response to its actions. RL is inspired by natural learning mechanisms, where animals adjust their actions based on reward and punishment stimuli received from the environment (Mendel and MacLaren, 1970; Busoniu *et al.*, 2009; Doya *et al.*, 2001).

The actor–critic structures shown in Figure 1.4 (Werbos, 1991; Bertsekas and Tsitsiklis, 1996; Sutton and Barto, 1998; Barto *et al.*, 1983; Cao, 2007) are one type of RL system. These structures give algorithms that are implemented in real time where an actor component applies an action, or control policy, to the

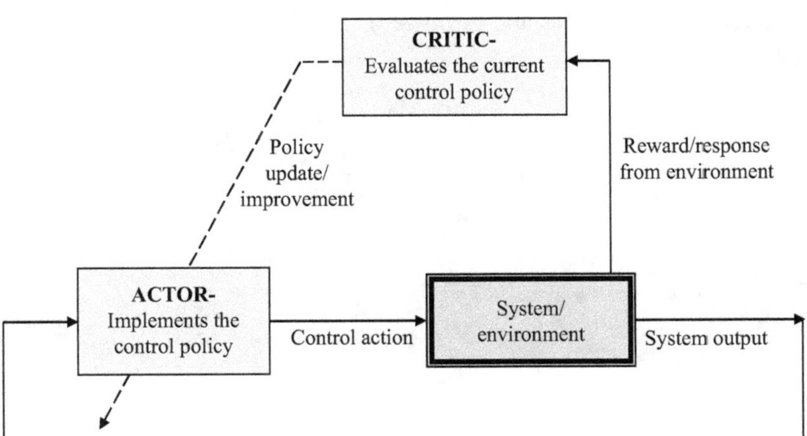

Figure 1.4 Reinforcement learning with an actor–critic structure. This structure provides methods for learning optimal control solutions online based on data measured along the system trajectories

environment and a critic component assesses the value of that action. The learning mechanism supported by the actor–critic structure has two steps, namely, policy evaluation, executed by the critic, followed by policy improvement, performed by the actor. The policy evaluation step is performed by observing from the environment the results of applying current actions, and determining how close to optimal the current action is. Based on the assessment of the performance, one of several schemes can then be used to modify or improve the control policy in the sense that the new policy yields a value that is improved relative to the previous value.

Note that the actor–critic RL structure is fundamentally different from the adaptive control structures in Figures 1.1–1.3, both in structure and in principle.

1.4 Optimal adaptive control

In this book, we show how to use RL techniques to unify optimal control and adaptive control. By this we mean that a novel class of adaptive control structures will be developed that learn the solutions of optimal control problems in real time by measuring data along the system trajectories online. We call these optimal adaptive controllers. These optimal adaptive controllers have structures based on the actor–critic learning architecture in Figure 1.4.

The main contributions of this book are to develop RL control systems for continuous-time dynamical systems and for multiplayer games. Previously, RL had generally only been applied for the control of discrete-time dynamical systems, through the family of ADP controllers of Paul Werbos (Werbos, 1989, 1991, 1992, 2009). The results in this book are from the work of Abu-Khalaf (Abu-Khalaf and Lewis, 2005, 2008; Abu-Khalaf *et al.*, 2006), Vrabie (Vrabie, 2009; Vrabie *et al.*, 2008, 2009) and Vamvoudakis (Vamvoudakis, 2011; Vamvoudakis and Lewis, 2010a, 2010b; Vamvoudakis and Lewis, 2011).

Chapter 2

Reinforcement learning and optimal control of discrete-time systems: Using natural decision methods to design optimal adaptive controllers

This book will show how to use principles of reinforcement learning to design a new class of feedback controllers for continuous-time dynamical systems. Adaptive control and optimal control represent different philosophies for designing feedback controllers. Optimal controllers are normally designed offline by solving Hamilton–Jacobi–Bellman (HJB) equations, for example, the Riccati equation, using complete knowledge of the system dynamics. Determining optimal control policies for non-linear system requires the offline solution of non-linear HJB equations, which are often difficult or impossible to solve. By contrast, adaptive controllers learn online to control unknown systems using data measured in real time along the system trajectories. Adaptive controllers are not usually designed to be optimal in the sense of minimizing user-prescribed performance functions. Indirect adaptive controllers use system identification techniques to first identify the system parameters, then use the obtained model to solve optimal design equations (Ioannou and Fidan, 2006). Adaptive controllers may satisfy certain inverse optimality conditions, as shown in Li and Krstic (1997).

Reinforcement learning (RL) is a type of machine learning developed in the Computational Intelligence Community in computer science engineering. It has close connections to both optimal control and adaptive control. Reinforcement learning refers to a class of methods that allow the design of adaptive controllers that learn online, in real time, the solutions to user-prescribed optimal control problems. RL methods were used by Ivan Pavlov in the 1860s to train his dogs. In machine learning, reinforcement learning (Mendel and MacLaren, 1970; Powell, 2007; Sutton and Barto, 1998) is a method for solving optimization problems that involve an actor or agent that interacts with its environment and modifies its actions, or control policies, based on stimuli received in response to its actions. Reinforcement learning is inspired by natural learning mechanisms, where animals adjust their actions based on reward and punishment stimuli received from the environment (Mendel and MacLaren, 1970; Busoniu *et al.*, 2009; Doya *et al.*, 2001). Other reinforcement learning mechanisms operate in the human brain, where the dopamine neurotransmitter acts as a reinforcement informational signal that favors learning at the level of the neuron (Doya *et al.*, 2001; Schultz, 2004; Doya, 2000).

Reinforcement learning implies a cause and effect relationship between actions and reward or punishment. It implies goal directed behavior at least insofar as the agent has an understanding of reward versus lack of reward or punishment. The reinforcement learning algorithms are constructed on the idea that effective control decisions must be remembered, by means of a reinforcement signal, such that they become more likely to be used a second time. Reinforcement learning is based on real-time evaluative information from the environment and could be called *action-based learning*. Reinforcement learning is connected from a theoretical point of view with both adaptive control and optimal control methods.

The actor–critic structures shown in Figure 2.1 (Barto *et al.*, 1983) are one type of reinforcement learning algorithms. These structures give forward-in-time algorithms that are implemented in real time where an actor component applies an action, or control policy, to the environment, and a critic component assesses the value of that action. The learning mechanism supported by the actor–critic structure has two steps, namely, policy evaluation, executed by the critic, followed by policy improvement, performed by the actor. The policy evaluation step is performed by observing from the environment the results of applying current actions. These results are evaluated using a performance index (Werbos, 1991; Bertsekas and Tsitsiklis, 1996; Sutton and Barto, 1998; Cao, 2007), which quantifies how close to optimal the current action is. Performance can be defined in terms of optimality objectives such as minimum fuel, minimum energy, minimum risk or maximum reward. Based on the assessment of the performance, one of the several schemes can then be used to modify or improve the control policy in the sense that the new policy yields a value that is improved relative to the previous value. In this scheme, reinforcement learning is a means of learning optimal behaviors by observing the real-time responses from the environment to non-optimal control policies.

It is noted that in computational intelligence, the control action is applied to the system, which is interpreted to be the environment. By contrast, in control system engineering, the control action is interpreted as being applied to a system or plant

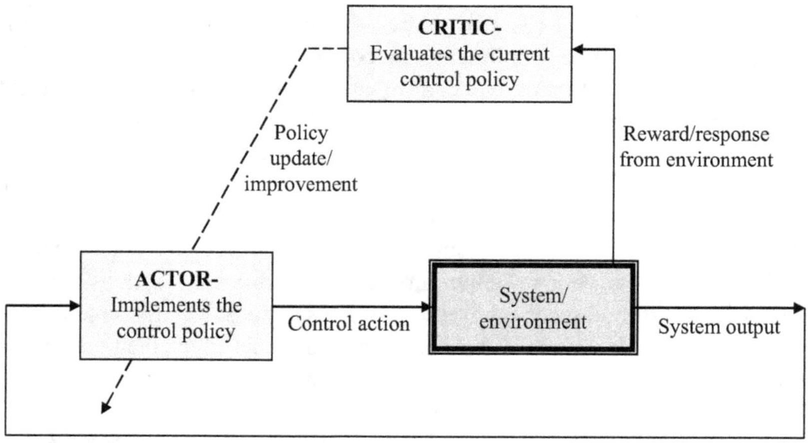

Figure 2.1 Reinforcement learning with an actor–critic structure. This structure provides methods for learning optimal control solutions online based on data measured along the system trajectories

that represents the vehicle, process or device being controlled. This difference captures the differences in philosophy between reinforcement learning and feedback control systems design.

One framework for studying reinforcement learning is based on Markov decision processes (MDPs). Many dynamical decision problems can be formulated as MDPs. Included are feedback control systems for human-engineered systems, feedback regulation mechanisms for population balance and survival of species (Darwin, 1859; Luenberger, 1979), decision making in multiplayer games and economic mechanisms for the regulation of global financial markets.

This chapter presents the main ideas and algorithms of reinforcement learning. We start from a discussion of MDP and then specifically focus on a family of techniques known as approximate (or adaptive) dynamic programming (ADP) or neurodynamic programming. These methods are suitable for control of dynamical systems, which is our main interest in this book. Bertsekas and Tsitsiklis developed RL methods for discrete-time dynamical systems in Bertsekas and Tsitsiklis (1996). This approach, known as neurodynamic programming, used offline solution methods.

Werbos (1989, 1991, 1992, 2009) presented RL techniques for feedback control of discrete-time dynamical systems that learn optimal policies online in real time using data measured along the system trajectories. These methods, known as approximate dynamic programming (ADP) or adaptive dynamic programming, comprised a family of four learning methods. The ADP controllers are actor–critic structures with one learning network for the control action and one learning network for the critic. Surveys of ADP are given in Si *et al.* (2004), Wang *et al.* (2009), Lewis and Vrabie (2009), Balakrishnan *et al.* (2008).

The use of reinforcement learning techniques provides optimal control solutions for linear or non-linear systems using online learning techniques. This chapter reviews current technology, showing that for discrete-time dynamical systems, reinforcement learning methods allow the solution of HJB design equations online, forward in time and without knowing the full system dynamics. In the discrete-time linear quadratic case, these methods determine the solution to the algebraic Riccati equation online, without explicitly solving the equation and without knowing the system *dynamics*.

The application of reinforcement learning methods for continuous-time systems is significantly more involved and forms the subject for the remainder of this book.

2.1 Markov decision processes

Markov decision processes (MDPs) provide a framework for studying reinforcement learning. In this section, we provide a review of MDP (Bertsekas and Tsitsiklis, 1996; Sutton and Barto, 1998; Busoniu *et al.*, 2009). We start by defining optimal sequential decision problems, where decisions are made at stages of a process evolving through time. Dynamic programming is next presented, which gives methods for solving optimal decision problems by working backward through time. Dynamic programming is an offline solution technique that cannot be implemented online in a forward-in-time fashion. Reinforcement learning and

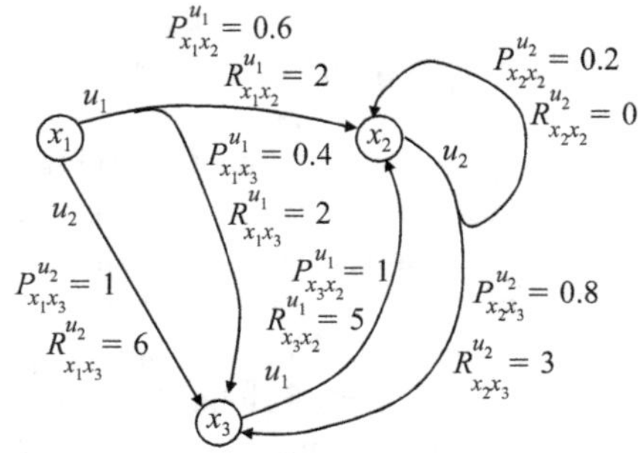

Figure 2.2 MDP shown as a finite-state machine with controlled state transitions and costs associated with each transition

adaptive control are concerned with determining control solutions in real time and forward in time. The key to this is provided by the Bellman equation, which is developed next. In the subsequent section, we discuss methods known as policy iteration and value iteration that give algorithms based on the Bellman equation for solving optimal decision problems in real-time forward-in-time fashion based on data measured along the system trajectories. Finally, the important notion of the Q function is introduced.

Consider the Markov decision process (MDP) (X, U, P, R), where X is a set of states and U is a set of actions or controls (see Figure 2.2). The transition probabilities $P : X \times U \times X \rightarrow [0,1]$ give for each state $x \in X$ and action $u \in U$ the conditional probability $P^u_{x,x'} = \Pr\{x'|x, u\}$ of transitioning to state $x' \in X$ given the MDP is in state x and takes action u. The cost function $R : X \times U \times X \rightarrow \mathbb{R}$ gives the expected immediate cost $R^u_{xx'}$ paid after transition to state $x' \in X$ given the MDP starts in state $x \in X$ and takes action $u \in U$. The Markov property refers to the fact that transition probabilities $P^u_{x,x'}$ depend only on the current state x and not on the history of how the MDP attained that state.

The basic problem for MDP is to find a mapping $\pi : X \times U \rightarrow [0,1]$ that gives for each state x and action u the conditional probability $\pi(x, u) = \Pr\{u|x\}$ of taking action u given the MDP is in state x. Such a mapping is termed a closed-loop control or action *strategy* or *policy*. The strategy or policy $\pi(x, u) = \Pr\{u|x\}$ is called *stochastic* or *mixed* if there is a non-zero probability of selecting more than one control when in state x. We can view mixed strategies as probability distribution vectors having as component i the probability of selecting the ith control action while in state $x \in X$. If the mapping $\pi : X \times U \rightarrow [0,1]$ admits only one control, with probability 1, when in every state x, the mapping is called a *deterministic* policy. Then, $\pi(x, u) = \Pr\{u|x\}$ corresponds to a function mapping states into controls $\mu(x) : X \rightarrow U$.

MDP that have finite state and action spaces are termed finite MDP.

2.1.1 Optimal sequential decision problems

Dynamical systems evolve causally through time. Therefore, we consider sequential decision problems and impose a discrete stage index k such that the MDP takes an action and changes states at non-negative integer stage values k. The stages may correspond to time or more generally to sequences of events. We refer to the stage value as the time. Denote state values and actions at time k by x_k, u_k. MDP evolve in discrete time.

It is often desirable for human-engineered systems to be optimal in terms of conserving resources such as cost, time, fuel and energy. Thus, the notion of optimality should be captured in selecting control policies for MDP. Define, therefore, a *stage cost* at time k by $r_k = r_k(x_k, u_k, x_{k+1})$. Then $R^u_{xx'} = E\{r_k | x_k = x, u_k = u, x_{k+1} = x'\}$, with $E\{\cdot\}$ the expected value operator. Define a performance index as the sum of future costs over time interval $[k,\ k+T]$

$$J_{k,T} = \sum_{i=0}^{T} \gamma^i r_{k+i} = \sum_{i=k}^{k+T} \gamma^{i-k} r_i \tag{2.1}$$

where $0 \le \gamma < 1$ is a discount factor that reduces the weight of costs incurred further in the future.

Usage of MDP in the fields of computational intelligence and economics usually consider r_k as a reward incurred at time k, also known as *utility*, and $J_{k,T}$ as a discounted return, also known as strategic reward. We refer instead to stage costs and discounted future costs to be consistent with objectives in the control of dynamical systems. For convenience we may call r_k the utility.

Consider that an agent selects a control policy $\pi_k(x_k, u_k)$ and uses it at each stage k of the MDP. We are primarily interested in *stationary policies*, where the conditional probabilities $\pi_k(x_k, u_k)$ are independent of k. Then $\pi_k(x, u) = \pi(x, u) = \Pr\{u|x\}$, *for all k*. Non-stationary deterministic policies have the form $\pi = \{\mu_0, \mu_1, \ldots\}$, where each entry is a function $\mu_k(x) : X \to U; k = 0, 1, \ldots$. Stationary deterministic policies are independent of time so that $\pi = \{\mu, \mu, \ldots\}$.

Select a fixed stationary policy $\pi(x, u) = \Pr\{u|x\}$. Then the 'closed-loop' MDP reduces to a Markov chain with state space X. That is, the transition probabilities between states are fixed with no further freedom of choice of actions. The transition probabilities of this Markov chain are given by

$$P_{x,x'} \equiv P^{\pi}_{x,x'} = \sum_u \Pr\{x'|x,u\}\Pr\{u|x\} = \sum_u \pi(x,u) P^u_{x,x'} \tag{2.2}$$

where the Chapman–Kolmogorov identity is used.

Under the assumption that the Markov chain corresponding to each policy, with transition probabilities as given in (2.2), is ergodic, it can be shown that every MDP has a stationary deterministic optimal policy (Bertsekas and Tsitsiklis, 1996; Wheeler and Narendra, 1986). A Markov chain is *ergodic* if all states are positive recurrent and aperiodic. Then, for a given policy there exists a stationary distribution $P_\pi(x)$ over X that gives the steady-state probability that the Markov chain is in state x.

The *value* of a policy is defined as the conditional expected value of future cost when starting in state x at time k and following policy $\pi(x,u)$ thereafter

$$V_k^\pi(x) = E_\pi\{J_{k,T}|x_k = x\} = E_\pi\left\{\sum_{i=k}^{k+T} \gamma^{i-k} r_i | x_k = x\right\} \tag{2.3}$$

Here, $E_\pi\{\cdot\}$ is the expected value given that the agent follows policy $\pi(x,u)$. $V^\pi(x)$ is known as the *value function* for policy $\pi(x,u)$. It tells the value of being in state x given that the policy is $\pi(x,u)$.

A main objective of MDP is to determine a policy $\pi(x,u)$ to minimize the expected future cost

$$\pi^*(x,u) = \arg\min_\pi V_k^\pi(x) = \arg\min_\pi E_\pi\left\{\sum_{i=k}^{k+T} \gamma^{i-k} r_i | x_k = x\right\} \tag{2.4}$$

This policy is termed the *optimal policy*, and the corresponding *optimal value* is given as

$$V_k^*(x) = \min_\pi V_k^\pi(x) = \min_\pi E_\pi\left\{\sum_{i=k}^{k+T} \gamma^{i-k} r_i | x_k = x\right\} \tag{2.5}$$

In computational intelligence and economics the interest is in utilities and rewards, and there we are interested in *maximizing* the expected performance index.

2.1.2 A backward recursion for the value

By using the Chapman–Kolmogorov identity and the Markov property we can write the value of policy $\pi(x,u)$ as

$$V_k^\pi(x) = E_\pi\{J_k|x_k = x\} = E_\pi\left\{\sum_{i=k}^{k+T} \gamma^{i-k} r_i | x_k = x\right\} \tag{2.6}$$

$$V_k^\pi(x) = E_\pi\left\{r_k + \gamma \sum_{i=k+1}^{k+T} \gamma^{i-(k+1)} r_i | x_k = x\right\} \tag{2.7}$$

$$V_k^\pi(x) = \sum_u \pi(x,u) \sum_{x'} P_{xx'}^u \left[R_{xx'}^u + \gamma E_\pi\left\{\sum_{i=k+1}^{k+T} \gamma^{i-(k+1)} r_i | x_{k+1} = x'\right\}\right] \tag{2.8}$$

Therefore, the value function for policy $\pi(x,u)$ satisfies

$$V_k^\pi(x) = \sum_u \pi(x,u) \sum_{x'} P_{xx'}^u \left[R_{xx'}^u + \gamma V_{k+1}^\pi(x')\right] \tag{2.9}$$

This equation provides a backward recursion for the value at time k in terms of the value at time $k+1$.

2.1.3 Dynamic programming

The optimal cost can be written as

$$V_k^*(x) = \min_{\pi} V_k^{\pi}(x) = \min_{\pi} \sum_u \pi(x, u) \sum_{x'} P_{xx'}^u \left[R_{xx'}^u + \gamma V_{k+1}^{\pi}(x') \right] \tag{2.10}$$

Bellman's optimality principle (Bellman, 1957) states that "An optimal policy has the property that no matter what the previous control actions have been, the remaining controls constitute an optimal policy with regard to the state resulting from those previous controls." Therefore, we can write

$$V_k^*(x) = \min_{\pi} \sum_u \pi(x, u) \sum_{x'} P_{xx'}^u \left[R_{xx'}^u + \gamma V_{k+1}^*(x') \right] \tag{2.11}$$

Suppose an arbitrary control u is now applied at time k and the optimal policy is applied from time $k+1$ on. Then Bellman's optimality principle says that the optimal control at time k is given by

$$\pi^*(x_k = x, u) = \arg\min_{\pi} \sum_u \pi(x, u) \sum_{x'} P_{xx'}^u \left[R_{xx'}^u + \gamma V_{k+1}^*(x') \right] \tag{2.12}$$

Under the assumption that the Markov chain corresponding to each policy, with transition probabilities as given in (2.2), is ergodic, every MDP has a stationary deterministic optimal policy. Then we can equivalently minimize the conditional expectation over all actions u in state x. Therefore

$$V_k^*(x) = \min_u \sum_{x'} P_{xx'}^u \left[R_{xx'}^u + \gamma V_{k+1}^*(x') \right] \tag{2.13}$$

$$u_k^* = \arg\min_u \sum_{x'} P_{xx'}^u \left[R_{xx'}^u + \gamma V_{k+1}^*(x') \right] \tag{2.14}$$

The backward recursion (2.11), (2.13) forms the basis for dynamic programming (DP) (Bellman, 1957), which gives offline methods for working *backward in time* to determine optimal policies (Lewis *et al.*, 2012). DP is an offline procedure for finding the optimal value and optimal policies that requires knowledge of the complete system dynamics in the form of transition probabilities $P_{x,x'}^u = \Pr\{x'|x, u\}$ and expected costs $R_{xx'}^u = E\{r_k | x_k = x, u_k = u, x_{k+1} = x'\}$.

2.1.4 Bellman equation and Bellman optimality equation

Dynamic programming is a backward-in-time method for finding the optimal value and policy. By contrast, reinforcement learning is concerned with finding optimal policies based on causal experience by executing sequential decisions that improve control actions based on the observed results of using a current policy.

This procedure requires the derivation of methods for finding optimal values and optimal policies that can be executed forward in time. The key to this is the Bellman equation, which we now develop. References for this section include Werbos (1992), Powell (2007), Busoniu *et al.* (2009), Barto *et al.* (1983).

To derive forward-in-time methods for finding optimal values and optimal policies, set now the time horizon T to infinity and define the infinite-horizon cost

$$J_k = \sum_{i=0}^{\infty} \gamma^i r_{k+1} = \sum_{i=k}^{\infty} \gamma^{i-k} r_i \tag{2.15}$$

The associated infinite-horizon value function for polgicy $\pi(x, u)$ is

$$V^{\pi}(x) = E_{\pi}\{J_k | x_k = x\} = E_{\pi}\left\{ \sum_{i=k}^{\infty} \gamma^{i-k} r_i | x_k = x \right\} \tag{2.16}$$

By using (2.8) with $T = \infty$ it can be seen that the value function for policy $\pi(x, u)$ satisfies the *Bellman equation*

$$V^{\pi}(x) = \sum_u \pi(x, u) \sum_{x'} P^u_{xx'} \left[R^u_{xx'} + \gamma V^{\pi}(x') \right] \tag{2.17}$$

The importance of this equation is that the same value function appears on both sides, which is due to the fact that the infinite-horizon cost is used. Therefore, the Bellman equation (2.17) can be interpreted as a consistency equation that must be satisfied by the value function at each time stage. It expresses a relation between the current value of being in state x and the value of being in next state x' given that policy $\pi(x, u)$ is used.

The Bellman equation (2.17) is the starting point for developing a family of reinforcement learning algorithms for finding optimal policies by using causal experiences received stagewise forward in time. The Bellman optimality equation (2.11) involves the 'minimum' operator, and so does not contain any specific policy $\pi(x, u)$. Its solution relies on knowing the dynamics, in the form of transition probabilities. By contrast, the form of the Bellman equation is simpler than that of the optimality equation, and it is easier to solve. The solution to the Bellman equation yields the value function of a specific policy $\pi(x, u)$. As such, the Bellman equation is well suited to the actor–critic method of reinforcement learning shown in Figure 2.1. It is shown subsequently that the Bellman equation provides methods for implementing the critic in Figure 2.1, which is responsible for evaluating the performance of the specific current policy. Two key ingredients remain to be put in place. First, it is shown that methods known as policy iteration and value iteration use the Bellman equation to solve optimal control problems forward in time. Second, by approximating the value function in (2.17) by a parametric structure, these methods can be implemented online using standard adaptive control system identification algorithms such as recursive least-squares.

In the context of using the Bellman equation (2.17) for reinforcement learning, $V^{\pi}(x)$ may be considered as a predicted performance, $\sum_u \pi(x, u) \sum_{x'} P^u_{xx'} R^u_{xx'}$ the observed one-step reward and $V^{\pi}(x')$ a current estimate of future behavior. Such notions can be capitalized on in the subsequent discussion of temporal difference learning, which uses them to develop adaptive control algorithms that can learn optimal behavior online in real-time applications.

If the MDP is finite and has N states, then the Bellman equation (2.17) is a system of N simultaneous linear equations for the value $V^{\pi}(x)$ of being in each state x given the current policy $\pi(x, u)$.

The optimal infinite-horizon value satisfies

$$V^*(x) = \min_{\pi} V^{\pi}(x) = \min_{\pi} \sum_{u} \pi(x, u) \sum_{x'} P^{u}_{xx'} \left[R^{u}_{xx'} + \gamma V^{\pi}(x') \right] \tag{2.18}$$

Bellman's optimality principle then yields the *Bellman optimality equation*

$$V^*(x) = \min_{\pi} V^{\pi}(x) = \min_{\pi} \sum_{u} \pi(x, u) \sum_{x'} P^{u}_{xx'} \left[R^{u}_{xx'} + \gamma V^{*}(x') \right] \tag{2.19}$$

Equivalently under the ergodicity assumption on the Markov chains corresponding to each policy, the Bellman optimality equation can be written as

$$V^*(x) = \min_{u} \sum_{x'} P^{u}_{xx'} \left[R^{u}_{xx'} + \gamma V^{*}(x') \right] \tag{2.20}$$

This equation is known as the Hamilton–Jacobi–Bellman (HJB) equation in control systems. If the MDP is finite and has N states, then the Bellman optimality equation is a system of N non-linear equations for the optimal value $V^*(x)$ of being in each state. The optimal control is given by

$$u^* = \arg\min_{u} \sum_{x'} P^{u}_{xx'} \left[R^{u}_{xx'} + \gamma V^{*}(x') \right] \tag{2.21}$$

The next example places these ideas into the context of dynamical systems. It is shown that, for the discrete-time linear quadratic regulator (LQR), the Bellman equation becomes a Lyapunov equation and the Bellman optimality equation becomes an algebraic Riccati equation (Lewis *et al.*, 2012).

Example 2.1. Bellman equation for the discrete-time LQR, the Lyapunov equation

This example shows the meaning of several of the ideas just discussed in terms of linear discrete-time dynamical systems.

a. MDP dynamics for deterministic discrete-time systems

Consider the discrete-time linear quadratic regulator (LQR) problem (Lewis *et al.*, 2012), where the MDP is deterministic and satisfies the state transition equation

$$x_{k+1} = Ax_k + Bu_k \tag{2.22}$$

with k the discrete-time index. The associated infinite-horizon performance index has deterministic stage costs and is

$$J_k = \frac{1}{2}\sum_{i=k}^{\infty} r_i = \frac{1}{2}\sum_{i=k}^{\infty}(x_i^T Q x_i + u_i^T R u_i) \tag{2.23}$$

where the cost weighting matrices satisfy $Q = Q^T \geq 0, R = R^T > 0$. In this example, the state space $X = R^n$ and action space $U = R^m$ are infinite and continuous.

b. Bellman equation for discrete-time LQR, the Lyapunov equation

Select a policy $u_k = \mu(x_k)$ and write the associated value function as

$$V(x_k) = \frac{1}{2}\sum_{i=k}^{\infty} r_i = \frac{1}{2}\sum_{i=k}^{\infty}(x_i^T Q x_i + u_i^T R u_i) \tag{2.24}$$

A difference equation equivalent is given by

$$\begin{aligned} V(x_k) &= \frac{1}{2}(x_k^T Q x_k + u_k^T R u_k) + \frac{1}{2}\sum_{i=k+1}^{\infty}(x_i^T Q x_i + u_i^T R u_i) \\ &= \frac{1}{2}(x_k^T Q x_k + u_k^T R u_k) + V(x_{k+1}) \end{aligned} \tag{2.25}$$

That is, the solution $V(x_k)$ to this equation that satisfies $V(0) = 0$ is the value given by (2.24). Equation (2.25) is exactly the Bellman equation (2.17) for the LQR.

Assuming the value is quadratic in the state so that

$$V_k(x_k) = \frac{1}{2} x_k^T P x_k \tag{2.26}$$

for some kernel matrix $P = P^T > 0$ yields the Bellman equation form

$$2V(x_k) = x_k^T P x_k = x_k^T Q x_k + u_k^T R u_k + x_{k+1}^T P x_{k+1} \tag{2.27}$$

which, using the state equation, can be written as

$$2V(x_k) = x_k^T Q x_k + u_k^T R u_k + (Ax_k + Bu_k)^T P(Ax_k + Bu_k) \tag{2.28}$$

Assuming a constant, that is stationary, state-feedback policy $u_k = \mu(x_k) = -Kx_k$ for some stabilizing gain K, write

$$2V(x_k) = x_k^T P x_k = x_k^T Q x_k + x_k^T K^T R K x_k + x_k^T (A - BK)^T P(A - BK) x_k \tag{2.29}$$

Since this equation holds for all state trajectories, we have

$$(A - BK)^T P(A - BK) - P + Q + K^T RK = 0 \tag{2.30}$$

which is a Lyapunov equation. That is, the Bellman equation (2.17) for the discrete-time LQR is equivalent to a Lyapunov equation. Since the performance index is undiscounted, that is $\gamma = 1$, a stabilizing gain K, that is a stabilizing policy, must be selected.

The formulations (2.25), (2.27), (2.29) and (2.30) for the Bellman equation are all equivalent. Note that forms (2.25) and (2.27) do not involve the system dynamics (A, B). On the other hand, note that the Lyapunov equation (2.30) can only be used if the state dynamics (A, B) are known. Optimal control design using the Lyapunov equation is the standard procedure in control systems theory. Unfortunately, by assuming that (2.29) holds for all trajectories and going to (2.30), we lose all possibility of applying any sort of reinforcement learning algorithms to solve for the optimal control and value online by observing data along the system trajectories. By contrast, we show that by employing the form (2.25) or (2.27) for the Bellman equation, reinforcement learning algorithms for learning optimal solutions online can be devised by using temporal difference methods. That is, reinforcement learning allows the Lyapunov equation to be solved online without knowing A *or* B.

c. Bellman optimality equation for discrete-time LQR, the algebraic Riccati equation

The discrete-time LQR Hamiltonian function is

$$H(x_k, u_k) = x_k^T Qx_k + u_k^T Ru_k + (Ax_k + Bu_k)^T P(Ax_k + Bu_k) - x_k^T Px_k \tag{2.31}$$

The Hamiltonian is equivalent to the *temporal difference error* (Section 2.4) in MDP. A necessary condition for optimality is the stationarity condition $\partial H(x_k, u_k)/\partial u_k = 0$, which is equivalent to (2.21). Solving this equation yields the optimal control

$$u_k = -Kx_k = -(B^T PB + R)^{-1} B^T PAx_k$$

Putting this equation into (2.29) yields the discrete-time algebraic Riccati equation (ARE)

$$A^T PA - P + Q - A^T PB(B^T PB + R)^{-1} B^T PA = 0 \tag{2.32}$$

ARE is exactly the Bellman optimality equation (2.19) for the discrete-time LQR. ■

2.2 Policy evaluation and policy improvement

Given a current policy $\pi(x, u)$, its value (2.16) can be determined by solving the Bellman equation (2.17). This procedure is known as *policy evaluation*.

Moreover, given the value for some policy $\pi(x,u)$, we can always use it to find another policy that is better, or at least no worse. This step is known as *policy improvement*. Specifically, suppose $V^\pi(x)$ satisfies (2.17). Then define a new policy $\pi'(x,u)$ by

$$\pi'(x, u) = \arg\min_{\pi(x,\cdot)} \sum_{x'} P^u_{xx'}[R^u_{xx'} + \gamma V^\pi(x')] \tag{2.33}$$

Then it can be shown that $V^{\pi'}(x) \le V^\pi(x)$ (Bertsekas and Tsitsiklis, 1996; Sutton and Barto, 1998). The policy determined as in (2.33) is said to be *greedy* with respect to value function $V^\pi(x)$.

In the special case that $V^{\pi'}(x) = V^\pi(x)$ in (2.33), then $V^{\pi'}(x), \pi'(x,u)$ satisfy (2.20) and (2.21); therefore, $\pi'(x,u) = \pi(x,u)$ is the optimal policy and $V^{\pi'}(x) = V^\pi(x)$ the optimal value. That is, an optimal policy, and only an optimal policy, is greedy with respect to its own value. In computational intelligence, *greedy* refers to quantities determined by optimizing over short or one-step horizons, regardless of potential impacts far into the future.

Now let us consider algorithms that repeatedly interleave the following two procedures.

Policy evaluation by Bellman equation

$$V^\pi(x) = \sum_u \pi(x, u) \sum_{x'} P^u_{xx'}\left[R^u_{xx'} + \gamma V^\pi(x')\right], \quad \text{for all } x \in S \subseteq X \tag{2.34}$$

Policy improvement

$$\pi'(x, u) = \arg\min_{\pi(x,\cdot)} \sum_{x'} P^u_{xx'}\left[R^u_{xx'} + \gamma V^\pi(x')\right], \quad \text{for all } x \in S \subseteq X \tag{2.35}$$

where S is a suitably selected subspace of the state space, to be discussed later. We call an application of (2.34) followed by an application of (2.35) one *step*. This terminology is in contrast to the decision time stage k defined above.

At each step of such algorithms, we obtain a policy that is no worse than the previous policy. Therefore, it is not difficult to prove convergence under fairly mild conditions to the optimal value and optimal policy. Most such proofs are based on the Banach fixed-point theorem. Note that (2.20) is a fixed-point equation for $V^*(\cdot)$. Then the two equations (2.34) and (2.35) define an associated map that can be shown under mild conditions to be a contraction map (Bertsekas and Tsitsiklis, 1996; Powell, 2007; Mehta and Meyn, 2009), which converges to the solution of (2.20).

A large family of algorithms is available that implement the policy evaluation and policy improvement procedures in different ways, or interleave them differently, or select subspace $S \subseteq X$ in different ways, to determine the optimal value and optimal policy. We soon outline some of them.

The relevance of this discussion for feedback control systems is that these two procedures can be implemented for dynamical systems online in real time by observing data measured along the system trajectories. This is shown subsequently. The result is a family of adaptive control algorithms that converge to optimal control solutions. Such algorithms are of the *actor–critic* class of reinforcement learning systems, shown in Figure 2.1. There, a critic agent evaluates the current control policy using methods based on (2.34). After this evaluation is completed, the action is updated by an actor agent based on (2.35).

2.2.1 Policy iteration

One method of reinforcement learning for using (2.34) and (2.35) to find the optimal value and optimal policy is *policy iteration.*

Algorithm 2.1. Policy iteration

Select an admissible initial policy $\pi_0(x,u)$.
Do for $j=0$ until convergence.

Policy evaluation (*value update*)

$$V_j(x) = \sum_{n} \pi_j(x, u) \sum_{x'} P^u_{xx'} \left[R^u_{xx'} + \gamma V_j(x')\right], \quad \text{for all } x \in X \tag{2.36}$$

Policy improvement (*policy update*)

$$\pi_{j+1}(x, u) = \underset{\pi(x,\cdot)}{\arg\min} \sum_{x'} P^u_{xx'} \left[R^u_{xx'} + \gamma V_j(x')\right], \quad \text{for all } x \in X \tag{2.37}$$

■

At each step j the policy iteration algorithm determines the solution of the Bellman equation (2.36) to compute the value $V_j(x)$ of using the current policy $\pi_j(x,u)$. This value corresponds to the infinite sum (2.16) for the current policy. Then the policy is improved using (2.37). The steps are continued until there is no change in the value or the policy.

Note that j is not the time or stage index k, but a policy iteration step iteration index. As detailed in the next sections, policy iteration can be implemented for dynamical systems online in real time by observing data measured along the system trajectories. Data for multiple times k are needed to solve the Bellman equation (2.36) at each step j.

If the MDP is finite and has N states, then the policy evaluation equation (2.36) is a system of N simultaneous linear equations, one for each state. The policy iteration algorithm must be suitably initialized to converge. The initial policy $\pi_0(x,u)$ and value V_0 must be selected so that $V_1 \leq V_0$. Initial policies that guarantee this are termed admissible. Then, for finite Markov chains with N states, policy iteration converges in a finite number of steps, less than or equal to N, because there are only a finite number of policies (Bertsekas and Tsitsiklis, 1996).

2.2.2 Iterative policy iteration

The Bellman equation (2.36) is a system of simultaneous equations. Instead of directly solving the Bellman equation, it can be solved by an iterative policy evaluation procedure. Note that (2.36) is a fixed-point equation for $V_j(\cdot)$. It defines the *iterative policy evaluation* map

$$V_j^{i+1}(x) = \sum_u \pi_j(x, u) \sum_{x'} P_{xx'}^u \left[R_{xx'}^u + \gamma V_j^i(x') \right], \quad i = 1, 2, \ldots \tag{2.38}$$

which can be shown to be a contraction map under rather mild conditions. By the Banach fixed-point theorem the iteration can be initialized at any non-negative value of $V_j^1(\cdot)$ and the iteration converges to the solution of (2.36). Under certain conditions, this solution is unique. A suitable initial value choice is the value function $V_{j-1}(\cdot)$ from the previous step $j-1$. On close enough convergence, we set $V_j(\cdot) = V_j^i(\cdot)$ and proceed to apply (2.37).

Index j in (2.38) refers to the step number of the policy iteration algorithm. By contrast i is an iteration index. Iterative policy evaluation (2.38) should be compared to the backward-in-time recursion (2.9) for the finite-horizon value. In (2.9), k is the time index. By contrast, in (2.38), i is an iteration index. Dynamic programming is based on (2.9) and proceeds backward in time. The methods for online optimal adaptive control described in this chapter proceed forward in time and are based on policy iteration and similar algorithms.

2.2.3 *Value iteration*

A second method for using (2.34) and (2.35) in reinforcement learning is *value iteration*.

Algorithm 2.2. Value iteration

Select an initial policy $\pi_0(x, u)$.
Do for $j = 0$ until convergence.

Value update

$$V_{j+1}(x) = \sum_u \pi_j(x, u) \sum_{x'} P_{xx'}^u \left[R_{xx'}^u + \gamma V_j(x') \right], \quad \text{for all } x \in S_j \subseteq X \tag{2.39}$$

Policy improvement

$$\pi_{j+1}(x, u) = \arg\min_{\pi(x,\cdot)} \sum_{x'} P_{xx'}^u \left[R_{xx'}^u + \gamma V_j(x') \right], \quad \text{for all } x \in S_j \subseteq X \tag{2.40}$$

■

We can combine the value update and policy improvement into one equation to obtain the equivalent form for value iteration

$$V_{j+1}(x) = \min_\pi \sum_u \pi(x, u) \sum_{x'} P_{xx'}^u \left[R_{xx'}^u + \gamma V_j(x') \right], \quad \text{for all } x \in S_j \subseteq X \tag{2.41}$$

or, equivalently under the ergodicity assumption, in terms of deterministic policies

$$V_{j+1}(x) = \min_u \sum_{x'} P^u_{xx'} \left[R^u_{xx'} + \gamma V_j(x') \right], \quad \text{for all } x \in S_j \subseteq X \tag{2.42}$$

Note that now, (2.39) is a simple one-step recursion, not a system of linear equations as is (2.36) in the policy iteration algorithm. In fact, value iteration uses one iteration of (2.38) in its value update step. It does not find the value corresponding to the current policy, but takes only one iteration toward that value. Again, j is not the time index, but the value iteration step index.

Compare Value Iteration (2.41) to Dynamic Programming (2.11). DP is a backwards-in-time procedure for finding optimal control policies, and as such cannot be implemented online in real-time. By contrast, in subsequent sections we show how to implement value iteration for dynamical systems online in real time by observing data measured along the system trajectories. Data for multiple times k are needed to solve the update (2.39) for each step j.

Standard value iteration takes the update set as $S_j = X$, for all j. That is, the value and policy are updated for all states simultaneously. *Asynchronous* value iteration methods perform the updates on only a subset of the states at each step. In the extreme case, updates can be performed on only one state at each step.

It is shown in Bertsekas and Tsitsikils (1996) that standard value iteration, which has $S_j = X$, for all j, converges for finite MDP for all initial conditions when the discount factor satisfies $0 < \gamma < 1$. When $S_j = X$, for all j and $\gamma = 1$ an absorbing state is added and a 'properness' assumption is needed to guarantee convergence to the optimal value. When a single state is selected for value and policy updates at each step, the algorithm converges, for all choices of initial value, to the optimal cost and policy if each state is selected for update infinitely often. More general algorithms result if value update (2.39) is performed multiple times for different choices of S_j prior to a policy improvement. Then, it is required that updates (2.39) and (2.40) be performed infinitely often for each state, and a monotonicity assumption must be satisfied by the initial starting value.

Considering (2.19) as a fixed-point equation, value iteration is based on the associated iterative map (2.39) and (2.40), which can be shown under certain conditions to be a contraction map. In contrast to policy iteration, which converges under certain conditions in a finite number of steps, value iteration usually takes an infinite number of steps to converge (Bertsekas and Tsitsiklis, 1996). Consider finite MDP and the transition probability graph having probabilities (2.2) for the Markov chain corresponding to an optimal policy $\pi^*(x,u)$. If this graph is acyclic for some $\pi^*(x,u)$, then value iteration converges in at most N steps when initialized with a large value.

Having in mind the dynamic programming equation (2.10) and examining the value iteration value update (2.41), $V_j(x')$ can be interpreted as an approximation or estimate for the future stage cost-to-go from the future state x', Those algorithms wherein the future cost estimate are themselves costs or values for some policy are called *rollout algorithms* in Bertsekas and Tsitsiklis (1996). Such policies are forward looking and self-correcting. It is shown that these methods can be used to derive adaptive learning algorithms for receding horizon control (Zhang *et al.*, 2009).

MDP, policy iteration and value iteration are closely tied to optimal and adaptive control. The next example shows that for the discrete-time LQR, policy iteration and value iteration can be used to derive algorithms for solution of the optimal control problem that are quite common in the feedback control systems, including Hewer's algorithm (Hewer, 1971).

Example 2.2. Policy iteration and value iteration for the discrete-time LQR

The Bellman equation (2.17) for the discrete-time LQR is equivalent to all the formulations (2.25), (2.27), (2.29) and (2.30). Any of these can be used to implement policy iteration and value iteration. Form (2.30) is a Lyapunov equation.

a. Policy iteration, Hewer's algorithm

With step index j, and using superscripts to denote algorithm steps and subscripts to denote the time k, the policy evaluation step (2.36) applied on (2.25) yields

$$V^{j+1}(x_k) = \frac{1}{2}(x_k^T Q x_k + u_k^T R u_k) + V^{j+1}(x_{k+1}) \tag{2.43}$$

Policy iteration applied on (2.27) yields

$$x_k^T P^{j+1} x_k = x_k^T Q x_k + u_k^T R u_k + x_{k+1}^T P^{j+1} x_{k+1} \tag{2.44}$$

and policy iteration on (2.30) yields the Lyapunov equation

$$0 = (A - BK^j)^T P^{j+1}(A - BK^j) - P^{j+1} + Q + (K^j)^T R K^j \tag{2.45}$$

In all cases the policy improvement step is

$$\mu^{j+1}(x_k) = K^{j+1} x_k = \arg\min(x_k^T Q x_k + u_k^T R u_k + x_{k+1}^T P^{j+1} x_{k+1}) \tag{2.46}$$

which can be written explicitly as

$$K^{j+1} = -(B^T P^{j+1} B + R)^{-1} B^T P^{j+1} A \tag{2.47}$$

Policy iteration algorithm format (2.45) and (2.47) relies on repeated solutions of Lyapunov equations at each step, and is Hewer's algorithm. This algorithm is proven to converge in Hewer (1971) to the solution of the Riccati equation (2.32). Hewer's algorithm is an offline algorithm that requires complete knowledge of the system dynamics (A, B) to find the optimal value and control. The algorithm requires that the initial gain K^0 be stabilizing.

b. Value iteration, Lyapunov recursions

Applying value iteration (2.39) to Bellman equation format (2.27) yields

$$x_k^T P^{j+1} x_k = x_k^T Q x_k + u_k^T R u_k + x_{k+1}^T P^j x_{k+1} \tag{2.48}$$

And applying on format (2.30) in Example 5 yields the Lyapunov recursion

$$P^{j+1} = (A - BK^j)^T P^j (A - BK^j) + Q + (K^j)^T R K^j \tag{2.49}$$

In both cases the policy improvement step is still given by (2.46) and (2.47).

Value iteration algorithm format (2.49) and (2.47) is a Lyapunov recursion, which is easy to implement and does not, in contrast to policy iteration, require Lyapunov equation solutions. This algorithm is shown to converge in Lancaster and Rodman (1995) to the solution of the Riccati equation (2.32). Lyapunov recursion is an offline algorithm that requires complete knowledge of the system dynamics (A, B) to find the optimal value and control. It *does not* require that the initial gain K^0 be stabilizing, and can be initialized with any feedback gain.

c. Online solution of the Riccati equation without knowing plant matrix A

Hewer's algorithm and the Lyapunov recursion algorithm are both offline methods of solving the algebraic Riccati equation (2.32). Full knowledge of the plant dynamics (A, B) is needed to implement these algorithms. By contrast, we shall show subsequently that both policy iteration algorithm format (2.44) and (2.46) and value iteration algorithm format (2.48) and (2.46) can be implemented *online* to determine the optimal value and control *in real time* using data measured along the system trajectories, and without knowing the system A matrix. This aim is accomplished through the *temporal difference method* to be presented. That is, reinforcement learning allows the solution of the algebraic Riccati equation online without knowing the system A matrix.

d. Iterative policy evaluation

Given a fixed policy K, the iterative policy evaluation procedure (2.38) becomes

$$P^{j+1} = (A - BK)^T P^j (A - BK) + Q + K^T RK \tag{2.50}$$

This recursion converges to the solution to the Lyapunov equation $P = (A - BK)^T P(A - BK) + Q + K^T RK$ if $(A - BK)$ is stable, for any choice of initial value P^0. ■

2.2.4 Generalized policy iteration

In policy iteration the system of linear equations (2.36) is completely solved at each step to compute the value (2.16) of using the current policy $\pi_j(x, u)$. This solution can be accomplished by running iterations (2.38) until convergence. By contrast, in value iteration one takes only one iteration of (2.38) in the value update step (2.39). *Generalized* policy iteration algorithms make several iterations (2.38) in their value update step.

Usually, policy iteration converges to the optimal value in fewer steps j since it does more work in solving equations at each step. On the other hand, value iteration is the easiest to implement as it takes only one iteration of a recursion as per (2.39). Generalized policy iteration provides a suitable compromise between computational complexity and convergence speed. Generalized policy iteration is a modified case of the value iteration algorithm given above, where we select $S_j = X$, for all j and perform value update (2.39) multiple times before each policy update (2.40).

2.2.5 Q function

The conditional expected value in (2.13)

$$Q_k^*(x, u) = \sum_{x'} P_{xx'}^u \left[R_{xx'}^u + \gamma V_{k+1}^*(x')\right] = E_\pi\{r_k + \gamma V_{k+1}^*(x')|x_k = x, u_k = u\} \tag{2.51}$$

is known as the *optimal Q function* (Watkins, 1989; Watkins and Dayan, 1992). The name comes from 'quality function'. The Q function is also called the action-value function (Sutton and Barto, 1998). The Q function is equal to the expected return for taking an arbitrary action u at time k in state x and thereafter following an optimal policy. The Q function is a function of the current state x and also the action u.

In terms of the Q function, the Bellman optimality equation has the particularly simple form

$$V_k^*(x) = \min_u Q_k^*(x, u) \tag{2.52}$$

$$u_k^* = \arg\min_u Q_k^*(x, u) \tag{2.53}$$

Given some fixed policy $\pi(x,u)$ define the Q function for that policy as

$$Q_k^\pi(x, u) = E_\pi\{r_k + \gamma V_{k+1}^\pi(x')|x_k = x, u_k = u\} = \sum_{x'} P_{xx'}^u \left[R_{xx'}^u + \gamma V_{k+1}^\pi(x')\right], \tag{2.54}$$

where (2.9) is used. This function is equal to the expected return for taking an arbitrary action u at time k in state x and thereafter following the existing policy $\pi(x,u)$. The meaning of the Q function is elucidated in Example 2.3.

Note that $V_k^\pi(x) = Q_k^\pi(x, \pi(x,u))$, hence (2.54) can be written as the backward recursion in the Q function

$$Q_k^\pi(x, u) = \sum_{x'} P_{xx'}^u [R_{xx'}^u + \gamma Q_{k+1}^\pi(x', \pi(x', u'))]. \tag{2.55}$$

The Q function is a two-dimensional (2D) function of both the current state x and the action u. By contrast the value function is a one-dimensional function of the state. For finite MDP, the Q function can be stored as a 2D lookup table at each state/action pair. Note that direct minimization in (2.11) and (2.12) requires knowledge of the state transition probabilities, which correspond to the system dynamics, and costs. By contrast, the minimization in (2.52) and (2.53) requires knowledge of only the Q function and not the system dynamics.

The utility of the Q function is twofold. First, it contains information about control actions in every state. As such, the best control in each state can be selected using (2.53) by knowing only the Q function. Second, the Q function can be estimated *online in real time* directly from data observed along the system trajectories, without knowing the system dynamics information, that is the transition probabilities. We see how this is accomplished later.

The infinite-horizon Q function for a prescribed fixed policy is given by

$$Q^{\pi}(x, u) = \sum_{x'} P^{u}_{xx'}[R^{u}_{xx'} + \gamma V^{\pi}(x')] \tag{2.56}$$

The Q function also satisfies a Bellman equation. Note that given a fixed policy $\pi(x, u)$

$$V^{\pi}(x) = Q^{\pi}(x, \pi(x, u)) \tag{2.57}$$

when according to (2.56) the Q function satisfies the Bellman equation

$$Q^{\pi}(x, u) = \sum_{x'} P^{u}_{xx'}[R^{u}_{xx'} + \gamma Q^{\pi}(x', \pi(x', u'))] \tag{2.58}$$

The Bellman optimality equation for the infinite-horizon Q function is

$$Q^{*}(x, u) = \sum_{x'} P^{u}_{xx'}[R^{u}_{xx'} + \gamma Q^{*}(x', \pi^{*}(x', u'))] \tag{2.59}$$

$$Q^{*}(x, u) = \sum_{x'} P^{u}_{xx'}[R^{u}_{xx'} + \gamma \min_{u'} Q^{*}(x', u')] \tag{2.60}$$

Compare (2.20) and (2.60), where the minimum operator and the expected value operator are interchanged.

Policy iteration and value iteration are especially easy to implement in terms of the Q function (2.54), as follows.

Algorithm 2.3. Policy iteration using Q function

Policy evaluation (value update)

$$Q_j(x, u) = \sum_{x'} P^{u}_{xx'}\left[R^{u}_{xx'} + \gamma Q_j(x', \pi(x', u'))\right], \quad \text{for all } x \in X \tag{2.61}$$

Policy improvement

$$\pi_{j+1}(x, u) = \arg\min_{u} Q_j(x, u), \quad \text{for all } x \in X \tag{2.62}$$

■

Algorithm 2.4. Value iteration using Q function

Value update

$$Q_{j+1}(x, u) = \sum_{x'} P^{u}_{xx'}\left[R^{u}_{xx'} + \gamma Q_j(x', \pi(x', u'))\right], \quad \text{for all } x \in S_j \subseteq X \tag{2.63}$$

Policy improvement

$$\pi_{j+1}(x, u) = \arg\min_{u} Q_{j+1}(x, u), \quad \text{for all } x \in S_j \subseteq X \tag{2.64}$$

■

Combining both steps of value iteration yields the form

$$Q_{j+1}(x, u) = \sum_{x'} P^u_{xx'} \left[R^u_{xx'} + \gamma \min_{u'} Q_j(x', u') \right], \quad \text{for all } x \in S_j \subseteq X \tag{2.65}$$

which may be compared to (2.42).

As we shall show, the utility of the Q function is that these algorithms can be implemented online in real time, without knowing the system dynamics, by measuring data along the system trajectories. They yield optimal adaptive control algorithms, that is adaptive control algorithms that converge online to optimal control solutions.

Example 2.3. Q function for discrete-time LQR

The Q function following a given policy $u_k = \mu(x_k)$ is defined in (2.54). For the discrete-time LQR in Example 2.1 the Q function is

$$Q(x_k, u_k) = \frac{1}{2}\left(x_k^T Q x_k + u_k^T R u_k\right) + V(x_{k+1}) \tag{2.66}$$

where the control u_k is arbitrary and the policy $u_k = \mu(x_k)$ is followed for $k+1$ and subsequent times. Writing

$$Q(x_k, u_k) = x_k^T Q x_k + u_k^T R u_k + (Ax_k + Bu_k)^T P (Ax_k + Bu_k) \tag{2.67}$$

with P the Riccati solution yields the Q function for the discrete-time LQR

$$Q(x_k, u_k) = \frac{1}{2} \begin{bmatrix} x_k \\ u_k \end{bmatrix}^T \begin{bmatrix} A^T PA + Q & A^T PB \\ B^T PA & B^T PB + R \end{bmatrix} \begin{bmatrix} x_k \\ u_k \end{bmatrix} \tag{2.68}$$

Define

$$Q(x_k, u_k) \equiv \frac{1}{2} \begin{bmatrix} x_k \\ u_k \end{bmatrix}^T S \begin{bmatrix} x_k \\ u_k \end{bmatrix} = \frac{1}{2} \begin{bmatrix} x_k \\ u_k \end{bmatrix}^T \begin{bmatrix} S_{xx} & S_{xu} \\ S_{ux} & S_{uu} \end{bmatrix} \begin{bmatrix} x_k \\ u_k \end{bmatrix} \tag{2.69}$$

for some kernel matrix $S = S^T > 0$.

Applying $\partial Q(x_k, u_k)/\partial u_k = 0$ to (2.69) yields

$$u_k = -S_{uu}^{-1} S_{ux} x_k \tag{2.70}$$

and to (2.68) yields

$$u_k = -(B^T PB + R)^{-1} B^T PA x_k \tag{2.71}$$

The latter equation requires knowledge of the system dynamics (A, B) to perform the policy improvement step of either policy iteration or value iteration. On the other hand, (2.70) requires knowledge only of the Q function matrix kernel S. It will be shown that these equations allow the use of reinforcement learning temporal difference methods to determine the kernel matrix S online in real time, without knowing the system dynamics (A, B), using data measured along the system trajectories. This procedure provides a family of Q learning algorithms that can solve the algebraic Riccati equation online without knowing the system dynamics (A, B). ■

2.3 Methods for implementing policy iteration and value iteration

Different methods are available for performing the value and policy updates for policy iteration and value iteration (Bertsekas and Tsitsiklis, 1996; Powell, 2007; Sutton and Barto, 1998). The main three are exact computation, Monte Carlo methods and temporal difference learning. The last two methods can be implemented without knowledge of the system dynamics. Temporal difference (TD) learning is the means by which optimal adaptive control algorithms can be derived for dynamical systems. Therefore, TD is covered in the next section.

Exact computation. Policy iteration requires solution at each step of Bellman equation (2.36) for the value update. For a finite MDP with N states, this is a set of linear equations in N unknowns, namely, the values of each state. Value iteration requires performing the one-step recursive update (2.39) at each step for the value update. Both of these can be accomplished exactly if we know the transition probabilities $P^u_{x,x'} = \Pr\{x'|x, u\}$ and costs $R^u_{xx'}$ of the MDP, which corresponds to knowing full system dynamics information. Likewise, the policy improvements (2.37) and (2.40) can be explicitly computed if the dynamics are known. It is shown in Example 2.1 that, for the discrete-time LQR, the exact computation method for computing the optimal control yields the Riccati equation solution approach. Policy iteration and value iteration boil down to repetitive solutions of Lyapunov equations or Lyapunov recursions. In fact, policy iteration becomes Hewer's method (Hewer, 1971), and value iteration becomes the Lyapunov recursion scheme that is shown to converge in Lancaster and Rodman (1995). These techniques are offline methods relying on matrix equation solutions and requiring complete knowledge of the system dynamics.

Monte Carlo learning. This is based on the definition (2.16) for the value function, and uses repeated measurements of data to approximate the expected value. The expected values are approximated by averaging repeated results along sample paths. An assumption on the ergodicity of the Markov chain with transition probabilities (2.2) for the given policy being evaluated is implicit. This assumption is suitable for *episodic tasks*, with experience divided into episodes (Sutton and Barto, 1998), namely, processes that start in an initial state and run until termination, and are then restarted at a new initial state. For finite MDP, Monte Carlo methods converge to the true value function if all states are visited infinitely often.

Therefore, in order to ensure accurate approximations of value functions, the episode sample paths must go through all the states $x \in X$ many times. This issue is called the problem of *maintaining exploration*. Several methods are available to ensure this, one of which is to use '*exploring starts*', in which every state has nonzero probability of being selected as the initial state of an episode.

Monte Carlo techniques are useful for dynamic systems control because the episode sample paths can be interpreted as system trajectories beginning in a prescribed initial state. However, no updates to the value function estimate or the control policy are made until after an episode terminates. In fact, Monte Carlo learning methods are closely related to repetitive or iterative learning control (Moore, 1993). They do not learn in real time along a trajectory, but learn as trajectories are repeated.

2.4 Temporal difference learning

It is now shown that the temporal difference method (Sutton and Barto, 1998) for solving Bellman equations leads to a family of optimal adaptive controllers, that is adaptive controllers that learn online the solutions to optimal control problems without knowing the full system dynamics. Temporal difference learning is true online reinforcement learning, wherein control actions are improved in real time based on estimating their value functions by observing data measured along the system trajectories.

Policy iteration requires solution at each step of N linear equations (2.36). Value iteration requires performing the recursion (2.39) at each step. Temporal difference reinforcement learning methods are based on the Bellman equation, and solve equations such as (2.36) and (2.39) without using system dynamics knowledge, but using data observed along a single trajectory of the system. Therefore, temporal difference learning is applicable for feedback control applications. Temporal difference updates the value at each time step as observations of data are made along a trajectory. Periodically, the new value is used to update the policy. Temporal difference methods are related to adaptive control in that they adjust values and actions online in real time along system trajectories.

Temporal difference methods can be considered to be *stochastic approximation* techniques whereby the Bellman equation (2.17), or its variants (2.36) and (2.39), is replaced by its evaluation along a single sample path of the MDP. Then, the Bellman equation becomes a deterministic equation that allows the definition of a *temporal difference error*.

Equation (2.9) was used to write the Bellman equation (2.17) for the infinite-horizon value (2.16). According to (2.7)–(2.9), an alternative form for the Bellman equation is

$$V^{\pi}(x_k) = E_{\pi}\{r_k|x_k\} + \gamma E_{\pi}\{V^{\pi}(x_{k+1})|x_k\} \tag{2.72}$$

This equation forms the basis for temporal difference learning.

Temporal difference reinforcement learning uses one sample path, namely, the current system trajectory, to update the value. Then, (2.72) is replaced by the *deterministic Bellman equation*

$$V^{\pi}(x_k) = r_k + \gamma V^{\pi}(x_{k+1}) \tag{2.73}$$

which holds for each observed data experience set (x_k, x_{k+1}, r_k) at each time stage k. This set consists of the current state x_k, the observed cost incurred r_k and the next state x_{k+1}. The *temporal difference error* is defined as

$$e_k = -V^{\pi}(x_k) + r_k + \gamma V^{\pi}(x_{k+1}) \tag{2.74}$$

and the current estimate for the value function is updated to make the temporal difference error small.

In the context of temporal difference learning, the interpretation of the Bellman equation is shown in Figure 2.3, where $V^{\pi}(x_k)$ may be considered as a predicted performance or value, r_k as the observed one-step reward and $\gamma V^{\pi}(x_{k+1})$ as a current estimate of future value. The Bellman equation can be interpreted as a consistency equation that holds if the current estimate for the predicted value $V^{\pi}(x_k)$ is correct. Temporal difference methods update the predicted value estimate $\hat{V}^{\pi}(x_k)$ to make the temporal difference error small. The idea, based on stochastic approximation, is that if we use the deterministic version of Bellman's equation repeatedly in policy iteration or value iteration, then on average these algorithms converge toward the solution of the stochastic Bellman equation.

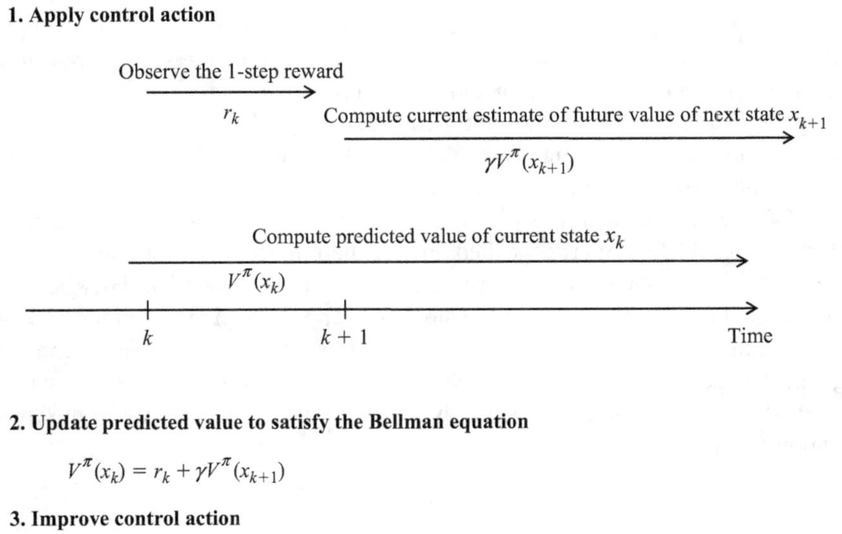

Figure 2.3 Temporal difference interpretation of Bellman equation. It shows how use of the Bellman equation captures the action, observation, evaluation and improvement mechanisms of reinforcement learning

2.5 Optimal adaptive control for discrete-time systems

A family of optimal adaptive control algorithms can now be developed for discrete-time dynamical systems. These algorithms determine the solutions to HJ design equations online in real time without knowing the system drift dynamics. In the LQR case, this means they solve the Riccati equation online without knowing the system A matrix. Physical analysis of dynamical systems using Lagrangian mechanics or Hamiltonian mechanics produces system descriptions in terms of non-linear ordinary differential equations. Discretization yields non-linear difference equations. Most research in reinforcement learning is conducted for systems that operate in discrete time. Therefore, we cover discrete-time dynamical systems here. The application of reinforcement learning techniques to continuous-time systems is significantly more involved and is the topic of the remainder of the book.

RL policy iteration and value iteration methods have been used for many years to provide methods for solving the optimal control problem for discrete-time (DT) dynamical systems. Bertsekas and Tsitsiklis developed RL methods based on policy iteration and value iteration for infinite-state discrete-time dynamical systems in Bertsekas and Tsitsiklis (1996). This approach, known as neurodynamic programming, used value function approximation to approximately solve the Bellman equation using iterative techniques. Offline solution methods were developed in Bertsekas and Tsitsiklis (1996). Werbos (1989, 1991, 1992, 2009) presented RL techniques based on value iteration for feedback control of discrete-time dynamical systems using value function approximation. These methods, known as approximate dynamic programming (ADP) or adaptive dynamic programming, are suitable for online learning of optimal control techniques for DT systems online in real time. As such, they are true adaptive learning techniques that converge to optimal control solutions by observing data measured along the system trajectories in real time. A family of four methods was presented under the aegis of ADP, which allowed learning of the value function and its gradient (e.g. the costate), and the Q function and its gradient. The ADP controllers are actor–critic structures with one learning network for the control action and one learning network for the critic. The ADP method for learning the value function is known as heuristic dynamic programming (HDP). Werbos called his method for online learning of the Q function for infinite-state DT dynamical systems 'action-dependent HDP'.

Temporal difference learning is a stochastic approximation technique based on the deterministic Bellman's equation (2.73). Therefore, we lose little by considering deterministic systems here. Therefore, consider a class of discrete-time systems described by deterministic non-linear dynamics in the affine state space difference equation form

$$x_{k+1} = f(x_k) + g(x_k)u_k, \tag{2.75}$$

with state $x_k \in R^n$ and control input $u_k \in R^m$. We use this affine form because its analysis is convenient. The following development can be generalized to the sampled data form $x_{k+1} = F(x_k, u_k)$.

A deterministic control policy is defined as a function from state space to control space $h(\cdot) : R^n \to R^m$. That is, for every state x_k, the policy defines a control action

$$u_k = h(x_k) \tag{2.76}$$

That is, a policy is a feedback controller.

Define a deterministic cost function that yields the value function

$$V^h(x_k) = \sum_{i=k}^{\infty} \gamma^{i-k} r(x_i, u_i) = \sum_{i=k}^{\infty} \gamma^{i-k} (Q(x_i) + u_i^T R u_i) \tag{2.77}$$

with $0 < \gamma \leq 1$ a discount factor, $Q(x_k) > 0$, $R = R^T > 0$ and $u_k = h(x_k)$ a prescribed feedback control policy. That is, the stage cost is

$$r(x_k, u_k) = Q(x_k) + u_k^T R u_k \tag{2.78}$$

The stage cost is taken quadratic in u_k to simplify developments, but can be any positive definite function of the control. We assume the system is stabilizable on some set $\Omega \in R^n$, that is there exists a control policy $u_k = h(x_k)$ such that the closed-loop system $x_{k+1} = f(x_k) + g(x_k)h(x_k)$ is asymptotically stable on Ω. A control policy $u_k = h(x_k)$ is said to be *admissible* if it is stabilizing, continuous and yields a finite cost $V^h(x_k)$.

For the deterministic value (2.77), the optimal value is given by Bellman's optimality equation

$$V^*(x_k) = \min_{h(\cdot)}(r(x_k, h(x_k)) + \gamma V^*(x_{k+1})) \tag{2.79}$$

which is known as the discrete-time Hamilton–Jacobi–Bellman (HJB) equation. The optimal policy is then given as

$$h^*(x_k) = \arg\min_{h(\cdot)}(r(x_k, h(x_k)) + \gamma V^*(x_{k+1})) \tag{2.80}$$

In this setup, Bellman's equation (2.17) is

$$V^h(x_k) = r(x_k, u_k) + \gamma V^h(x_{k+1}) = Q(x_k) + u_k^T R u_k + \gamma V^h(x_{k+1}), \quad V^h(0) = 0 \tag{2.81}$$

which is the same as (2.73). This is a difference equation equivalent of the value (2.77). That is, instead of evaluating the infinite sum (2.77), the difference equation (2.81) can be solved, with boundary condition $V(0) = 0$, to obtain the value of using a current policy $u_k = h(x_k)$.

The discrete-time Hamiltonian function can be defined as

$$H(x_k, h(x_K), \Delta V_k) = r(x_k, h(x_k)) + \gamma V^h(x_{k+1}) - V^h(x_k) \tag{2.82}$$

where $\Delta V_k = \gamma V_h(x_{k+1}) - V_h(x_k)$ is the forward difference operator. The Hamiltonian function captures the energy content along the trajectories of a system as reflected in the desired optimal performance. In fact, the Hamiltonian is the temporal difference error (2.74). The Bellman equation requires that the Hamiltonian be equal to zero for the value associated with a prescribed policy.

For the discrete-time linear quadratic regulator case we have

$$x_{k+1} = Ax_k + Bu_k \tag{2.83}$$

$$V^h(x_k) = \frac{1}{2}\sum_{i=k}^{\infty} \gamma^{i-k}(x_i^T Q x_i + u_i^T R u_i) \tag{2.84}$$

and the Bellman equation is written in several ways as seen in Example 2.1.

2.5.1 Policy iteration and value iteration for discrete-time dynamical systems

Two forms of reinforcement learning are based on policy iteration and value iteration. For temporal difference learning, policy iteration is written as follows in terms of the deterministic Bellman equation. Here, where $\nabla V(x) = \partial V(x)/\partial x$ is the gradient of the value function, interpreted as a column vector.

Algorithm 2.5. Policy iteration using temporal difference learning

Initialize. Select some admissible control policy $h_0(x_k)$.
Do for $j = 0$ until convergence:

Policy evaluation (value update)

$$V_{j+1}(x_k) = r(x_k, h_j(x_k)) + \gamma V_{j+1}(x_{k+1}) \tag{2.85}$$

Policy improvement

$$h_{j+1}(x_k) = \arg\min_{h(\cdot)}(r(x_k, h(x_k)) + \gamma V_{j+1}(x_{k+1})) \tag{2.86}$$

or

$$h_{j+1}(x_k) = -\frac{\gamma}{2}R^{-1}g^T(x_k)\,\nabla V_{j+1}(x_{k+1}) \tag{2.87}$$

■

Value iteration is similar, but the policy evaluation procedure is performed as follows.

Value iteration using temporal difference learning

Update the value using

Value update step

$$V_{j+1}(x_k) = r(x_k, h_j(x_k)) + \gamma V_j(x_{k+1}) \tag{2.88}$$

In value iteration we can select any initial control policy $h_0(x_k)$, not necessarily admissible or stabilizing.

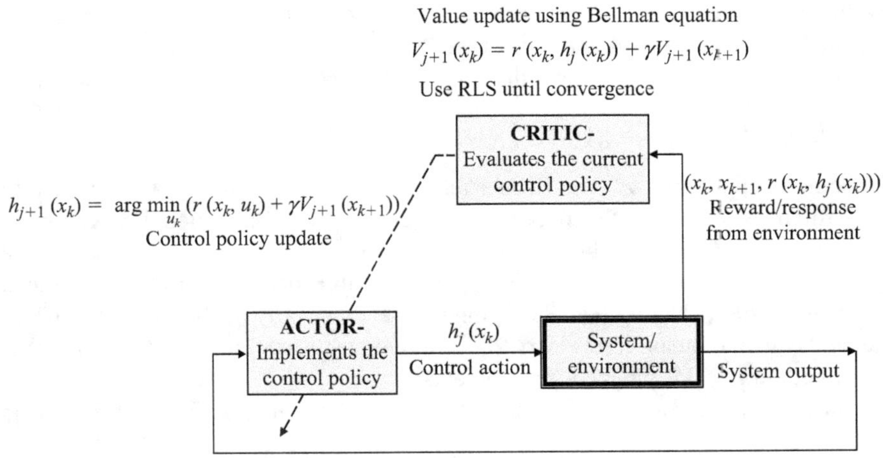

Figure 2.4 Temporal difference learning using policy iteration. At each time one observes the current state, the next state and the cost incurred. This is used to update the value estimate in the critic. Based on the new value, the control action is updated

These algorithms are illustrated in Figure 2.4. They are actor–critic formulations as is evident in Figure 2.1.

It is shown in Example 2.1 that, for the discrete-time LQR, the Bellman equation (2.81) is a linear Lyapunov equation and that (2.79) yields the discrete-time algebraic Riccati equation (ARE). For the discrete-time LQR, the policy evaluation step (2.85) in policy iteration is a Lyapunov equation and policy iteration exactly corresponds to Hewer's algorithm (Hewer, 1971) for solving the discrete-time ARE. Hewer proved that the algorithm converges under stabilizability and detectability assumptions. For the discrete-time LQR, value iteration is a Lyapunov recursion that is shown to converge to the solution to the discrete-time ARE under the stated assumptions by Lancaster and Rodman (1995) (see Example 2.2). These methods are offline design methods that require knowledge of the discrete-time dynamics (A, B). By contrast, we next desire to determine online methods for implementing policy iteration and value iteration that do not require full dynamics information.

2.5.2 Value function approximation

Policy iteration and value iteration can be implemented for finite MDP by storing and updating lookup tables. The key to implementing policy iteration and value iteration online for dynamical systems with infinite state and action spaces is to approximate the value function by a suitable approximator structure in terms of unknown parameters. Then, the unknown parameters are tuned online exactly as in system identification (Ljung, 1999). This idea of *value function approximation* (VFA) is used by Werbos (1989, 1991, 1992, 2009) for control of discrete-time dynamical systems and called approximate dynamic programming (ADP) or adaptive dynamic programming. The approach is used by Bertsekas and Tsitsiklis (1996) and is called neurodynamic programming (see Powell, 2007; Busoniu *et al.*, 2009).

In the LQR case it is known that the value is quadratic in the state, therefore

$$V(x_k) = \frac{1}{2}x_k^T P x_k = \frac{1}{2}(\text{vec}(P))^T(x_k \otimes x_k) \equiv \bar{p}^T \bar{x}_k \equiv \bar{p}^T \phi(x_k) \tag{2.89}$$

for some kernel matrix P. The Kronecker product $\otimes$ allows this quadratic form to be written as linear in the parameter vector $\bar{p} = \text{vec}(P)$, which is formed by stacking the columns of the P matrix. The vector $\phi(x_k) = \bar{x}_k = x_k \otimes x_k$ is the quadratic polynomial vector containing all possible pairwise products of the n components of x_k. Noting that P is symmetric and has only $n(n+1)/2$ independent elements, the redundant terms in $x_k \otimes x_k$ are removed to define a quadratic basis set $\phi(x_k)$ with $n(n+1)/2$ independent elements.

In illustration, for second-order systems the value function for the DT LQR is

$$V(x) = \begin{bmatrix} x_1 \\ x_2 \end{bmatrix}^T \begin{bmatrix} p_{11} & p_{12} \\ p_{21} & p_{22} \end{bmatrix} \begin{bmatrix} x_1 \\ x_2 \end{bmatrix} \tag{2.90}$$

This can be written as

$$V(x) = \begin{bmatrix} p_{11} & 2p_{12} & p_{22} \end{bmatrix} \begin{bmatrix} x_1^2 \\ x_1 x_2 \\ x_2^2 \end{bmatrix} \tag{2.91}$$

which is linear in the unknown parameters p_{11}, p_{12}, p_{22}.

For non-linear systems (2.75) the value function contains higher-order non-linearities. Then, we assume the Bellman equation (2.81) has a local smooth solution (Van Der Schaft, 1992). Then, according to the Weierstrass Higher-Order Approximation Theorem (Abu-Khalaf *et al.*, 2006; Finlayson, 1990), there exists a dense basis set $\{\varphi_i(x)\}$ such that

$$V(x) = \sum_{i=1}^{\infty} w_i \varphi_i(x) = \sum_{i=1}^{L} w_i \varphi_i(x) + \sum_{i=L+1}^{\infty} w_i \varphi_i(x) \equiv W^T \phi(x) + \varepsilon_L(x) \tag{2.92}$$

where basis vector $\phi(x) = [\varphi_1(x) \quad \varphi_2(x) \quad \dots \quad \varphi_L(x)] : R^n \to R^L$ and $\varepsilon_L(x)$ converges uniformly to zero as the number of terms retained $L \to \infty$. In the Weierstrass Theorem, standard usage takes a polynomial basis set. In the neural-network research, approximation results are shown for other basis sets including sigmoid, hyperbolic tangent, Gaussian radial basis functions and others (Hornik *et al.*, 1990; Sandberg, 1998). There, standard results show that the neural-network approximation error $\varepsilon_L(x)$ is bounded by a constant on a compact set. L is referred to as the number of hidden-layer neurons, $\varphi_i(x)$ as the neural-network activation functions, and w_i as the neural-network weights.

2.5.3 Optimal adaptive control algorithms for discrete-time systems

We are now in a position to present several adaptive control algorithms for discrete-time systems based on temporal difference reinforcement learning that converge online to the optimal control solution.

The VFA parameters in $\bar{p}$ in (2.89) or W in (2.92) are unknown. Substituting the value function approximation into the value update (2.85) in policy iteration the following algorithm is obtained.

Algorithm 2.6. Optimal adaptive control using policy iteration

Initialize. Select some admissible control policy $h_0(x_k)$:
Do for $j=0$ until convergence.

Policy evaluation step. Determine the least-squares solution W_{j+1} to

$$W_{j+1}^T(\phi(x_k) - \gamma\phi(x_{k+1})) = r(x_k, h_j(x_k)) = Q(x_k) + h_j^T(x_k)Rh_j(x_k) \tag{2.93}$$

Policy improvement step. Determine an improved policy using

$$h_{j+1}(x_k) = -\frac{\gamma}{2}R^{-1}g^T(x_k)\,\nabla\phi^T(x_{k+1})W_{j+1} \tag{2.94}$$

■

This algorithm is easily implemented online by standard system identification techniques (Ljung, 1999). In fact, note that (2.93) is a scalar equation, whereas the unknown parameter vector $W_{j+1} \in R^L$ has L elements. Therefore, data from multiple time steps are needed for its solution. At time $k+1$ we measure the previous state x_k, the control $u_k = h_j(x_k)$, the next state x_{k+1} and compute the resulting utility $r(x_k, h_j(x_k))$. This data give one scalar equation. This procedure is repeated for subsequent times using the same policy $h_j(\cdot)$ until we have at least L equations, at which point the least-squares solution W_{j+1} can be found. Batch least-squares can be used for this.

Alternatively, note that equations of the form (2.93) are exactly those solved by recursive least-squares (RLS) techniques (Ljung, 1999). Therefore, RLS can be run online until convergence. Write (2.93) as

$$W_{j+1}^T\Phi(k) \equiv W_{j+1}^T(\phi(x_k) - \gamma\phi(x_{k+1})) = r(x_k, h_j(x_k)) \tag{2.95}$$

where $\Phi(k) \equiv (\phi(x_k) - \gamma\phi(x_{k+1}))$ is a regression vector. At step j of the policy iteration algorithm, the control policy is fixed at $u = h_j(x)$. Then, at each time k the data set $(x_k, x_{k+1}, r(x_k, h_j(x_k)))$ is measured. One step of RLS is then performed. This procedure is repeated for subsequent times until convergence to the parameters corresponding to the value $V_{j+1}(x) = W_{j+1}^T\phi(x)$. Note that for RLS to converge, the regression vector $\Phi(k) \equiv (\phi(x_k) - \gamma\phi(x_{k+1}))$ must be persistently exciting.

As an alternative to RLS, we could use a gradient descent tuning method such as

$$W_{j+1}^{j+1} = W_{j+1}^i - \alpha\Phi(k)((W_{j+1}^i)^T\Phi(k) - r(x_k, h_j(x_k))) \tag{2.96}$$

where $\alpha > 0$ is a tuning parameter. The step index j is held fixed, and index i is incremented at each increment of the time index k. Note that the quantity inside the large brackets is just the temporal difference error.

Once the value parameters have converged, the control policy is updated according to (2.94). Then, the procedure is repeated for step $j+1$. This entire procedure is repeated until convergence to the optimal control solution. This method provides an online reinforcement learning algorithm for solving the optimal control problem using policy iteration by measuring data along the system trajectories.

Likewise, an online reinforcement learning algorithm can be given based on value iteration. Substituting the value function approximation into the value update (2.88) in value iteration the following algorithm is obtained.

Algorithm 2.7. Optimal adaptive control using value iteration

Initialize. Select some control policy $h_0(x_k)$, not necessarily admissible or stabilizing.
Do for $j = 0$ until convergence.

Value update step. Determine the least-squares solution W_{j+1} to

$$W_{j+1}^T \phi(x_k) = r(x_k, h_j(x_k)) + W_j^T \gamma \phi(x_{k+1}) \tag{2.97}$$

Policy improvement step. Determine an improved policy using (2.94). ∎

To solve (2.97) in real time we can use batch least-squares, RLS or gradient-based methods based on data $(x_k, x_{k+1}, r(x_k, h_j(x_k)))$ measured at each time along the system trajectories. Then the policy is improved using (2.94). Note that the old weight parameters are on the right-hand side of (2.97). Thus, the regression vector is now $\phi(x_k)$, which must be persistently exciting for convergence of RLS.

2.5.4 Introduction of a second 'Actor' neural network

Using value function approximation allows standard system identification techniques to be used to find the value function parameters that approximately solve the Bellman equation. The approximator structure just described that is used for approximation of the value function is known as the critic neural network, as it determines the value of using the current policy. Using VFA, the policy iteration reinforcement learning algorithm solves a Bellman equation during the value update portion of each iteration step j by observing only the data set $(x_k, x_{k+1}, r(x_k, h_j(x_k)))$ at each time along the system trajectory and solving (2.93). In the case of value iteration, VFA is used to perform a value update using (2.97).

However, note that in the LQR case the policy update (2.94) is given by

$$K^{j+1} = -(B^T P^{j+1} B + R)^{-1} B^T P^{j+1} A \tag{2.98}$$

which requires full knowledge of the dynamics (A, B). Note further that the embodiment (2.94) cannot easily be implemented in the non-linear case because it is implicit in the control, since x_{k+1} depends on $h(\cdot)$ and is the argument of a nonlinear activation function.

These problems are both solved by introducing a *second neural network* for the control policy, known as the *actor neural network*. Actor–critic structures using two neural networks, one for approximation in the critic and one for approximating the control policy in the actor, were developed in approximate dynamic programming (ADP) by Werbos (1989, 1991, 1992, 2009).

Therefore, consider a parametric approximator structure for the control action

$$u_k = h(x_k) = U^T \sigma(x_k) \tag{2.99}$$

with $\sigma(x) : R^n \to R^M$ a vector of M basis or activation functions and $U \in R^{M \times m}$ a matrix of weights or unknown parameters.

After convergence of the critic neural-network parameters to W_{j+1} in policy iteration or value iteration, it is required to perform the policy update (2.94). To achieve this aim, we can use a gradient descent method for tuning the actor weights U such as

$$U_{j+1}^{i+1} = U_{j+1}^{i} - \beta\sigma(x_k)(2R(U_{j+1}^{i})^T \sigma(x_k) + \gamma g(x_k)^T \nabla\phi^T(x_{k+1})W_{j+1})^T \tag{2.100}$$

with $\beta > 0$ a tuning parameter. The tuning index i can be incremented with the time index k.

Note that the tuning of the actor neural network requires observations at each time k of the data set (x_k, x_{k+1}), that is the current state and the next state. However, as per the formulation (2.99), the actor neural network yields the control u_k at time k in terms of the state x_k at time k. The next state x_{k+1} is not needed in (2.99). Thus, after (2.100) converges, (2.99) is a legitimate feedback controller. Note also that, in the LQR case, the actor neural network (2.99) embodies the feedback gain computation (2.98). Equation (2.98) contains the state drift dynamics A, but (2.99) does not. Therefore, the A matrix is not needed to compute the feedback control. The reason is that, during the tuning or training phase, the actor neural network learns information about A in its weights, since (x_k, x_{k+1}) are used in its tuning.

Finally, note that only the input function $g(\cdot)$ or, in the LQR case, the B matrix, is needed in (2.100) to tune the actor neural network. Thus, introducing a second actor neural network completely avoids the need for knowledge of the state drift dynamics $f(\cdot)$, or A in the LQR case.

Example 2.4. Discrete-time optimal adaptive control of power system using value iteration

In this simulation it is shown how to use discrete-time value iteration to solve the discrete-time ARE online without knowing the system matrix A. We simulate the online value iteration algorithm (2.97) and (2.100) for load-frequency control of an electric power system. Power systems are complicated non-linear systems. However, during normal operation the system load, which produces the non-linearity, has only small variations. As such, a linear model can be used to represent the system dynamics around an operating point specified by a constant load value. A problem rises from the fact that in an actual plant the parameter values are not precisely known, reflected in an unknown system A matrix, yet an optimal control solution is sought.

The model of the system that is considered here is $\dot{x} = Ax + Bu$, where

$$A = \begin{bmatrix} -1/T_p & K_p/T_p & 0 & 0 \\ 0 & -1/T_T & 1/T_T & 0 \\ -1/RT_G & 0 & -1/T_G & -1/T_G \\ K_E & 0 & 0 & 0 \end{bmatrix}, \quad B = \begin{bmatrix} 0 \\ 0 \\ 1/T_G \\ 0 \end{bmatrix}$$

The system state is $x(t) = \begin{bmatrix} \Delta f(t) & \Delta P_g(t) & \Delta X_g(t) & \Delta E(t) \end{bmatrix}^T$, where $\Delta f(t)$ is incremental frequency deviation in Hz, $\Delta P_g(t)$ is incremental change in generator output (p.u. MW), $\Delta X_g(t)$ is incremental change in governor position in p.u. MW and $\Delta E(t)$ is incremental change in integral control. The system parameters are T_G, the governor time constant, T_T, turbine time constant, T_P, plant model time constant, K_P, plant model gain, R, speed regulation due to governor action and K_E, integral control gain.

The values of the continuous-time system parameters were randomly picked within specified ranges so that

$$A = \begin{bmatrix} -0.0665 & 8 & 0 & 0 \\ 0 & -3.663 & 3.663 & 0 \\ -6.86 & 0 & -13.736 & -13.736 \\ 0.6 & 0 & 0 & 0 \end{bmatrix},$$

$$B = \begin{bmatrix} 0 & 0 & 13.7355 & 0 \end{bmatrix}$$

The discrete-time dynamics is obtained using the zero-order hold method with sampling period of $T = 0.01$ s. The solution to the discrete-time ARE

$$A^T PA - P + Q - A^T PB(B^T PB + R)^{-1} B^T PA = 0$$

with cost function weights $Q = I$, $R = I$ and $\gamma = 1$ is

$$P_{DARE} = \begin{bmatrix} 0.4750 & 0.4766 & 0.0601 & 0.4751 \\ 0.4766 & 0.7831 & 0.1237 & 0.3829 \\ 0.0601 & 0.1237 & 0.0513 & 0.0298 \\ 0.4751 & 0.3829 & 0.0298 & 2.3370 \end{bmatrix}$$

In this simulation, only the time constant T_G of the governor, which appears in the B matrix, is considered to be known, while the values for all the other parameters appearing in the system A matrix are not known. That is, the A matrix is needed only to simulate the system and obtain the data, and is not needed by the control algorithm.

For the discrete-time LQR, the value is quadratic in the states $V(x) = \frac{1}{2} x^T Px$ as in (2.89). Therefore, the basic functions for the critic neural network in (2.92) are selected as the quadratic polynomial vector in the state components. Since there are $n = 4$ states, this vector has $n(n+1)/2 = 10$ components. The control is linear in the states $u = -Kx$; hence, the basic functions for the actor neural network (2.99) are taken as the state components.

The online implementation of value iteration can be done by setting up a batch least-squares problem to solve for the 10 critic neural-network parameters, that is the Riccati solution entries $\overline{p}_{j+1} \equiv W_{j+1}$ in (2.97) for each step j. In this simulation the matrix P^{j+1} is determined after collecting 15 points of data $(x_k, x_{k+1}, r(x_k, u_k))$ for each least-squares problem. Therefore, a least-squares problem for the critic weights is solved each 0.15 s. Then the actor neural-network parameters, that is the feedback gain matrix entries, are updated using (2.100). The simulations were performed over a time interval of 60 s.

The system states trajectories are given in Figure 2.5, which shows that the states are regulated to zero as desired. The convergence of the Riccati matrix parameters is shown in Figure 2.6. The final values of the critic neural-network parameter estimates are

$$P_{critic\ NN} = \begin{bmatrix} 0.4802 & 0.4768 & 0.0603 & 0.4754 \\ 0.4768 & 0.7887 & 0.1239 & 0.3834 \\ 0.0603 & 0.1239 & 0.0567 & 0.0300 \\ 0.4754 & 0.3843 & 0.0300 & 2.3433 \end{bmatrix}$$

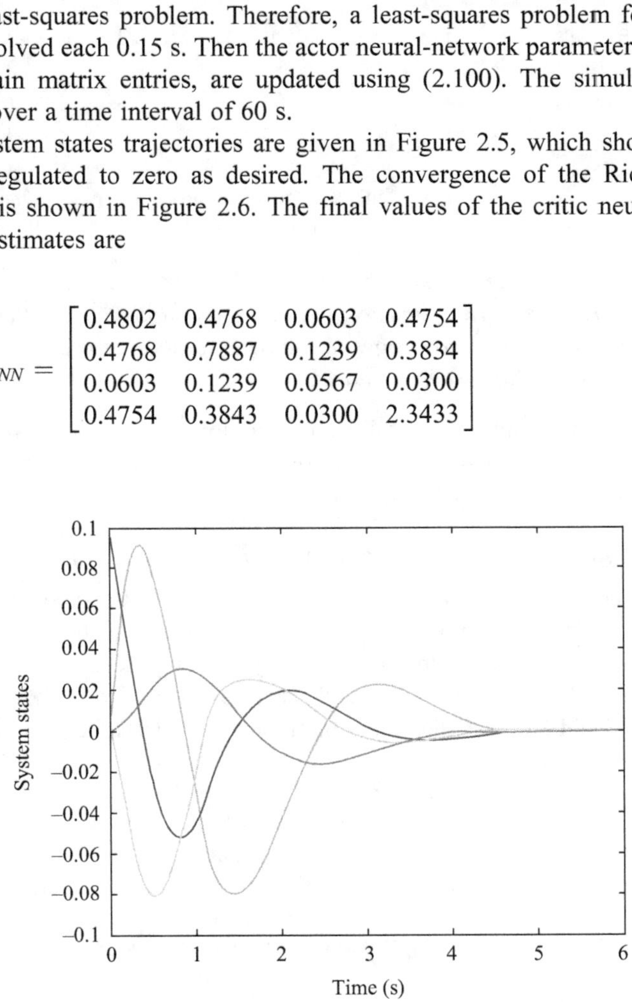

Figure 2.5 System states during the first 6 s. This figure shows that even though the A matrix of the power system is unknown, the adaptive controller based on value iteration keeps the states stable and regulates them to zero

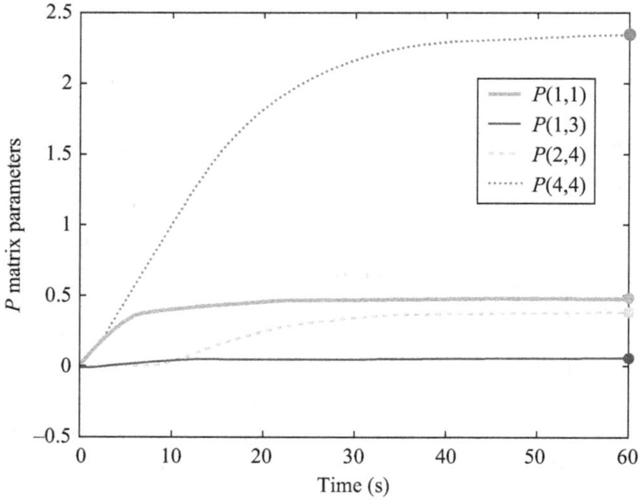

Figure 2.6 Convergence of selected algebraic Riccati equation solution parameters. This figure shows that the adaptive controller based on value iteration converges to the ARE solution in real time without knowing the system A matrix

Thus, the optimal adaptive control value iteration algorithm converges to the optimal control solution as given by the ARE solution. This solution is performed in real time without knowing the system A matrix. ■

2.5.5 Online solution of Lyapunov and Riccati equations

Note that the policy iteration and value iteration adaptive optimal control algorithms just given solve the Bellman equation (2.81) and the HJB equation (2.79) online in real time by using data measured along the system trajectories. The system drift function $f(x_k)$, or the A matrix in the LQR case, is not needed in these algorithms. That is, in the DT LQR case these algorithms solve the Riccati equation

$$A^TPA - P + Q - A^TPB(B^TPB + R)^{-1}B^TPA = 0 \tag{2.101}$$

online in real time without knowledge of the A matrix.

Moreover, at each step of policy iteration, the Lyapunov equation

$$0 = (A - BK^j)^T P^{j+1}(A - BK^j) - P^{j+1} + Q + (K^j)^T RK^j \tag{2.102}$$

is solved without knowing matrices A or B. At each step of value iteration the Lyapunov recursion

$$P^{j+1} = (A - BK^j)^T P^j(A - BK^j) + Q + (K^j)^T RK^j \tag{2.103}$$

is solved without knowing either A or B.

2.5.6 *Actor–critic implementation of discrete-time optimal adaptive control*

Two algorithms for optimal adaptive control of discrete-time systems based on reinforcement learning have just been described. They are implemented by using two approximators to approximate respectively the value and the control action. The implementation of reinforcement learning using two neural networks, one as a critic and one as an actor, yields the actor–critic reinforcement learning structure shown in Figure 2.4. In this two-loop control system, the critic and the actor are tuned online using the observed data $(x_k, x_{k+1}, r(x_k, h_j(x_k)))$ along the system trajectory. The critic and actor are tuned sequentially in both the policy iteration and the value iteration algorithms. That is, the weights of one neural network are held constant while the weights of the other are tuned until convergence. This procedure is repeated until both neural networks have converged. Thus, this learning controller learns the optimal control solution online. This procedure amounts to an *online adaptive optimal control system* wherein the value function parameters are tuned online and the convergence is to the optimal value and control. The convergence of value iteration using two neural networks for the discrete-time non-linear system (2.75) is proven in Al-Tamimi *et al.* (2008).

2.5.7 *Q learning for optimal adaptive control*

It has just been seen that actor–critic implementations of policy iteration and value iteration based on value function approximation yield adaptive control methods that converge in real time to optimal control solutions by measuring data along the system trajectories. The system drift dynamics $f(x_k)$ or A is not needed, but the input-coupling function $g(x_k)$ or B must be known. It is now shown that the Q learning–reinforcement learning method gives an adaptive control algorithm that converges online to the optimal control solution for completely unknown systems. That is, it solves the Bellman equation (2.81) and the HJB equation (2.79) online in real time by using data measured along the system trajectories, without any knowledge of the dynamics $f(x_k)$, $g(x_k)$.

Q learning was developed by Watkins (1989), Watkins and Dayan (1992) for MDP and by Werbos (1991, 1992, 2009) for discrete-time dynamical systems. It is a simple method for reinforcement learning that works for systems with completely unknown dynamics. Q learning is called action-dependent heuristic dynamic programming (ADHDP) by Werbos as the Q function depends on the control input. Q learning learns the Q function (2.56) using temporal difference methods by performing an action u_k and measuring at each time stage the resulting data experience set (x_k, x_{k+1}, r_k) consisting of the current state, the next state, and the resulting stage cost. Writing the Q function Bellman equation (2.58) along a sample path gives

$$Q^{\pi}(x_k, u_k) = r(x_k, u_k) + \gamma Q^{\pi}(x_{k+1}, h(x_{k+1})) \tag{2.104}$$

which defines a temporal difference error

$$e_k = -Q^{\pi}(x_k, u_k) + r(x_k, u_k) + \gamma Q^{\pi}(x_{k+1}, h(x_{k+1})) \tag{2.105}$$

The value iteration algorithm using Q function is given as (2.65). Based on this, the Q function is updated using the algorithm

$$Q_k(x_k, u_k) = Q_{k-1}(x_k, u_k) + \alpha_k\left[r(x_k, u_k) + \gamma \min_u Q_{k-1}(x_{k+1}, u) - Q_{k-1}(x_k, u_k)\right] \quad (2.106)$$

This algorithm is developed for finite MDP and the convergence proven by Watkins (1989) using stochastic approximation methods. It is shown the algorithm converges for finite MDP provided that all state–action pairs are visited infinitely often and

$$\sum_{k=1}^{\infty} \alpha_k = \infty, \qquad \sum_{k=1}^{\infty} \alpha_k^2 < \infty \quad (2.107)$$

that is standard stochastic approximation conditions. On convergence, the temporal difference error is approximately equal to zero.

The requirement that all state-action pairs are visited infinitely often translates to the problem of maintaining sufficient *exploration* during learning.

The Q learning algorithm (2.106) is similar to stochastic approximation methods of adaptive control or parameter estimation used in control systems. Let us now derive methods for Q learning for dynamical systems that yield adaptive control algorithms that converge to optimal control solutions.

Policy iteration and value iteration algorithms are given using the Q function in (2.61)–(2.65). A Q learning algorithm is easily developed for discrete-time dynamical systems using Q function approximation (Werbos, 1989, 1991, 1992; Bradtke *et al.*, 1994). It is shown in Example 2.3 that, for the discrete-time LQR the Q function is a quadratic form in terms of $z_k \equiv [x_k^T \quad u_k^T]^T$. Assume therefore that for non-linear systems the Q function is parameterized as

$$Q(x, u) = W^T\phi(z) \quad (2.108)$$

for some unknown parameter vector W and basis set vector $\phi(z)$. Substituting the Q function approximation into the temporal difference error (2.105) yields

$$e_k = -W^T\phi(z_k) + r(x_k, u_k) + \gamma W^T\phi(z_{k+1}) \quad (2.109)$$

on which either policy iteration or value iteration algorithms can be based. Considering the policy iteration algorithm (2.61), (2.62) yields the Q function evaluation step

$$W_{j+1}^T(\phi(z_k) - \gamma\phi(z_{k+1})) = r(x_k, h_j(x_k)) \quad (2.110)$$

and the policy improvement step

$$h_{j+1}(x_k) = \arg\min_u(W_{j+1}^T\phi(x_k, u)), \quad \text{for all } x \in X \quad (2.111)$$

Q learning using value iteration (2.63) is given by

$$W_{j+1}^T \phi(z_k) = r(x_k, h_j(x_k)) + \gamma W_j^T \phi(z_{k+1}) \tag{2.112}$$

and (2.111). These equations do not require knowledge of the dynamics $f(\cdot)$, $g(\cdot)$. For instance, it is seen in Example 2.3 that for the discrete-time LQR case the control can be updated knowing the Q function without knowing A, B.

For online implementation, batch least-squares or RLS can be used to solve (2.110) for the parameter vector W_{j+1} given the regression vector $(\phi(z_k) - \gamma\phi(z_{k+1}))$, or (2.112) using regression vector $\phi(z_k)$. The observed data at each time instant are $(z_k, z_{k+1}, r(x_k, u_k))$ with $z_k \equiv [x_k^T \quad u_k^T]^T$. Vector $z_{k+1} \equiv [x_{k+1}^T \quad u_{k+1}^T]^T$ is computed using $u_{k+1} = h_j(x_{k+1})$ with $h_j(\cdot)$ the current policy. Probing noise must be added to the control input to obtain persistence of excitation. On convergence, the action update (2.111) is performed. This update is easily accomplished without knowing the system dynamics due to the fact that the Q function contains u_k as an argument, therefore $\partial(W_{j+1}^T \phi(x_k, u))/\partial u$ can be explicitly computed. This is illustrated for the DT LQR in Example 2.3.

Due to the simple form of action update (2.111), the actor neural network is not needed for Q learning; Q learning can be implemented using only one critic neural network for Q function approximation.

Example 2.5. Adaptive controller for online solution of discrete-time LQR using Q learning

This example presents an adaptive control algorithm based on Q learning that converges online to the solution to the discrete-time LQR problem. This is accomplished by solving the algebraic Riccati equation in real time without knowing the system dynamics (A, B) by using data measured along the system trajectories.

Q learning is implemented by repeatedly performing the iterations (2.110) and (2.111). In Example 2.3, it is seen that the LQR Q function is quadratic in the states and inputs so that $Q(x_k, u_k) = Q(z_k) \equiv \frac{1}{2} z_k^T S z_k$, where $z_k = \begin{bmatrix} x_k^T & u_k^T \end{bmatrix}^T$. The kernel matrix S is explicitly given by (2.68) in terms of the system parameters A and B. However, matrix S can be estimated online without knowing A and B by using system identification techniques. Specifically, write the Q function in parametric form as

$$Q(x, u) = Q(z) = W^T(z \otimes z) = W^T \phi(z) \tag{2.113}$$

with W the vector of the elements of S and $\otimes$ the Kronecker product. Function $\phi(z) = (z \otimes z)$ is the quadratic polynomial basis set in terms of elements of z, which contains state and input components. Redundant entries are removed so that W is composed of the $(n+m)(n+m+1)/2$ elements in the upper half of S, with $x_k \in R^n, u_k \in R^m$

Now, for the LQR, the Q learning Bellman equation (2.110) can be written as

$$W_{j+1}^T (\phi(z_k) - \phi(z_{k+1})) = \frac{1}{2}(x_k^T Q x_k + u_k^T R u_k) \tag{2.114}$$

Note that the Q matrix here is the state weighting matrix in the performance index; it should not be confused with the Q function $Q(x_k, u_k)$. This equation must be solved at each step j of the Q learning process. Note that (2.114) is one equation in $(n+m)(n+m+1)/2$ unknowns, namely, the entries of vector W. This is exactly the sort of equation encountered in system identification, and is solved online using methods from adaptive control such as recursive least-squares (RLS).

Therefore, Q learning is implemented as follows.

Algorithm 2.8. Optimal adaptive control using Q learning

Initialize

Select an initial feedback policy $u_k = -K^0 x_k$ at $j = 0$. The initial gain matrix need not be stabilizing and can be selected equal to zero.

Step *j*

Identify the Q function using RLS

At time k, apply control u_k based on the current policy $u_k = -K^j x_k$ and measure the data set $(x_k, u_k, x_{k+1}, u_{k+1})$, where u_{k+1} is computed using $u_{k+1} = -K^j x_{k+1}$. Compute the quadratic basis sets $\phi(z_k), \phi(z_{k+1})$. Now perform a one-step update in the parameter vector W using RLS on equation (2.114). Repeat at the next time $k+1$ and continue until RLS converges and the new parameter vector W_{j+1} is found.

Update the control policy

Unpack vector W_{j+1} into the kernel matrix

$$Q(x_k, u_k) \equiv \frac{1}{2}\begin{bmatrix} x_k \\ u_k \end{bmatrix}^T S \begin{bmatrix} x_k \\ u_k \end{bmatrix} = \frac{1}{2}\begin{bmatrix} x_k \\ u_k \end{bmatrix}^T \begin{bmatrix} S_{xx} & S_{xu} \\ S_{ux} & S_{uu} \end{bmatrix} \begin{bmatrix} x_k \\ u_k \end{bmatrix} \tag{2.115}$$

Perform the control update using (2.111), which is

$$u_k = -S_{uu}^{-1} S_{ux} x_k \tag{2.116}$$

Set $j = j+1$. Go to Step *j*

Termination. This algorithm is terminated when there are no further updates to the Q function or the control policy at each step.

This is an adaptive control algorithm implemented using Q function identification by RLS techniques. No knowledge of the system dynamics A, B is needed for its implementation. The algorithm effectively solves the algebraic Riccati equation online in real time using data $(x_k, u_k, x_{k+1}, u_{k+1})$ measured in real time at each time stage k. It is necessary to add probing noise to the control input to guarantee persistence of excitation to solve (2.114) using RLS. ■

2.6 Reinforcement learning for continuous-time systems

Reinforcement learning is considerably more difficult for continuous-time (CT) systems than for discrete-time systems, and fewer results are available. Therefore, RL adaptive learning techniques for CT systems have not been developed until recently.

To see the problem with formulating policy iterations and value iterations for CT systems. Consider the time-invariant affine-in-the-input dynamical system given by

$$\dot{x}(t) = f(x(t)) + g(x(t))u(x(t)); \quad x(0) = x_0 \tag{2.117}$$

with state $x(t) \in R^n$, drift dynamics $f(x(t)) \in R^n$, control input function $g(x(t)) \in R^{n \times m}$ and control input $u(t) \in R^m$. Given a stabilizing control policy define the infinite-horizon integral cost

$$V^u(x(t)) = \int_t^{\infty} r(x(\tau), u(\tau))\, d\tau \tag{2.118}$$

Using Leibniz's formula, the infinitesimal version of (2.118) is found to be

$$0 = r(x, \mu(x)) + (\nabla V_x^{\mu})^T (f(x) + g(x)\mu(x)), \quad V^{\mu}(0) = 0 \tag{2.119}$$

where ∇V_x^{μ} (a column vector) denotes the gradient of the cost function V^{μ} with respect to x.

In analogy with the development for discrete-time systems in Section 2.5, (2.119) should be considered as a Bellman equation for CT systems. Unfortunately, this CT Bellman equation does not share any of the beneficial properties of the DT Bellman equation (2.81). Specifically, the dynamics $(f(\cdot), g(\cdot))$ do not appear in the DT Bellman equation, whereas they do appear in the CT Bellman equation. This makes it difficult to formulate algorithms such as Q learning, which do not require knowledge of the system dynamics. Moreover, in the DT Bellman equation there are two occurrences of the value function, evaluated at different times k and $k+1$. This allows the formulation of value iteration, or heuristic dynamic programming, for DT systems. However, with only one occurrence of the value in the CT Bellman equation, it is not at all clear how to formulate any sort of value iteration procedure. Several studies have been made about reinforcement learning and ADP for CT systems, including those of Baird (1994), Doya (2000), Hanselmann *et al.* (2007), Murray *et al.* (2002) and Mehta and Meyn (2009). These involve either approximation of derivatives by Euler's method, integration on an infinite horizon or manipulations of partial derivatives of the value function.

In the remainder of this book we shall show how to apply reinforcement learning methods for optimal adaptive control of continuous-time systems. See Abu-Khalaf *et al.* (2006), Vamvoudakis and Lewis (2010a), Vrabie and Lewis (2009) for the development of a policy iteration method for continuous-time systems. Using a method known as *integral reinforcement learning* (IRL) (Vrabie *et al.*, 2008, 2009; Vrabie, 2009; Vrabie and Lewis, 2009) allows the application of reinforcement learning to formulate online optimal adaptive control methods for continuous-time systems. These methods find solutions to optimal HJ design equations and Riccati equations online in real time without knowing the system drift dynamics, that is, in the LQR case without knowing the A matrix.

Part I
Optimal adaptive control using reinforcement learning structures

This book shows how to use reinforcement learning (RL) methods to design adaptive controllers of novel structure that learn optimal control solutions for continuous-time systems. We call these *optimal adaptive controllers*. They stand in contrast to standard adaptive control systems in the control systems literature, which do not normally converge to optimal solutions in terms of solving a Hamilton–Jacobi–Bellman equation.

RL is a powerful technique for online learning in a complex decision-making system that is based on emulating naturally occurring learning systems in nature. RL is based on an agent selecting a control action, observing the consequences of this action, evaluating the resulting performance and using that evaluation to update the action so as to improve its performance. RL has been used for sequential decisions in complicated stochastic systems, and it has been applied with great effect in the online real-time control of discrete-time dynamical systems. Chapter 2 provides a background on RL and its applications in optimal adaptive control design for discrete-time systems. The applications of RL to continuous-time systems have lagged due to the inconvenient form of the continuous-time Bellman equation, which contains all the system dynamics. In this book, we show how to apply RL to continuous-time systems to learn optimal control solutions online in real time using adaptive tuning techniques.

In Part I of the book we lay the foundations for RL applications in continuous-time systems based on a form of the continuous-time Bellman equation known as the integral reinforcement learning (IRL) Bellman equation, as developed in Vrabie *et al.* (2008, 2009), Vrabie (2009), Vrabie and Lewis (2009). It is shown how to use IRL to formulate policy iterations and value iterations for continuous-time systems. The result is a family of online adaptive learning systems that converge to optimal solutions in real time by measuring data along the system trajectories.

The optimal adaptive controllers in Part I are based on the standard RL actor–critic structure, with a critic network to evaluate the performance base on a selected control policy, and a second actor network to update the policy so as to improve the performance. In these controllers, the two networks learn sequentially. That is, while the critic is performing the value update for the current policy, that policy is not changed. Policy updates are performed only after the critic converges to the value update solution. The proofs of performance in Part I are based on the methods in general use from the perspective of RL.

The controllers developed in Part I learn in the usual RL manner of updating only one of the two learning networks at a time. This strategy seems a bit odd for the adaptive feedback control systems' practitioner. These two-loop sequential reinforcement learning structures are not like standard adaptive control systems currently used in feedback control. They are hybrid controllers with a continuous inner action control loop and an outer learning critic loop that operates on a discrete-time scale.

In Part II of the book, we adopt a philosophy more akin to that of adaptive controllers in the feedback control literature (Ioannou and Fidan, 2006; Astrom and Wittenmark, 1995; Tao, 2003). To learn optimal control solutions, we develop novel structures of adaptive controllers based on RL precepts. The resulting controllers have multi-loop learning networks, yet they are continuous-time controllers that are more in keeping with adaptive methods such as direct and indirect adaptive control. The controllers of Part II operate as normally expected in adaptive control in that the control loops are not tuned sequentially as in Part I, but the parameter tuning in all control loops is performed simultaneously through time. We call this *synchronous tuning* of the critic and actor networks. In contrast to Part I that uses proof techniques standard in RL, in Part II the convergence proofs are carried out using methods standard in adaptive control, namely, Lyapunov energy-based techniques.

In Part III of the book we develop adaptive controllers that learn optimal solutions in real time for several differential game problems, including zero-sum and multiplayer non–zero-sum games. The design procedure is to first formulate RL policy iteration algorithms for these problems, then use the structure of policy iteration to motivate novel multi-loop adaptive controller structures. Then, tuning laws for these novel adaptive controllers are determined by adaptive control Lyapunov techniques or, in Chapter 11, by RL techniques.

Chapter 3

Optimal adaptive control using integral reinforcement learning for linear systems

This chapter presents a new algorithm based on policy iterations that provide an online solution procedure for the optimal control problem for continuous-time (CT), linear, time-invariant systems having the state-space model $\dot{x}(t) = Ax(t) + Bu(t)$. This is an adaptive learning algorithm based on reinforcement learning (RL) that converges to the optimal control solution to the linear quadratic regulator problem. We term this an *optimal adaptive controller*. The algorithm is partially model-free in the sense that it does not require full knowledge of the system dynamics. Specifically, the drift dynamics or system matrix A is not required, but the input-coupling matrix B must be known. It is well known that solving the optimal control problem for these systems is equivalent to finding the unique positive definite solution of the underlying algebraic Riccati equation (ARE). The algorithm in this chapter provides an online optimal adaptive learning algorithm that solves the ARE online in real time without knowing the A matrix by measuring state and input data $(x(t), u(t))$ along the system trajectories. The algorithm is based on policy iterations and as such has an actor–critic structure consisting of two interacting adaptive learning structures.

Considerable effort has been devoted to solving ARE, including the following approaches:

- Backward integration of the differential Riccati equation or Chandrasekhar equations (Kailath, 1973),
- Eigenvector-based algorithms (MacFarlane, 1963; Potter, 1966) and the numerically advantageous Schur vector-based modification (Laub, 1979),
- Matrix sign-based algorithms (Balzer, 1980; Byers, 1987; Hasan *et al.*, 1999), and
- Newton's method (Kleinman, 1968; Gajic and Li, 1988; Moris and Navasca, 2006; Banks and Ito, 1991).

All of these methods, and their more numerically efficient variants, are offline procedures that have been proved to converge to the solution of the ARE. However, all of these techniques require exact knowledge of the state-space description (A, B) of the system to be controlled, since they either operate on the Hamiltonian matrix associated with the ARE (eigenvector and matrix sign-based algorithms) or require solving Lyapunov equations (Newton's method). In either case a model of the system is required. For unknown systems, this means that a preliminary system identification procedure is necessary. Furthermore, even if a model is available, the state-feedback controller obtained based on it will only be optimal for the model approximating the actual system dynamics.

Reinforcement learning for discrete-time systems. RL policy iteration and value iteration methods have been used for many years to provide methods for solving the optimal control problem for discrete-time (DT) systems. These methods are outlined in Chapter 2. Methods were developed by Watkins for Q learning for finite-state, discrete-time systems (Watkins and Dayan, 1992). Bertsekas and Tsitsiklis developed RL methods based on policy iteration and value iteration for infinite-state discrete-time dynamical systems in Bertsekas and Tsitsiklis (1996). This approach, known as neurodynamic programming, used value function approximation to approximately solve the Bellman equation using iterative techniques. Offline solution methods were developed in Bertsekas and Tsitsiklis (1996). Werbos (1989, 1991, 1992, 2009) presented RL techniques based on value iteration for feedback control of discrete-time dynamical systems using value function approximation. These methods, known as approximate dynamic programming (ADP) or adaptive dynamic programming, are suitable for online learning of optimal control techniques for DT systems online in real time. As such, they are true adaptive learning techniques that converge to optimal control solutions by observing data measured along the system trajectories in real time. A family of four methods was presented under the aegis of ADP, which allowed learning of the value function and its gradient (e.g. the costate), and the Q function and its gradient. The ADP controllers are actor–critic structures with one learning network for the control action and one learning network for the critic. The ADP method for learning the value function is known as heuristic dynamic programming (HDP). Werbos called his method of online learning of the Q function for infinite-state DT dynamical systems 'action-dependent HDP'.

Reinforcement learning for continuous-time systems. Applications of RL in feedback control for continuous-time dynamical systems $\dot{x} = f(x) + g(x)u$ have lagged. This is due to the fact that the Bellman equation for DT systems $V(x_k) = r(x_k, u_k) + \gamma V(x_{k+1})$ does not depend on the system dynamics, whereas the Bellman equation $0 = r(x,u) + (\nabla V)^T(f(x) + g(x)u)$ for CT systems does depend on the system dynamics $f(x)$, $g(x)$. See the discussion in Section 2.6. In this chapter, a new method known as *integral reinforcement learning* (IRL) presented in Vrabie *et al.* (2008, 2009), Vrabie (2009), Vrabie and Lewis (2009) is used to circumvent this problem and formulate meaningful policy iteration algorithms for CT systems.

The integral reinforcement learning policy iteration technique proposed in this chapter solves the linear quadratic regulator problem for continuous-time systems online in real time, using only partial knowledge about the system dynamics (i.e. the drift dynamics A of the system need not be known), and without requiring measurements of the state derivative. This is in effect a direct (i.e. no system identification procedure is employed) adaptive control scheme for partially unknown linear systems that converges to the optimal control solution. It will be shown that the optimal adaptive control scheme based on IRL is a dynamic controller with an actor–critic structure that has a memory whose state is given by the cost or value function.

The IRL method for continuous-time policy iteration for linear time-invariant systems (Vrabie *et al.*, 2008, 2009; Vrabie, 2009; Vrabie and Lewis, 2009) is given in Section 3.1. Equivalence with iterations on underlying Lyapunov equations is proved. It is shown that IRL policy iteration is actually a Newton method for solving the Riccati equation, so that convergence to the optimal control is established. In Section 3.2, an online optimal adaptive control algorithm is developed that implements IRL in real time, without knowing the plant matrix A, to find the optimal controller. Extensive discussions are given about the dynamical nature and structure of the IRL optimal adaptive control algorithm. To demonstrate the capabilities of the proposed IRL policy iteration scheme, in Section 3.3 are presented simulation results of applying the algorithm to find the optimal load-frequency controller for a power plant (Wang *et al.*, 1993). It is shown that IRL optimal adaptive control solves the algebraic Riccati equation online in real time, without knowing the plant matrix A, by measuring data $(x(t), u(t))$ along the system trajectories.

3.1 Continuous-time adaptive critic solution for the linear quadratic regulator

This section develops the Integral Reinforcement Learning (IRL) approach to solving online the linear quadratic regulator (LQR) problem (Lewis *et al.*, 2012) without using knowledge of the system matrix A. IRL solves the algebraic Riccati equation online in real time, without knowing the system A matrix, by measuring data $(x(t), u(t))$ along the system trajectories.

Consider the linear time-invariant dynamical system described by

$$\dot{x}(t) = Ax(t) + Bu(t) \tag{3.1}$$

with state $x(t) \in \mathbb{R}^n$, control input $u(t) \in \mathbb{R}^m$ and (A, B) stabilizable. To this system associate the infinite-horizon quadratic cost function

$$V(x(t_0), t_0) = \int_{t_0}^{\infty} (x^T(\tau)Qx(\tau) + u^T(\tau)Ru(\tau))\, d\tau \tag{3.2}$$

with $Q \geq 0, R > 0$ such that $(Q^{1/2}, A)$ is detectable. The LQR optimal control problem requires finding the control policy that minimizes the cost

$$u^*(t) = \arg \min_{\substack{u(t) \\ t_0 \leq t \leq \infty}} V(t_0, x(t_0), u(t)) \tag{3.3}$$

The solution of this optimal control problem, determined by Bellman's optimality principle, is given by the state feedback $u(t) = -Kx(t)$ given by

$$K = R^{-1}B^T P \tag{3.4}$$

where the matrix P is the unique positive definite solution of the algebraic Riccati equation (ARE)

$$A^T P + PA - PBR^{-1}B^T P + Q = 0 \tag{3.5}$$

Under the detectability condition for $(Q^{1/2}, A)$ the unique positive semidefinite solution of the ARE determines a stabilizing closed-loop controller given by (3.4).

It is known that the solution of the infinite-horizon optimization problem can be obtained using the dynamic programming method. This amounts to solving backward in time a finite-horizon optimization problem while extending the horizon to infinity. The following Riccati differential equation must be solved

$$\begin{aligned} -\dot{P} &= A^T P + PA - PBR^{-1}B^T P + Q \\ P(t_f) &= P_{tf} \end{aligned} \tag{3.6}$$

Its solution will converge to the solution of the ARE as $t_f \to \infty$.

It is important to note that, in order to solve (3.5), complete knowledge of the model of the system is needed, that is both the system matrix A and the control input matrix B must be known. Thus, a system identification procedure is required prior to solving the optimal control problem, a procedure that most often ends with finding only an approximate model of the system. For this reason, developing algorithms that converge to the solution of the optimization problem without performing prior system identification or using explicit models of the system dynamics is of particular interest from the control systems point of view.

3.1.1 Policy iteration algorithm using integral reinforcement

This section presents a new policy iteration algorithm that solves online for the optimal control gain (3.4) without using knowledge of the system matrix A (Vrabie *et al.*, 2008, 2009; Vrabie, 2009; Vrabie and Lewis, 2009). The result is an adaptive controller that converges to the state-feedback optimal controller. The algorithm is based on an actor–critic structure and consists of a two-step iteration, namely, the critic update and the actor update. The update of the critic structure results in calculating the infinite-horizon cost associated with a given stabilizing controller. The actor parameters (i.e. the controller feedback gain matrix K) are then updated in the sense of reducing the cost compared to the present control policy. The derivation of the algorithm is given in Section 3.1.1. An analysis is done and proof of convergence is provided in Section 3.1.2.

Let K be a stabilizing state-feedback gain for (3.1) such that $\dot{x} = (A - BK)x$ is a stable closed-loop system. Then the corresponding infinite-horizon quadratic cost or *value* is given by

$$V(x(t)) = \int_t^{\infty} x^T(\tau)(Q + K^T RK)x(\tau)\, d\tau = x^T(t)Px(t) \tag{3.7}$$

where P is the real symmetric positive definite solution of the Lyapunov matrix equation

$$(A - BK)^T P + P(A - BK) = -(K^T RK + Q) \tag{3.8}$$

Then, $V(x(t))$ serves as a Lyapunov function for (3.1) with controller gain K. The value function (3.7) can be written in the following form.

Integral reinforcement form of value function: IRL Bellman equation

$$V(x(t)) = \int_t^{t+T} x^T(\tau)(Q + K^T RK)x(\tau)\, d\tau + V(x(t+T)) \tag{3.9}$$

This is a Bellman equation for the LQR problem of the same form as the DT Bellman equation in Section 2.5. Using the IRL Bellman equation, one can circumvent the problems noted in Section 2.6 of applying RL to continuous-time systems.

Denote $x(t)$ by x_t and write the value function as $V(x_t) = x_t^T P x_t$. Then, based on the IRL Bellman equation (3.9) one can write the following RL algorithm.

Algorithm 3.1. Integral reinforcement learning policy iteration algorithm

$$x_t^T P_i x_t = \int_t^{t+T} x_\tau^T (Q + K_i^T RK_i) x_\tau\, d\tau + x_{t+T}^T P_i x_{t+T} \tag{3.10}$$

$$K_i{+}1 = R^{-1} B^T P_i \tag{3.11}$$

Equations (3.10) and (3.11) formulate a new policy iteration algorithm for continuous-time systems. An initial stabilizing control gain K_1 is required. Note that implementing this algorithm *does not involve the plant matrix A.*

Writing the cost function as in (3.9) is the key to the optimal adaptive control method developed in this chapter. This equation has the same form as the Bellman equation for discrete-time systems discussed in Section 2.5. In fact, it is a Bellman equation for CT systems that can be used instead of the Bellman equation given in Section 2.6 in terms of the Hamiltonian function. We call

$$\rho(x(t), t, T) \equiv \int_t^{t+T} x^T(\tau)(Q + K^T RK)x(\tau)d\tau \tag{3.12}$$

the *integral reinforcement*, and (3.9) the *integral reinforcement* form of the value function. Then, (3.10), (3.11) is the IRL form of policy iterations for CT systems. ■

3.1.2 *Proof of convergence*

The next results establish the convergence of the IRL algorithm (3,10), (3.11).

Lemma 3.1. Assuming that the system $x = A_i x$, with $A_i = A - BK_i$, is stable, solving for P_i in (3.10) is equivalent to finding the solution of the underlying Lyapunov equation

$$A_i^T P_i + P_i A_i = -(K_i^T Rk_i + Q) \tag{3.13}$$

Proof: Since A_i is a stable matrix and $K_i^T RK_i + Q > 0$ then there exists a unique solution of the Lyapunov equation (3.13), $P_i > 0$. Also, since $V_i(x_t) = x_t^T P_i x_t$, $\forall x_t$ is a Lyapunov function for the system $x = A_i x$ and

$$\frac{d(x_t^T P_i x_t)}{dt} = x_t^T (A_i^T P_i + P_i A_i) x_t = x_t^T (K_t^T RK_i + Q) x_t \tag{3.14}$$

then, $\forall T > 0$ the unique solution of the Lyapunov equation satisfies

$$\int_t^{t+T} x_\tau^T (Q + K_i^T RK_i) x_\tau d\tau = -\int_t^{t+T} \frac{d(x_\tau^T P_i x_\tau)}{d\tau} d\tau = x_t^T P_i x_i - x_{t+T}^T P_i x_t + T$$

that is (3.10). That is, provided that the system $\dot{x} = A_i x$ is asymptotically stable, the solution of (3.10) is the unique solution of (3.13). ■

Remark 3.1. Although the same solution is obtained whether solving (3.13) or (3.10), (3.10) can be solved without using any knowledge of the system matrix A.

From Lemma 3.1 it follows that the iterative algorithm on (3.10), (3.11) is equivalent to iterating between (3.13), (3.11), without using knowledge of the system drift dynamics, if $\dot{x} = A_i x$ is stable at each iteration. The algorithm (3.13), (3.11) is the same as Kleinman's algorithm, whose convergence was proven in Kleinman (1968). ■

Lemma 3.2. Assume that the control policy K_i is stabilizing at iteration i with $V_i(x_t) = x_t{}^T P_i x_t$ the associated value. Then, if (3.11) is used to update the control policy, the new control policy K_{i+1} is stabilizing.

Proof: Take the positive definite cost function $V_i(x_t)$ as a Lyapunov function candidate for the state trajectories generated while using the controller K_{i+1}. Taking the derivative of $V_i(x_t)$ along the trajectories generated by K_{i+1} one obtains

$$\begin{aligned} V_i(x_t) &= x_t^T [P_i(A - BK_{i+1}) + (A - BK_{i+1})^T P_i] x_t \\ &= x_t^T [P_i(A - BK_i) + (A - BK_i)^T P_i] x_t \\ &\quad + x_t^T [P_i B(K_i - k_{i+1}) + (K_i - k_{i+1})^T B^T P_i] x_t \end{aligned} \tag{3.15}$$

The second term, using the update given by (3.11) and completing the squares, can be written as

$$x_t^T \left[K_{i+1}{}^T R(k_i - K_{i+1}) + (K_i - K_{i+1})^T R K_{i+1} \right] x_t$$
$$= x_t^T \left[-(K_i - K_{i+1})^T R(K_i - K_{i+1}) - K_{i+1}{}^T R K_{i+1} + K_t^T R K_i \right] x_t$$

Using (3.13) the first term in (3.15) can be written as $-x_t^T[K_i^T R K_i + Q]x_t$ and summing up the two terms one obtains

$$\dot{V}_i(x_t) = -x_t^T[(K_i - K_{i+1})^T R(K_i - K_{i+1})]x_t - x_t^T[Q + K_{i+1}^T R K_{i-1}]x_t \tag{3.16}$$

Thus, under the initial assumptions from the problem setup $Q \geq 0$, $R > 0$, $V_i(x_t)$ is a Lyapunov function proving that the updated control policy $u = -K_{i+1}x$, with K_{i+1} given by (3.11), is stabilizing. ■

Remark 3.2. Based on Lemma 3.2, one can conclude that if the initial control policy given by K_1 is stabilizing, then all policies obtained using the iteration (3.10)–(3.11) are stabilizing for each iteration i.

Denote by $Ric(P_i)$ the matrix-valued function defined as

$$Ric(P_i) = A^T P_i + P_i A + Q - P_i B R^{-1} B^T P_i \tag{3.17}$$

and let Ric'_{P_i} denote the Fréchet derivative of $Ric(P_i)$ taken with respect to P_i. The matrix function Ric'_{P_i} evaluated at a given matrix M is given by

$$Ric'_{p_i}(M) = (A - BR^{-}1B^T P_i)^T M + M(A - BR^{-1}B^T P_i)$$

■

Lemma 3.3. The iteration (3.10), (3.11) is equivalent to the Newton's method

$$P_i = P_i - 1 - (Ric'_{P_{i-1}})^{-1} Ric(P_{i-1}) \tag{3.18}$$

Proof: Equations (3.13) and (3.11) can be compactly written as

$$A_i^T P_i + P_i A_i = -(P_{i-1} B R^{-1} B^T P_{i-1} + Q) \tag{3.19}$$

Subtracting $A_t^T P_{i-1} + P_{i-1} A_i$ on both sides gives

$$\begin{aligned} A^T(P_i - P_{i-1}) + (P_i - P_{i-1})A_i = &- (P_{i-1}A + A^T P_{i-1} \\ &- P_{i-1} B R^{-1} B^T P_{i-1} + Q) \end{aligned} \tag{3.20}$$

which, making use of the introduced notations $Ric(P_i)$ and Ric'_{P_i}, is the Newton method formulation (3.18). ■

Theorem 3.1. (Convergence) Assume stabilizability of (A, B) and detectability of $(Q^{1/2}, A)$. Let the initial controller K_1 be stabilizing. Then the policy iteration (3.10), (3.11) converges to the optimal control solution given by (3.4) where the matrix P satisfies the ARE (3.5).

Proof: In Kleinman (1968) it has been shown that Newton's method, that is the iteration (3.13) and (3.11), conditioned by an initial stabilizing policy will converge to the solution of the ARE. Also, if the initial policy is stabilizing, all the subsequent control policies will be stabilizing (as by Lemma 3.2). Based on the proven equivalence between (3.13) and (3.11), and (3.10) and (3.11), we can conclude that the proposed new online policy iteration algorithm will converge to the solution of the optimal control problem (3.2) with the infinite-horizon quadratic cost (3.3) – without using knowledge of the drift dynamics of the controlled system (3.1). ■

Note that the only requirement for convergence of IRL (3.10), (3.11) to the optimal controller is that the initial policy be stabilizing. This guarantees a finite value for the cost $V_1(x_t) = x_t^T P_1 x_t$. Under the assumption that the system is stabilizable, it is reasonable to assume that a stabilizing (though not optimal) state-feedback controller is available to begin the IRL iteration (Kleinman, 1968; Moris and Navasca, 2006). In fact in many cases the system to be controlled is itself stable; then, the initial control gain can be chosen as zero.

3.2 Online implementation of IRL adaptive optimal control

In this section we present an online adaptive learning algorithm to implement the IRL policy iteration scheme (3.10), (3.11) in real time. This algorithm performs the IRL iterations in real time by measuring at times t the state $x(t)$ and the next state $x(t+T)$, and measuring or computing the control $u(t)$. For this procedure, one only requires knowledge of the B matrix because it explicitly appears in the policy update (3.11). The system A matrix does not appear in IRL and so need not be known. This is because information regarding the system A matrix is embedded in the measured states $x(t)$ and $x(t+T)$, which are observed online.

3.2.1 Adaptive online implementation of IRL algorithm

The parameters of the value function $V_i(x(t)) = x^T(t)P_i x(t)$ at iteration i of IRL are the elements of the symmetric kernel matrix P_i. These must be found at each iteration i by measuring the data $(x(t), x(t+T), u(t))$ at times t along the system trajectories. To compute these parameters, the term $x^T(t)P_i x(t)$ is written as

$$x^T(t)P_i x(t) = \bar{p}^T \bar{x}(t) \tag{3.21}$$

where $\bar{x}(t)$ denotes the Kronecker product quadratic polynomial basis vector having elements $\{x_i(t)x_j(t)\}_{i=1,n;j=i,n}$. The parameter vector $\bar{p}_i$ contains the elements of the matrix P_i ordered by columns and with the redundant elements removed. Removing the elements of P_i below the diagonal, for instance, $\bar{p}_i$ is obtained by stacking the elements of the diagonal and upper triangular part of the symmetric matrix P_i into a vector where the off-diagonal elements are taken as $2P_{ij}$ (see Brewer (1978)). Using (3.21), (3.10) is rewritten as

$$\bar{p}_i^T(\bar{x}(t) - \bar{x}(t+T)) = \int_t^{t+T} x^T(\tau)(Q + K_i^T R K_i)x(\tau)\, d\tau \tag{3.22}$$

Here $\bar{p}_i$ is the vector of unknown parameters and $\bar{x}(t) - \bar{x}(t+T)$ acts as a regression vector. The right-hand side

$$d(\bar{x}(t), K_i) = \int_t^{t+T} x^T(\tau)(Q + K_i{}^T R K_i)x(\tau)\, d\tau$$

is a desired value or target function to which $\bar{p}_i{}^T(\bar{x}(t) - \bar{x}(t+T))$ is equal when the parameter vector $\bar{p}_i$ contains the correct parameters.

Note that $d(\bar{x}(t), K_i)$ is the integral reinforcement (3.12) on the time interval $[t, t+T]$. To compute it efficiently, define a controller state $V(t)$ and add the state equation

$$\dot{V}(t) = x^T(t)Qx(t) + u^T(t)Ru(t) \tag{3.23}$$

to the controller dynamics. By measuring $V(t)$ along the system trajectory, the value of $d(\bar{x}(t), K_i)$ can be computed by using $d(\bar{x}(t), K_i) = V(t+T) - V(t)$. This new state signal $V(t)$ is simply the output of an analog integration block having as input the quadratic terms $x^T(t)Qx(t)$ and $u^T(t)Ru(t)$ that can also be obtained using an analog processing unit.

Equation (3.22) is a scalar equation involving an unknown parameter vector. As such, it is a standard form encountered in adaptive control and can be solved using methods such a recursive least-squares (RLS) or gradient descent. Then, a persistence of excitation condition is required.

A batch solution method for (3.22) can also be used. At each iteration step i, after a sufficient number of state-trajectory points are collected using the same control policy K_i, a least squares method can be employed to solve for the parameters $\bar{p}_i$ of the function $V_t(x_t)$ (i.e. the critic), which are the elements of matrix P_i. The parameter vector $\bar{p}_i$ is found by minimizing, in the least-squares sense, the error between the target function, $d(\bar{x}(t), K_i)$, and the parameterized left-hand side of (3.22). Matrix P_i has $n(n+1)/2$ independent elements.

Therefore, the right-hand side of (3.22) must be computed at $N \geq n(n+1)/2$ points $\bar{x}^i$ in the state space, over time intervals T. Then, the batch least-squares solution is obtained as

$$\bar{p}_i = (XX^T)^{-1}XY \tag{3.24}$$

where

$$X = [\bar{x}_\Delta^1 \quad \bar{x}_\Delta^2 \quad \ldots \quad \bar{x}_\Delta^N]$$

$$\bar{x}_\Delta^i = \bar{x}^i(t) - \bar{x}^i(t+T)$$

$$Y = [d(\bar{x}^i, K_i) \quad d(\bar{x}^2, K_i) \quad \ldots \quad d(\bar{x}^N, K_i)]^T$$

The least-squares problem can also be solved in real time after a sufficient number of data points are collected along a single state trajectory, under the presence of an excitation requirement.

A flowchart of this online adaptive IRL algorithm is presented in Figure 3.1.

Implementation issues. Concerning the convergence speed of this algorithm, it has been proven in Kleinman (1968) that Newton's method has quadratic convergence. According to the equivalence proven in Theorem 3.1, the online adaptive IRL algorithm converges quadratically in the iteration step *i*. To capitalize on this, the value function (3.10) associated with the current control policy should be computed using a method such as the batch least squares described by (3.24). For the case in which the solution of (3.10) is obtained iteratively online using a method such as RLS or gradient descent, the convergence speed of the online algorithm proposed in this chapter will decrease. Such algorithms generally have exponential convergence. From this perspective one can resolve that the convergence speed of the online algorithm will depend on the technique selected for solving (3.10). Analyses along these lines are presented in detail in the adaptive control literature (see, e.g. Ioannou and Fidan, 2006).

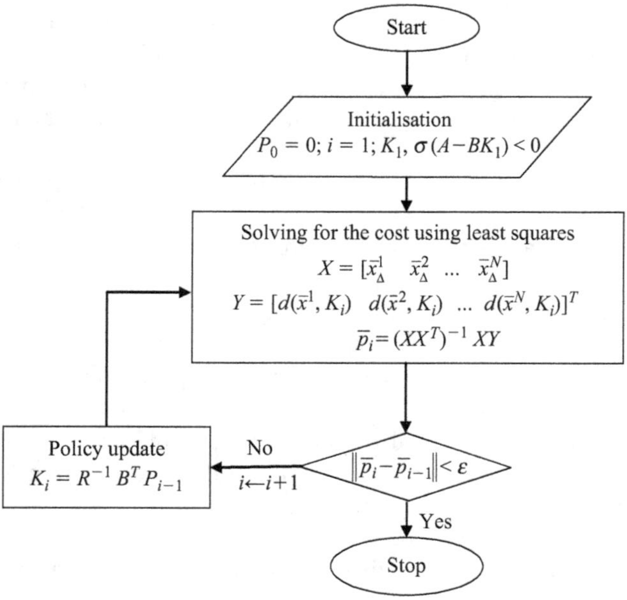

Figure 3.1 Flowchart for online IRL policy iteration algorithm for continuous-time linear systems

In relation with the choice of the value of the sample time T used for acquiring the data necessary in the iterations, it must be specified that this parameter does not affect in any way the convergence property of the online algorithm. It is, however, related to the excitation condition (see Section 3.2.2) necessary in the setup of a numerically well-posed least-squares problem to obtain the batch least-squares solution (3.24).

The RL integration period T is not a sample period in the normal sense used in sampled data systems. In fact, T can change at each measurement time. The data acquired to set up the batch least squares could be obtained by using different values of the sample time T for each element in the vectors X and Y, as long as the information relating the target elements in the Y vector is consistent with the state samples used for obtaining the corresponding elements in the X vector.

The adaptive IRL policy iteration procedure requires only data measurements $(x(t), x(t+T))$ of the states at discrete moments in time, t and $t+T$, as well as knowledge of the observed value over the time interval $[t, t-T]$, which is $d(\overline{x}(t), K_i)$ and can be computed using the added controller state equation (3.23). Therefore, no knowledge about the system A matrix is needed in this algorithm. However, the B matrix is required for the update of the control policy using (3.11), and this makes the tuning algorithm partially model free.

3.2.2 Structure of the adaptive IRL algorithm

The structure of the IRL adaptive controller is presented in Figure 3.2. It is an actor–critic structure with the critic implemented by solving the IRL Bellman equation (3.10) and the actor implemented using the policy update (3.11) (see Vrabie *et al.*, 2008, 2009; Vrabie, 2009; Vrabie and Lewis, 2009). It is important to note that the system is augmented by an extra state $V(t)$, namely, the value defined as $\dot{V} = x^T Qx + u^T Ru$, in order to extract the information regarding the cost associated with the given policy. This newly introduced value state is part of the IRL controller, thus the control scheme is actually a dynamic controller with the state given by the cost function V. One can observe that the IRL adaptive optimal controller has a hybrid structure with a continuous-time internal state $V(t)$ followed by a sampler and discrete-time control policy update rule. The RL integration period

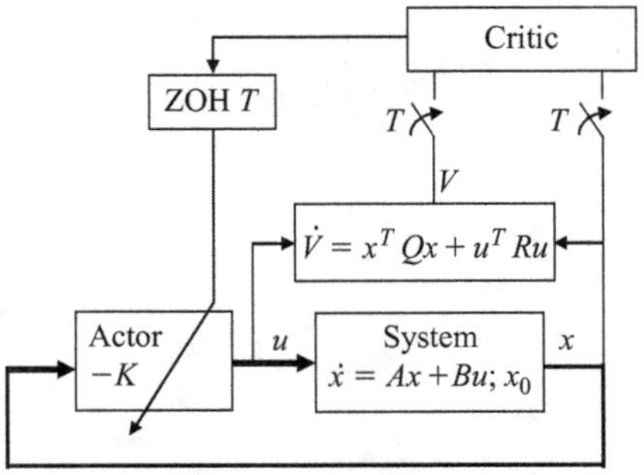

Figure 3.2 Hybrid actor–critic structure of the IRL optimal adaptive controller

T is not a sample period in the normal sense used in sampled data systems. In fact, T can change at each measurement time.

The IRL controller only requires data measurements ($x(t)$, $x(t+T)$) and information about the value $V(t)$ at discrete-time values. That is, the algorithm uses only the data samples $x(t)$, $x(t+T)$ and $V(t+T) - V(t)$ over several time samples. Nevertheless, the critic is able to evaluate the performance of the system associated with the given control policy. The control policy is improved after the solution given by (3.24) is obtained. In this way, by using state measurements over a single state trajectory the algorithm converges to the optimal control policy.

It is observed that the updates of both the actor and the critic are performed at discrete moments in time. However, the control action is a full-fledged continuous-time control, only that its constant gain is updated at certain points in time. Moreover, the critic update is based on the observations of the continuous-time cost over a finite sample interval. As a result, the algorithm converges to the solution of the continuous-time optimal control problem, as proven in Section 3.1.

The hybrid nature of the IRL optimal adaptive controller is illustrated in Figure 3.3. It is shown there that the feedback gain or policy is updated at discrete-times using (3.11) after the solution to (3.10) has been determined. On the other hand, the control input is a continuous-time signal depending on the state $x(t)$ at each time t.

Persistence of excitation. It is necessary that sufficient excitation exists to guarantee that the matrix XX^T in (3.24) is non-singular. Specifically, the difference signal $\phi(t) = \bar{x}(t) - \bar{x}(t+T)$ in (3.22) must be persistently exciting (PE). Then the matrix XX^T in (3.24) is invertible. The PE condition is that there exist constants $\beta_1, \beta_2 > 0$ such that

$$\beta_1 I \leq \int_t^{t+T} \phi(\tau)\phi^T(\tau)d\tau \leq \beta_2 I$$

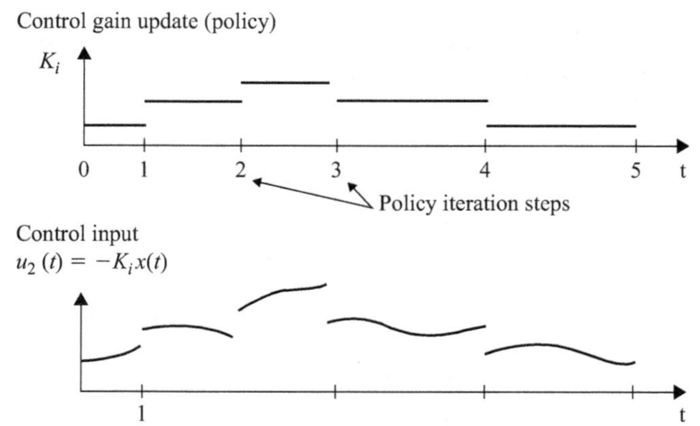

Figure 3.3 Hybrid nature of control signal. The control gains are updated at discrete times, but the control signal is piecewise continuous

This condition depends on the selection of the period of integration T. Generally, for larger T the constant β_1 can be selected larger.

This persistence of excitation condition is typical of adaptive controllers that require system identification procedures (Ioannou and Fidan, 2006). The algorithm iterates only on stabilizing policies, so that the state goes exponentially to zero. In the case that excitation is lost prior to obtaining the convergence of the algorithm, a new experiment needs to be conducted with a new non-zero initial state $x(0)$. The policy for this new run can be selected as the last policy from the previous experiment.

In contrast to indirect adaptive control methods that require identification of the full system dynamics (3.1), the adaptive IRL algorithm identifies the value function (3.2). As such, the PE condition on $\phi(t) = \bar{x}(t) - \bar{x}(t+T)$ is milder than the PE condition required for system identification.

Adaptive IRL for time-varying systems. The A matrix is not needed to implement the IRL algorithm. In fact, the algorithm works for time-varying systems. If the A matrix changes suddenly, as long as the current controller is stabilizing for the new A matrix, the algorithm will converge to the solution to the corresponding new ARE.

Online operation of the adaptive IRL algorithm. The next two figures provide a visual overview of the IRL online policy iteration algorithm. Figure 3.4 shows that over the time intervals $[T_i, T_{i+1}]$ the system is controlled using a state-feedback control policy that has a constant gain K_i. During this time interval a reinforcement learning procedure, which uses data measured from the system, is employed to determine the value associated with this controller. The value is described by the parametric structure P_i. Once the learning procedure results in convergence to the value P_i, this result is used for calculating a new gain for the state-feedback controller, namely, K_t. The length of the interval $[T_i, T_{i+1}]$ is given by the end of the learning procedure, in the sense that T_{i+1} is the time moment when convergence of the learning procedure has been obtained and the value P_i has been determined. In view of this fact, we must emphasize that the time intervals $[T_i, T_{i+1}]$ need not be equal with each other and their length is not a design parameter.

At every step in the iterative procedure it is guaranteed that the new controller K_{i+1} will result in a better performance, that is smaller associated cost, than the previous controller. This will result in a monotonically decreasing sequence of cost

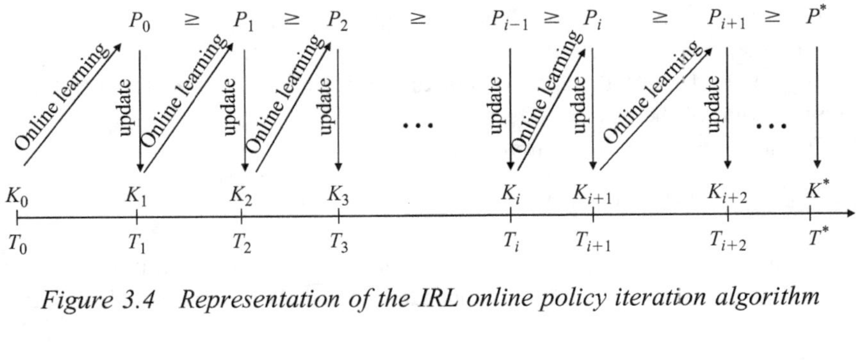

Figure 3.4 Representation of the IRL online policy iteration algorithm

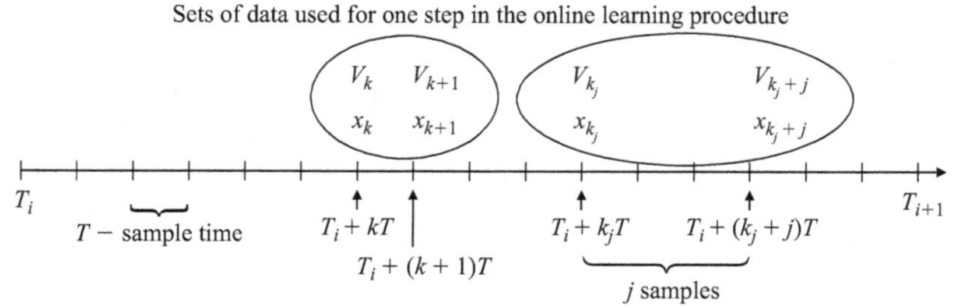

Figure 3.5 Data measurements used for learning the value described by P_i over the time interval $[T_i, T_{i+1}]$, while the state-feedback gain is K_i

functions, $\{P_i\}$, that converges to the smaller possible value, that is optimal cost P^*, associated with the optimal control policy K^*.

Figure 3.5 presents sets of data $(x(t), x(t+T))$ that are required for online learning of the value described by P_i. We denoted by T the smallest sampling time that can be used to make measurements of the state of the system. A data point that will be used for the online learning procedure is given, in a general notation, by the quadruple $(x_k, x_{k+j}, V_k, V_{k+j})$. Denoting with $d(x_k, x_{k+j}, K_i) = V_{k+j} - V_k$ the reinforcement over the time interval, where $J \in \mathbb{N}^*$, then $(x_k, x_{k+j}, d(x_k, x_{k+j}, K_i))$ is a data point of the sort required for setting up the solution given by (3.24). It is emphasized that the data that will be used by the learning procedure need not be collected at fixed sample time intervals.

3.3 Online IRL load-frequency controller design for a power system

In this section are presented the results that were obtained in simulation while finding the optimal controller for a power system. The plant is the linearized model of the power system presented in Wang *et al.* (1993).

Even though power systems are characterized by non-linearities, linear state-feedback control is regularly employed for load-frequency control at certain nominal operating points that are characterized by small variations of the system load around a constant value. Although this assumption seems to have simplified the design problem of a load-frequency controller, a new problem appears from the fact that the parameters of the actual plant are not precisely known and only the range of these parameters can be determined. For this reason it is particularly advantageous to apply model-free methods to obtain the optimal LQR controller for a given operating point of the power system.

The state vector of the system is

$$x = [\Delta f \quad \Delta P_g \quad \Delta X_g \quad \Delta E]^T$$

where the state components are the incremental frequency deviation Δf (Hz), incremental change in generator output ΔP_g (p.u. MW), incremental change in governor value position ΔX_g (p.u. MW) and the incremental change in integral control ΔE. The matrices of the *linearized nominal model* of the plant, used in Wang *et al.* (1993), are

$$A_{nom} = \begin{bmatrix} -0.0665 & 8 & 0 & 0 \\ 0 & -3.663 & 3.663 & 0 \\ -6.86 & 0 & -13.736 & -13.736 \\ 0.6 & 0 & 0 & 0 \end{bmatrix}$$
$$B = \begin{bmatrix} 0 & 0 & 13.736 & 0 \end{bmatrix}^T \tag{3.25}$$

Having the model of the system matrices one can easily calculate the LQR controller that is

$$K = \begin{bmatrix} 0.8267 & 1.7003 & 0.7049 & 0.4142 \end{bmatrix} \tag{3.26}$$

The iterative algorithm can be started while using this controller, which was calculated for the nominal model of the plant. The parameters of the controller will then be adapted in an online procedure, using reinforcement learning, to converge to the parameters of the optimal controller for the real plant.

For this simulation it was considered that the linear drift dynamics of the real plant is given by

$$A = \begin{bmatrix} -0.0665 & 11.5 & 0 & 0 \\ 0 & -2.5 & 2.5 & 0 \\ -9.5 & 0 & -13.736 & 13.736 \\ 0.6 & 0 & 0 & 0 \end{bmatrix} \tag{3.27}$$

Notice that the drift dynamics of the real plant, given by (3.27), differ from the nominal model used for calculation of the initial stabilizing controller, given in (3.25). In fact it is the purpose of the reinforcement learning adaptation scheme to find the optimal control policy for the real plant while starting from the 'optimal' controller corresponding to the nominal model of the plant.

The simulation was conducted using data obtained from the system at every 0.05 s. For the purpose of demonstrating the algorithm the closed-loop system was excited with an initial condition of 0.1 MW incremental change in generator output, the initial state of the system being $X_0 = \begin{bmatrix} 0 & 0.1 & 0 & 0 \end{bmatrix}$. The cost function parameters, namely, the Q and R matrices, were chosen to be identity matrices of appropriate dimensions.

To solve online for the values of the P matrix that parameterizes the cost function, before each iteration step a least-squares problem of the sort described in Section 3.2.1, with the solution given by (3.24), was setup. Since there are 10 independent elements in the symmetric matrix P the setup of the least-squares problem requires at least 10 measurements of the cost function associated with the given control policy and measurements of the system's states at the beginning and the end of each time interval, provided that there is enough excitation in the system.

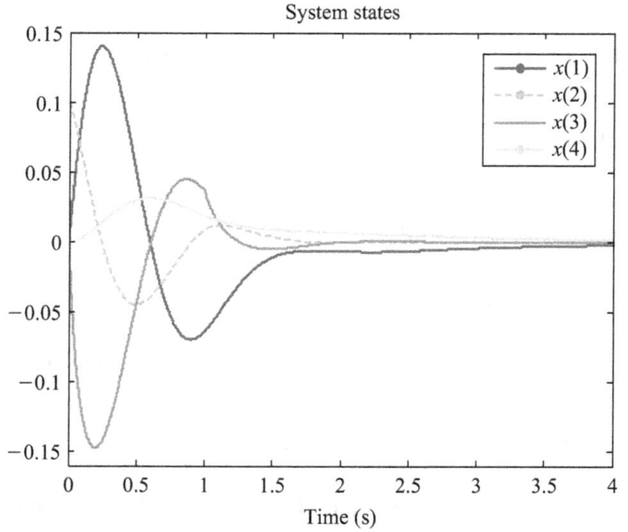

Figure 3.6 State trajectories of the linear closed-loop power system over the duration of the experiment

In this case, since the system states are not continuously excited and because resetting the state at each step is not an acceptable solution for online implementation, in order to have consistent data necessary to obtain the solution given by (3.24) one has to continue reading information from the system until the solution of the least-squares problem is feasible. A least-squares problem was solved after 20 sample data were acquired and thus the controller was updated every 1 s. The trajectory of the state of the system for the duration of the online experiment is presented in Figure 3.6.

The critic parameters (i.e. the elements of the value kernel matrix P) obtained from the IRL algorithm are presented in Figure 3.7. It is clear that the cost function (i.e. critic) parameters converged to the optimal ones – indicated in the figure with star-shaped points, which were placed for ease of comparison at $t = 5$ s. The values of the P matrix parameters at $t = 0$ s correspond to the solution of the Riccati equation that was solved, considering the approximate model of the system, to find the initial controller (3.26).

The P matrix obtained online using the adaptive critic algorithm, without knowing the plant internal dynamics, is

$$P = \begin{bmatrix} 0.4599 & 0.6910 & 0.0518 & 0.4641 \\ 0.6910 & 1.8665 & 0.2000 & 0.5798 \\ 0.0518 & 0.2000 & 0.0532 & 0.0300 \\ 0.4641 & 0.5798 & 0.0300 & 2.2105 \end{bmatrix} \tag{3.28}$$

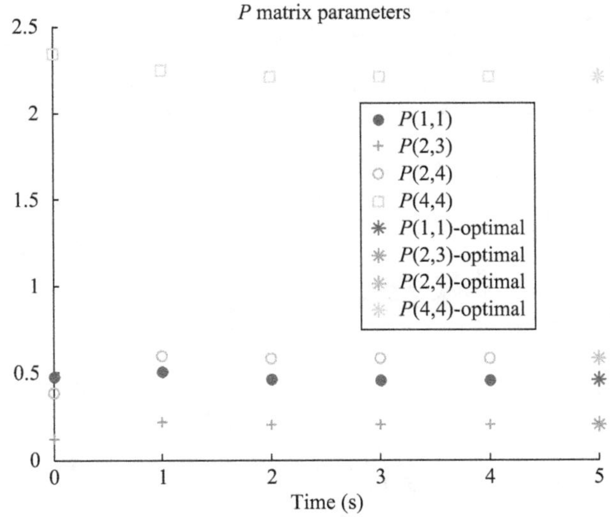

Figure 3.7 Evolution of the parameters of the P matrix for the duration of the experiment

The solution that was obtained by directly solving the algebraic Riccati equation considering the real plant internal dynamics (3.27) is

$$P = \begin{bmatrix} 0.4600 & 0.6911 & 0.0519 & 0.4642 \\ 0.6911 & 1.8668 & 0.2002 & 0.5800 \\ 0.0519 & 0.2002 & 0.0533 & 0.0302 \\ 0.4642 & 0.5800 & 0.0302 & 2.2106 \end{bmatrix} \tag{3.29}$$

One can see that the error difference between the parameters of the two matrices is in the range of 10^{-4}.

In practice, the convergence of the algorithm is considered to be achieved when the difference between the measured cost and the expected cost crosses below a designer-specified threshold value. It is important to note that after the convergence to the optimal controller was attained, the algorithm need not continue to be run and subsequent updates of the controller need not be performed.

Figure 3.8 presents a detail of the system state trajectories for the first two seconds of the simulation. The state values that were actually measured and subsequently used for the critic update computation are represented by the points on the state trajectories. Note that the control policy was updated at time $t = 1$ s.

Although in this case a nominal model of the system was available, and this allowed us to calculate a stabilizing controller to initialize the adaptive algorithm, it is important to point out that in the case when the system is itself stable this allows starting the iteration while using no controller (i.e. the initial controller is zero and no identification procedure needs to be performed).

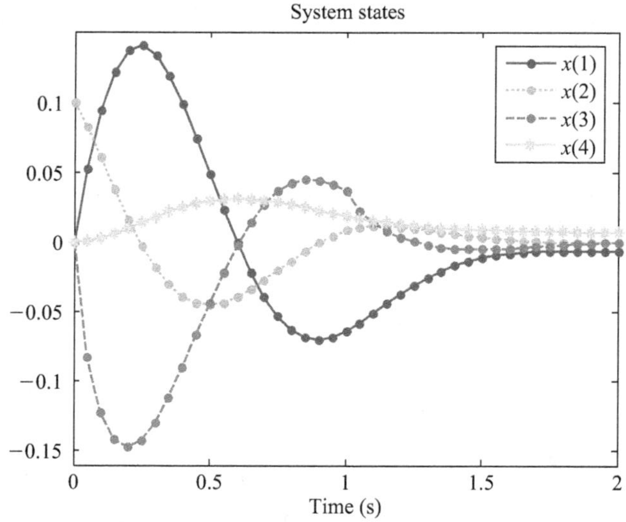

Figure 3.8 *System state trajectories (lines), and state information that was actually used for the critic update (dots on the state trajectories)*

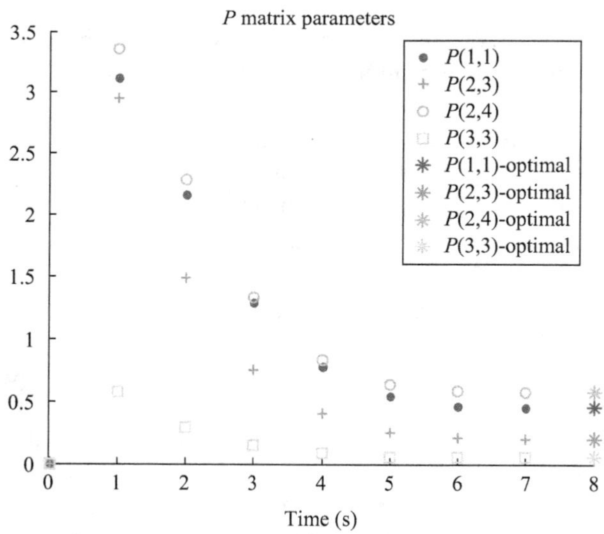

Figure 3.9 *Evolution of the parameters of the P matrix for the duration of the experiment when the adaptive algorithm was started without controller for the power system*

Figure 3.9 presents the parameter convergence result for the case the adaptive optimal control algorithm was initialized with no controller. The critic parameters converged to the optimal ones at time $t = 7$ s after seven updates of the controller parameters. The P matrix calculated with the adaptive algorithm is

$$P = \begin{bmatrix} 0.4601 & 0.6912 & 0.0519 & 0.4643 \\ 0.6912 & 1.8672 & 0.2003 & 0.5800 \\ 0.0519 & 0.2003 & 0.0533 & 0.0302 \\ 0.4643 & 0.5800 & 0.0302 & 2.2107 \end{bmatrix} \tag{3.30}$$

The error difference between the parameters of the solution (3.30) obtained iteratively and the optimal solution (3.29) is in the range of 10^{-4}.

3.4 Conclusion

This chapter presented a new policy iteration technique that solves the continuous-time LQR problem online without using knowledge about the system's internal dynamics (system matrix A). The algorithm was derived by writing the value function in integral reinforcement form to yield a new form of Bellman equation for CT systems. This allows the derivation of an integral reinforcement learning (IRL) algorithm, which is an adaptive controller that converges online to the solution of the optimal LQR controller. IRL is based on an adaptive critic scheme in which the actor performs continuous-time control while the critic incrementally corrects the actor's behavior at discrete moments in time until best performance is obtained. The critic evaluates the actor performance over a period of time and formulates it in a parameterized form. Based on the critic's evaluation the actor behavior policy is updated for improved control performance.

The result can be summarized as an algorithm that solves online in real time the algebraic Riccati equation associated with the optimal control problem without using knowledge of the system matrix A. Convergence to the solution of the optimal control problem, under the condition of initial stabilizing controller, was established by proving equivalence with the algorithm presented by Kleinman in Kleinman (1968). A simulation for load-frequency optimal control of a power system generator shows that the IRL algorithm learns the solution to the optimal control problem online in real time by measuring data along the system trajectories.

Chapter 4

Integral reinforcement learning (IRL) for non-linear continuous-time systems

This chapter presents an adaptive method based on actor–critic reinforcement learning (RL) for solving online the optimal control problem for non-linear continuous-time systems in the state space form $\dot{x}(t) = f(x) + g(x)u(t)$. The algorithm, first presented in Vrabie *et al.* (2008, 2009), Vrabie (2009), Vrabie and Lewis (2009), solves the optimal control problem without requiring knowledge of the drift dynamics $f(x)$. The method is based on policy iteration (PI), a RL algorithm that iterates between the steps of policy evaluation and policy improvement. The PI method starts by evaluating the cost of a given admissible initial policy and then uses this information to obtain a new control policy, which is improved in the sense of having a smaller associated cost compared with the previous policy. These two steps are repeated until the policy improvement step no longer changes the present policy, indicating that the optimal control behavior has been obtained.

Section 2.6 discusses the difficulties of implementing RL methods for continuous-time systems. These difficulties stem from the fact that the continuous-time Bellman equation $0 = r(x, u(x)) + (\nabla V_x)^T (f(x) + g(x)u(x))$, with ∇V_x the gradient and $r(x, u(x))$ the cost function integrand, contains the full system dynamics, unlike the discrete-time Bellman equation in Section 2.5. This chapter uses a new method known as *integral reinforcement learning* (IRL) to write an alternative form of the Bellman equation that has the same form as the discrete-time (DT) Bellman equation. The IRL form of the Bellman equation does not contain the system dynamics.

A PI algorithm was first developed for non-linear continuous-time systems by Leake and Liu (1967), but at that time the mathematical techniques required for real-time implementation had not been developed. Three decades later continuous-time PI was revisited and presented in Beard *et al.* (1997) as a feasible adaptive solution to the continuous-time (CT) optimal control problem. The main contribution of Beard *et al.* (1997) resides in the fact that the CT Bellman equation was solved using successive Galerkin approximation algorithms. A neural-network–based approach was developed for the cases of H2 and H-infinity control in Abu-Khalaf and Lewis (2005). Neural-network-based actor–critic structures with neural-network update laws are given in Hanselmann *et al.* (2007). All of these methods are offline solution methods. They require complete knowledge of the system dynamics since they are based on the CT Bellman equation $0 = r(x, u(x)) + (\nabla V_x)^T (f(x) + g(x)u(x))$.

This chapter gives a formulation of the PI algorithm for continuous-time non-linear systems, known as IRL, that does not depend on the system drift dynamics $f(x)$. The algorithm was first presented in Vrabie *et al.* (2008, 2009), Vrabie (2009), Vrabie and Lewis (2009). The structure of the IRL algorithm allows development of an online adaptive IRL algorithm that converges to the solution of the optimal control problem in real time by measuring data along the system trajectories. Knowledge of the system drift dynamics $f(x)$ is not required. We call this an *optimal adaptive control* algorithm, since it is an adaptive controller that converges to the optimal control solution. This adaptive IRL controller has an actor–critic structure and is a hybrid combination between a continuous-time controller and a learning structure that operates based on discrete sampled data from the system. The continuous-time controller is dynamic, and has a memory whose state is the value or cost. This hybrid continuous/sampled date structure is unlike any of the standard forms of controllers appearing in the literature.

Section 4.1 gives an overview of the optimal control problem for non-linear continuous-time systems. The Bellman equation and the Hamilton–Jacobi–Bellman (HJB) equation are derived. Section 4.2 presents an IRL policy iteration algorithm that solves the HJB equation without requiring knowledge of the drift dynamics $f(x)$. It is based on an alternative IRL form of the Bellman equation that does not contain the system dynamics. Convergence of the IRL algorithm is proved by showing equivalence with the general PI algorithm for non-linear systems. Section 4.3 presents a practical method to implement IRL policy iterations based on approximating the value function. Convergence of this approximate IRL PI algorithm, considering the error between the cost function and its approximation, is then proven. Section 4.4 discusses the online implementation on an actor–critic structure, while commenting also on the relations between the proposed online algorithm and certain learning mechanisms in the mammal brain. Section 4.5 presents simulation results considering two non-linear systems with quadratic and quartic cost functions.

4.1 Non-linear continuous-time optimal control

This section presents the formulation of the non-linear optimal control problem for continuous-time systems. Consider the time-invariant affine-in-the-input dynamical system given by

$$\dot{x}(t) = f(x(t)) + g(x(t))u(x(t)), \quad x(0) = x_0 \tag{4.1}$$

with state $x(t) \in \mathbb{R}^n$, $f(x(t)) \in \mathbb{R}^n$, $g(x(t)) \in \mathbb{R}^{n \times m}$ and control input $u(t) \in U \subset \mathbb{R}^m$. It is assumed that $f(0) = 0$, that $f(x) + g(x)u$ is Lipschitz continuous on a set $\Omega \subseteq \mathbb{R}^n$ that contains the origin, and that the dynamical system is stabilizable on Ω, that is there exists a continuous control function $u(t) \in U$ such that the closed-loop system is asymptotically stable on Ω.

We note here that although global asymptotic stability is guaranteed in a linear system case, it is generally difficult to guarantee in a general continuous-time non-linear system problem setting. This is due to the possible non-smooth nature of the non-linear system dynamics. At points where there exist discontinuities in $\dot{x}$, there will also exist discontinuities of the gradient of the cost function. For this reason, the discussion here is restricted to the case in which asymptotic stability is sought only in a region $\Omega \subseteq \mathbb{R}^n$ in which the cost function is continuously differentiable.

Define an infinite-horizon integral cost associated with the control input $\{u(\tau); \tau \geq t\}$. as

$$V(x(t)) = \int_t^{\infty} r(x(\tau), u(\tau))\, d\tau = \int_t^{\infty} (Q(x) + u^T Ru)\, d\tau \tag{4.2}$$

where $x(\tau)$ denotes the solution of (4.1) for initial condition $x(t) \in \Omega$ and input $\{u(\tau); \tau \geq t\}$. The cost integrand is taken as $r(x, u) = Q(x) + u^T Ru$ with $Q(x)$ positive definite (i.e. $\forall x \neq 0, Q(x) > 0$ and $x = 0 \Rightarrow Q(x) = 0$) and $R \in \mathbb{R}^{m \times m}$ a symmetric positive definite matrix.

Definition 4.1. Beard *et al.* (1997) (Admissible (stabilizing) policy) A control policy $u(t) = \mu(x)$ is defined as admissible with respect to (4.2) on Ω, denoted by $\mu \in \Psi(\Omega)$, if $\mu(x)$ is continuous on Ω, $\mu(0) = 0, \mu(x)$ stabilizes (4.1) on Ω and $V(x_0)$ is finite $\forall x_0 \in \Omega$.

The cost function or *value* associated with any admissible control policy $\mu(t) = \mu(x(t)) \in \Psi(\Omega)$ is

$$V^{\mu}(x(t)) = \int_t^{\infty} r(x(\tau), \mu(x(\tau)))\, d\tau \tag{4.3}$$

where $V^{\mu}(x)$ is C^1. Using Leibniz's formula, the infinitesimal version of (4.3) is found to be the following.

CT Bellman equation

$$0 = r(x, \mu(x)) + (\nabla Vx^{\mu})^T (f(x) + g(x)\mu(x)), \quad V^{\mu}(0) = 0 \tag{4.4}$$

Here, ∇V_x^{μ} (a column vector) denotes the gradient of the cost function V^{μ} with respect to x. That is $\nabla V_x^{\mu} = \partial V^{\mu}/\partial x$. Equation (4.4) is a Bellman equation for non-linear continuous-time (CT) systems, which, given the control policy $\mu(x) \in \Psi(\Omega)$, can be solved for the value $V^{\mu}(x)$ associated with it. Given that $\mu(x)$ is an admissible control policy, if $V^{\mu}(x)$ satisfies (4.4), with $r(x, \mu(x)) \geq 0$, then it can be shown that $V^{\mu}(x)$ is a Lyapunov function for the system (4.1) with control policy $\mu(x)$.

The optimal control problem can now be formulated: given the continuous-time system (4.1), the set $u \in \Psi(\Omega)$ of admissible control policies, and the infinite-horizon cost functional (4.2), find an admissible control policy such that the value (4.3) is minimized.

Defining the Hamiltonian

$$H(x, u, V_x) = r(x(t), u(t)) + (\nabla V_x)^T (f(x(t)) + g(x(t))u(t)) \tag{4.5}$$

the optimal cost function $V^*(x)$ satisfies the Hamilton–Jacobi–Bellman (HJB) equation

$$0 = \min_{u \in \Psi(\Omega)} [H(x, u, \nabla V_x^*)]$$

Assuming that the minimum on the right-hand side of this equation exists and is unique, then the optimal control function is

$$u^*(x) = -\frac{1}{2} R^{-1} g^T(x) \nabla V_x^* \tag{4.6}$$

Inserting this optimal control policy in the Hamiltonian we obtain the formulation of the HJB equation

$$0 = Q(x) + (\nabla V_x^*)^T f(x) - \frac{1}{4} (\nabla V_x^*)^T g(x) R^{-1} g^T(x) \nabla V_x^*, \quad V^*(0) = 0 \tag{4.7}$$

This is a necessary condition for the optimal cost function. For the linear system case with a quadratic cost functional, the equivalent of this HJB equation is the algebraic Riccati equation discussed in Chapter 3.

To find the optimal control solution for the problem one can solve the HJB equation (4.7) for the cost function and then substitute the solution in (4.6) to obtain the optimal control. However, solving the HJB equation is generally difficult, and analytic solutions may not exist. Explicitly solving the HJB also requires complete knowledge of the system dynamics $f(x), g(x)$.

4.2 Integral reinforcement learning policy iterations

The development of RL methods such as policy iteration and value iteration for CT systems has lagged their development for discrete-time systems. This is because the CT Bellman equation (4.4) does not share any of the beneficial properties of the DT Bellman equation in Section 2.5, which is as follows

DT Bellman equation

$$V(x_k) = r(x_k, u_k) + \gamma V(x_{k+1}) = Q(x_k) u_k^T R u_k + \gamma V(x_{k+1})$$

with k the discrete-time index and $r \leq 1$ the discount factor. Specifically, the dynamics ($f(\cdot)$, $g(\cdot)$) do not appear in the DT Bellman equation, whereas they do appear in the CT Bellman equation (4.4). This makes it difficult to formulate algorithms such as Q learning, which do not require knowledge of the system dynamics. Moreover, in the DT Bellman equation there are two occurrences of the value function, evaluated at different times k and $k+1$. This allows the formulation of value iteration, or heuristic dynamic programming, for DT systems. However, with only one occurrence of the value in the CT Bellman equation, it is not at all clear how to formulate any sort of value iteration procedure.

Based on the CT Bellman equation (4.4), one could write a policy iteration algorithm for CT systems as follows. Index i is the iteration step number.

Algorithm 4.1. Policy iteration algorithm for continuous-time systems

Select $\mu^{(0)}(x(t)) \in \Psi(\Omega)$ as an admissible policy.

1. (*Policy evaluation* step) Solve for the value $V^{\mu^{(i)}}(x(t))$ using the CT Bellman equation

$$0 = r(x, \mu^{(i)}(x)) + (\nabla V_x^{\mu^{(i)}})^T (f(x) + g(x)\mu^{(i)}(x)) \text{ with } V^{\mu^{(i)}}(0) = 0$$

2. (*Policy improvement* step) Update the control policy using

$$\mu^{(i+1)} = \arg \min_{u \in \Psi(\Omega)} [H(x, u, \nabla V_x^{\mu^{(i)}})]$$

■

The PI Algorithm 4.1 solves the non-linear HJB equation by iterations on equations linear in the value function gradient. The algorithm was proven to converge in Abu-Khalaf and Lewis (2005), Leake and Liu (1967), Beard *et al.* (1997). Algorithm 4.1 has an undesirable feature in that it requires full knowledge of the system dynamics ($f(\cdot)$, $g(\cdot)$) to solve the CT Bellman equation at each iteration step. This is fixed in Section 4.2.1.

Example 4.1. Policy iteration for continuous-time linear quadratic regulator

For the continuous-time linear quadratic regulator (LQR), policy iteration Algorithm 4.1 is a familiar algorithm in control theory. Consider the linear time-invariant dynamical system

$$\dot{x} = Ax + Bu$$

with state $x(t) \in \mathbb{R}^n$, control input $u(t) \in \mathbb{R}^m$ and (A, B) stabilizable. To this system associate the infinite-horizon quadratic cost function or value

$$V(x(t_0), t_0) = \int_{t_0}^{\infty} (x^T(\tau)Qx(\tau) + u^T(\tau)Ru(\tau))\, d\tau$$

with $Q \geq 0$, $R \geq 0$ such that $(Q^{1/2}, A)$ is detectable. For the LQR, the value is quadratic so that

$$V(x(t)) = x^T(t)Px(t)$$

Then, the gradient of the value is

$$\nabla V_x = \frac{\partial V}{\partial x} = 2Px$$

The HJB equation (4.7) therefore becomes the algebraic Riccati equation (ARE)

$$A^T P + PA - PBR^{-1}B^T P + Q = 0$$

The optimal control (4.6) is given as $u(t) = -Kx(t)$ with optimal feedback gain

$$K = R^{-1}B^T P$$

where the matrix P is the unique positive definite solution of the ARE. Under the detectability condition for $(Q^{1/2}, A)$ the unique positive semidefinite solution of the ARE determines a stabilizing closed-loop controller K.

For admissible feedback controls $u(t) = -Kx(t)$, the Bellman equation (4.4) is the Lyapunov equation

$$(A - BK)^T P + P(A - BK) + K^T RK + Q = 0$$

Policy iteration Algorithm 4.1 is then given as

$$A_i^T P_i + P_i A_i = -(K_i^T RK_i + Q)$$

$$K_{i+1} = R^{-1}B^T P_i$$

with $A_i = A - BK_i$. This is exactly Kleinman's Algorithm (Kleinman, 1968), known in feedback control theory since the 1960s. Kleinman proved that if the algorithm is started with an initial stabilizing gain K_0, then all subsequent gains are stabilizing and the algorithm converges to the solution to the ARE.

The importance of Kleinman's Algorithm is that it solves the quadratic ARE by iterations on solving Lyapunov equations, which are linear. In the early days, there were no good numerical algorithms for solving ARE, whereas the Lyapunov equation can be solved by linear equation methods.

Compare this discussion to the development in Chapter 3. ■

4.2.1 Integral reinforcement learning policy iteration algorithm

To improve upon Algorithm 4.1, we now present a new formulation of the CT Bellman equation based on integral reinforcement learning (IRL). The IRL form allows development of novel policy iteration algorithms for CT systems. This leads in turn to an online adaptive algorithm that solves the optimal control problem without using knowledge of the drift dynamics $f(x)$. Given an admissible policy and an integration time interval $T > 0$, write the value function (4.3) as the following equivalent form Vrabie *et al.* (2008, 2009), Vrabie (2009), Vrabie and Lewis (2009).

Integral reinforcement form of value function: IRL Bellman equation

$$V^{\mu}(x(t)) = \int_t^{t+T} r(x(\tau), \mu(x(\tau)))\, d\tau + V^{\mu}(x(t+T)) \tag{4.8}$$

Note that this form does not contain the system dynamics ($f(\cdot)$, $g(\cdot)$). Lemma 4.1 shows that this equation is equivalent to Bellman equation (4.4) in the sense that both equations have the same solution. Therefore, using the IRL Bellman equation (4.8) allows the formulation of continuous-time PI algorithms that share the beneficial features of discrete-time PI algorithms. The integrand

$$\rho(x(t), t, t+T) = \int_t^{t+T} r(x(\tau), \mu(x(\tau)))\, d\tau$$

is known as the *integral reinforcement* on the time interval $[t, t+T]$.

Let $\mu^{(0)}(x(t)) \in \Psi(\Omega)$ be an admissible policy, select $T>0$ such that, if $x(t) \in \Omega$, then also $x(t+T) \in \Omega$. The existence of such a time period $T>0$ is guaranteed by the admissibility of $\mu^{(0)}(\cdot)$ on Ω. Define the following PI algorithm based on the IRL Bellman equation.

Algorithm 4.2. Integral reinforcement learning policy iteration algorithm

Select $\mu^{(0)}(x(t)) \in \Psi(\Omega)$ as an admissible policy.

1. (*Policy evaluation* step) Solve for the value $V^{\mu^{(i)}}(x(t))$ using the IRL Bellman equation

$$V^{\mu^{(i)}}(x(t)) = \int_t^{t+T} r(x(s), \mu^{(i)}(x(s)))\, ds + V^{\mu^{(i)}}(x(t+T)) \text{ with } V^{\mu^{(i)}}(0) = 0 \tag{4.9}$$

2. (*Policy improvement* step) Update the control policy using

$$\mu^{(i+1)} = \arg\min_{u \in \Psi(\Omega)} [H(x, u, \nabla V_x^{\mu^{(i)}})] \tag{4.10}$$

which explicitly is

$$\mu^{(i+1)}(x) = -\frac{1}{2} R^{-1} g^T(x)\, \nabla V_x^{\mu^{(i)}} \tag{4.11}$$

■

Algorithm 4.2 gives a new formulation for the policy iteration algorithm that allows solution of the optimal control problem without requiring knowledge of the drift dynamics $f(x)$. This IRL PI algorithm is an online version of the offline algorithms proposed in Abu-Khalaf and Lewis (2005), Beard *et al.* (1997), and is motivated by the success of the online adaptive critic techniques proposed by computational intelligence researchers (Prokhorov and Wunsch, 1997; Bertsekas and Tsitsiklis, 1996; Murray *et al.*, 2002).

The IRL Bellman equation (4.9) is a discretized version of $V^{\mu^{(i)}}(x(t)) = \int_t^{\infty} r(x(\tau), \mu^{(i)}(x(\tau)))\, d\tau$ and it can be viewed as a Lyapunov equation for non-linear systems. In this chapter, we shall refer to it also as

$$LE(V^{\mu^{(i)}}(x(t))) \triangleq \int_t^{t+T} r(x(s), \mu^{(i)}(x(s)))\, ds + V^{\mu^{(i)}}(x(t+T)) - V^{\mu^{(i)}}(x(t))$$

with $V^{\mu^{(i)}}(0) = 0$.

The convergence of the new PI algorithm is proven in Section 4.2.2. The implementation of the algorithm using value function approximation structures is discussed in Section 4.3.

4.2.2 Convergence of IRL policy iteration

It is now shown that the IRL PI algorithm converges to the optimal control policy $\mu^* \in \Psi(\Omega)$ with corresponding cost $V^*(x_0) = \min_\mu \left(\int_0^\infty r(x(\tau), \mu(x(\tau)))\, d\tau\right)$.

It has been proven that if $\mu^{(i)} \in \Psi(\Omega)$ and $V^{\mu^{(i)}}(x(t)) \in C^1(\Omega)$ satisfy (4.9) then the new control policy $\mu^{(i+1)}$, determined based on (4.10), is also admissible for the system (4.1) (for proof see Abu-Khalaf and Lewis, 2005; Leake and Liu, 1967; Beard *et al.*, 1997).

The following result is required in order to prove the convergence of the IRL policy iteration algorithm.

Lemma 4.1. Solving for $V^{\mu^{(i)}}$ in (4.9) is equivalent to finding the solution of

$$0 = r(x, \mu^{(i)}(x)) + (\nabla V_x^{\mu^{(i)}})^T (f(x) + g(x)\mu^{(i)}(x)), \quad V^{\mu^{(i)}}(0) = 0 \tag{4.12}$$

Proof: See appendix. ■

Remark 4.1. Although the same solution is obtained solving either (4.9) or (4.12), solving (4.9) does not require any knowledge of the drift dynamics $f(x)$, which does appear explicitly in (4.12). From Lemma 4.1, it follows that Algorithm 4.2 is equivalent to Algorithm 4.1. ■

Theorem 4.1. (Convergence) The IRL policy iteration Algorithm 4.2 converges uniformly to the optimal control solution on trajectories originating in Ω, that is

$$\begin{aligned} &\forall \varepsilon > 0 \exists i_0 : \forall i \geq i_0 \\ &\sup_{x \in \Omega} |V^{\mu^{(i)}}(x) - V^*(x)| < \varepsilon, \sup_{x \in \Omega} |\mu^{(i)}(x) - u^*(x)| < \varepsilon \end{aligned} \tag{4.13}$$

Proof: In Abu-Khalaf and Lewis (2005), Beard *et al.* (1997) it was shown that in Algorithm 4.1, if the admissible policy $\mu^{(0)}(x)$ is admissible then all policies are admissible at each iteration. Moreover, the algorithm converges to the solution of the HJB equation, that is (4.13) is satisfied.

Based on the equivalence shown in Lemma 1 between the (4.9) and (4.12), one concludes that IRL Algorithm 4.2 converges to the solution of the optimal control problem (4.2), on Ω. ■

For the implementation of the IRL policy iteration Algorithm 4.2, one only needs knowledge of the input-to-state dynamics $g(x)$, which is required for the policy update in (4.11). One can see that knowledge of the drift dynamics $f(x)$ is not required. This is due to the fact that information regarding the system $f(x)$ matrix is embedded in the states $x(t)$ and $x(t+T)$ that are measured online and are required to solve (4.9). Thus, it will be seen that IRL policy iteration can be implemented in real time without requiring any system identification procedure.

4.3 Implementation of IRL policy iterations using value function approximation

This section shows how to implement IRL policy iterations in a practical fashion by using an approximator structure to approximate the value function. This is known as value function approximation (VFA).

4.3.1 Value function approximation and temporal difference error

The IRL Bellman equation (4.9) is difficult to solve, and may not have an analytic solution. It can be approximately solved by practical means by making use of a structure to approximate the value function solution for any $x \in \Omega$.

Consider an infinite set of linearly independent basis functions $\{\phi_j(x)\}_1^\infty$, such that $\phi_j(x) \in C^1(\Omega)$, $\phi_j(0) = 0$, $j = \overline{1, \infty}$. Let $\{\phi_j(x)\}_1^\infty$ satisfy the completeness property. That is, any function $f(x) \in C^1(\Omega), f(0) = 0$ can be represented as a linear combination of a finite subset of $\{\phi_j(x)\}_1^\infty$. Then the exact solution of Bellman equation (4.9) can be expressed as

$$V^{\mu^{(i)}}(x) = \sum_{j=1}^{\infty} c_j^{\mu^{(i)}} \phi_j(x) = (c_\infty^{\mu^{(i)}})^T \varphi_\infty(x) \tag{4.14}$$

where $\varphi_\infty(x)$ is the vector of basis functions and $c^{\infty\mu^{(i)}}$ denotes a weight vector.

Assume now that a linear combination of a finite set of basis functions can be determined that closely approximates the value function $V^{\mu^{(i)}}(x)$, for $x \in \Omega$. That is, assume the value function can be approximately represented as

$$V_L^{\mu^{(i)}}(x) = \sum_{j=1}^{L} w_j^{\mu^{(i)}} \phi_j(x) = (w_L{}^{\mu^{(i)}})^T \varphi_L(x) \tag{4.15}$$

where L is the number of retained basis functions, $\varphi_L(x)$ is the vector of corresponding basis function, and $w_L^{\mu^{(i)}}$ is a vector of unknown weights to be determined.

Using this VFA for the cost function, the IRL Bellman equation (4.9) can be written as

$$w_L^{\mu^{(i)}T} \varphi_L(x(t)) = \int_t^{t+T} r(x, \mu^{(i)}(x))\, d\tau + w_L^{\mu^{(i)}T} \varphi_L(x(t+T)) \tag{4.16}$$

Because the cost function is replaced by its approximation, (4.16) will have the residual error

$$\delta_L^{\mu^{(i)}}(x(t), T) = \int_t^{t+T} r(x, \mu^{(i)}(x))\, d\tau + w_L^{\mu^{(i)}T}[\varphi_L(x(t+T)) - \varphi_L(x(t))] \quad (4.17)$$

From the perspective of temporal difference learning methods (e.g. Doya, 2000; Baird, 1994; see Section 2.4) this error can be viewed as a temporal difference residual error for continuous-time systems.

To determine the parameters $w_L^{\mu^{(i)}}$ in the VFA that best approximate the cost function $V_L^{\mu^{(i)}}$ in the least-squares sense, we use the method of weighted residuals. Thus, the parameters $w_L^{\mu^{(i)}}$ of the cost function approximation $V_L^{\mu^{(i)}}$ are adapted such that to minimize the objective

$$S = \int_\Omega \delta_L^{\mu^{(i)}}(x, T)\delta_L^{\mu^{(i)}}(x, T)\, dx \quad (4.18)$$

This amounts to $\int_\Omega ((d\delta_L^{\mu^{(i)}}(x, T))/dw_L^{\mu^{(i)}})\delta_L^{\mu^{(i)}}(x, T)\, dx = 0$. Using the inner product notation for the Lebesgue integral one can write

$$\left\langle \frac{d\delta_L^{\mu^{(i)}}(x, T)}{dw_L^{\mu^{(i)}}} \delta_L^i(x, T) \right\rangle_\Omega = 0 \quad (4.19)$$

which is

$$\begin{aligned} &\left\langle [\varphi_L(x(t+T)) - \varphi_L(x(t))],\ [\varphi_L(x(t+T)) - \varphi_L(x(t))]^T \right\rangle_\Omega w_L^{\mu^{(i)}} \\ &+\left\langle [\varphi_L(x(t+T)) - \varphi_L(x(t))],\ \textstyle\int_t^{t+T} r(x(s), \mu^{(i)}(x(s)))\, ds \right\rangle_\Omega = 0 \end{aligned} \quad (4.20)$$

Conditioned on $\Phi = \langle [\varphi_L(x(t+T)) - \varphi_L(x(t))],\ [\varphi_L(x(t+T)) - \varphi_L(x(t))]^T \rangle_\Omega$ being invertible, then the solution is

$$w_L^{\mu^{(i)}} = -\Phi^{-1}\left\langle [\varphi_L(x(t+T)) - \varphi_L(x(t))], \int_t^{t+T} r(x(s), \mu^{(i)}(x(s)))\, ds \right\rangle_\Omega \quad (4.21)$$

To show that Φ can be inverted the following technical results are needed.

Definition 4.2. (Linearly independent set of functions) Kolmogorov and Fomin (1999) A set of functions $\{\phi_j\}_1^N$ is said to be linearly independent on a set Ω if $\sum_{j=1}^N c_j\phi_j(x) = 0$ a.e. on Ω implies that $c_1 = \cdots = c_N = 0$.

Lemma 4.2. If the set $\{\phi_j\}_1^N$ is linearly independent and $u \in \Psi(\Omega)$ then the set $\{\nabla\phi_j^T(f + gu)\}_1^N$ is also linearly independent.

The proof is given in Beard *et al.* (1997).

The next lemma shows that Φ can be inverted.

Lemma 4.3. Let $\mu(x) \in \Psi(\Omega)$ such that $f(x) + g(x)\mu(x)$ is asymptotically stable. Given that the set $\{\phi_j\}_1^N$ is linearly independent then $\exists T > 0$ such that $\forall x(t) \in \Omega - \{0\}$, the set $\{\bar{\phi}_j(x(t), T) = \phi_j(x(t+T)) - \phi_j(x(t))\}_1^N$ is also linearly independent.

Proof: See appendix. ■

IRL policy iteration with VFA. Based on Lemma 4.3, there exist values of T such that Φ is invertible and the parameters $w_L^{\mu^{(i)}}$ that solve (4.16) can be computed using (4.21). The value function $V_L^{\mu^{(i)}}$ at each step can be calculated using (4.15). Having solved (4.16) for the value function parameters $w_L^{\mu^{(i)}}$, the policy update step can be executed according to

$$\mu_L^{(i+1)}(x) = -\frac{1}{2}R^{-1}g^T(x)\,\nabla\varphi_L^T(x)w_L^{\mu^{(i)}} \tag{4.22}$$

Equation (4.22) gives the output of the actor structure. Note that in this implementation the controller (actor) can be seen as an approximation structure, which has the same weight parameters as the critic, but whose basis set of functions are the gradients of those in the critic.

In summary, the IRL policy iteration using VFA is given by the following.

Algorithm 4.3. IRL policy iteration algorithm using value function approximation

Select $\mu_L^{(0)}(x(t)) \in \Psi(\Omega)$ as an admissible policy.

1. (*Policy evaluation* step.) Solve for the weights $w_L^{\mu^{(i)}}$ approximating the value function using the approximate IRL Bellman equation

$$w_L^{\mu^{(i)}T}[\varphi_L(x(t)) - \varphi_L(x(t+T))] = \int_t^{t+T} r(x, \mu_L^{(i)}(x))\,d\tau$$

2. (*Policy improvement* step.) Update the control policy using

$$\mu_L^{(i+1)}(x) = -\frac{1}{2}R^{-1}g^T(x)\,\nabla\varphi_L^T(x)w_L^{\mu^{(i)}}$$

At each step one has $V_L^{\mu^{(i)}}(x) = (w_L^{\mu^{(i)}})^T\varphi_L(x)$. ■

4.3.2 Convergence of approximate value function to solution of the Bellman equation

This section shows that as the number of retained basis functions L becomes large, the solution to the VFA equation (4.16) converges to the solution to the IRL Bellman equation (4.9).

Definition 4.3. (Convergence in the mean.) A sequence of Lebesgue integrable functions on a set $\Omega, \{f_n(x)\} \in L_2(\Omega)$, is said to converge in the mean to $f(x) \in L_2(\Omega)$, if $\forall \varepsilon > 0$, $\exists N(\varepsilon)$ such that

$$\forall n > N(\varepsilon), \|f_n(x) - f(x)\|_{L_2(\Omega)} < \varepsilon, \text{ where } \|f(x)\|^2{}_{L_2(\Omega)} = \langle f, f \rangle$$

Equation (4.9) can be written using a linear operator A defined on the Hilbert space of continuous and differentiable functionals on Ω

$$\overbrace{V^{\mu}(x(t)) - V^{\mu}(x(t+T))}^{AV^{\mu}} = \overbrace{\int_t^{t+T} r(x(s), \mu(x(s)))\, ds}^{d(x,\mu(x),T)} \tag{4.23}$$

Function approximators, which are defined such that the basis functions are power series of order *m*, are differentiable and can uniformly approximate a continuous function with all its partial derivatives, up to order *m*, by differentiating the series term wise. This type of series is *m*-uniformly dense as shown in Lemma 4.4.

Lemma 4.4. (Higher-order Weierstrass approximation theorem (Hornik *et al.*, 1990)). Let $f(x) \in C^m(\Omega)$, then there exists a polynomial $P(x)$ such that it converges uniformly to $f(x)$, and all its partial derivatives up to order *m* converge uniformly.

The following facts hold under the stated standard conditions in optimal control.

Fact 4.1. The solution of (4.9) is positive definite. This is guaranteed when the system has stabilizable dynamics and when the performance functional satisfies zero-state observability (i.e. observability of the system state through the cost function) (Van Der Schaft 1992).

Fact 4.2. The system dynamics and the performance integrand $r(x(s), \mu(x(s)))$ are such that the solution of (4.9) is continuous and differentiable on Ω.

Fact 4.3. A complete set $\{\phi_j\}_1^{\infty} \in C^1(\Omega)$ can be chosen such that the solution $V \in C^1(\Omega)$ and ∇V can be uniformly approximated by the infinite series built based on $\{\phi_j\}_1^{\infty}$.

Fact 4.4. The sequence $\{\overline{\phi}_j(x(t), T) = \phi_j(x(t+T)) - \phi_j(x(t))\}_1^{\infty}$ is linearly independent and complete.

Proof: The linear independence results from Lemma 4.3, being conditioned by certain values of the sample time *T*. The completeness relies on the high-order Weierstrass approximation theorem.

$\forall V, \varepsilon\, \exists L, w_L$ such that $|V_L - V| < \varepsilon$. This implies that as $L \to \infty$ $\sup_{x\in\Omega} |AV_L - AV| \to 0 \Rightarrow \|AV_L - AV\|_{L_2(\Omega)} \to 0$ that proves completeness of $\{\overline{\phi}_j(x(t), T) = A\phi_j\}_1^{\infty}$.

The first three facts are taken as standard assumptions in optimal control, and we have proven the fourth. The next result is required.

Lemma 4.5. Given a set of N linearly independent functions $\{f_j(x)_1^N\}$ defined on Ω then

$$||\alpha_N^T f_N||_{L_2(\Omega)}^2 \to 0 \Leftrightarrow ||\alpha_N||_{l_2}^2 \to 0 \tag{4.24}$$

Proof: See Abu-Khalaf and Lewis (2005). ■

The next main result shows convergence in the mean of the VFA solutions of (4.16) to the solution to the IRL Bellman equation (4.9) as the number of retained basis functions L becomes large.

Theorem 4.2. Given that the Facts 4.1–4.4 hold, then approximate solutions exist for (4.9) using the method of least-squares and are unique for each L. In addition, as $L \to \infty$

$$\text{R1.}||LE(V_L^{\mu^{(i)}}(x)) - LE(V^{\mu^{(i)}}(x))||_{L_2(\Omega)} \to 0$$

where $LE(V(x))$ is defined in Section 4.2.1, and

$$\text{R2.}||V_L^{\mu^{(i)}}(x) - V^{\mu^{(i)}}(x)||_{L_2(\Omega)} \to 0$$

$$\text{R3.}||\nabla V_L^{\mu^{(i)}}(x) - \nabla V^{\mu^{(i)}}(x)||_{L_2(\Omega)} \to 0$$

$$\text{R4.}||\mu_L^{(i)}(x) - \mu^{(i)}(x)||_{L_2(\Omega)} \to 0$$

Proof: The least-squares solution $V_L^{\mu^{(i)}}$ of (4.9) is the solution of the minimization problem

$$||AV_L^{\mu^{(i)}} - d(x, \mu^{(i)}, T)||^2 = \min_{w_L} ||w_L^T \overline{\varphi}(x) - d(x, \mu^{(i)}, T)||^2 \tag{4.25}$$

The uniqueness of the solution follows directly from the linear independence of $\{\overline{\varphi}_j(x(t), T)\}_1^L$. R1 follows from the completeness of $\{\overline{\varphi}_j(x(t), T) = A\phi_j\}_1^\infty$.

R2 is next proved.

$$LE(V_L^{\mu^{(i)}}(x)) - LE(V^{\mu^{(i)}}(x)) = w_L^T \overline{\varphi}_L(x, T) - c_\infty^T \overline{\varphi}_\infty(x, T) = \varepsilon_L(x, T) \tag{4.26}$$

$$(w_L - c_L)^T \overline{\varphi}_L(x, T) = \varepsilon_L(x, T) + \sum_{j=L+1}^{\infty} c_j \overline{\varphi}_j(x, T) = \varepsilon_L(x, T) + e_L(x, T) \tag{4.27}$$

where $e_L(x,T)$ converges uniformly to zero due to the high-order Weierstrass approximation theorem (this implies convergence in the mean) and $\varepsilon_L(x,T)$ converges in the mean to zero due to R1. Then

$$\begin{aligned}\|(w_L-c_L)^T\overline{\varphi}_L(x,T)\|^2_{L_2(\Omega)} &= \|\varepsilon_L(x,T)+e_L(x,T)\|^2_{L_2(\Omega)}\\ &\le 2\|\varepsilon_L(x,T)\|^2_{L_2(\Omega)}+2\|e_L(x,T)\|^2_{L_2(\Omega)}\to 0 \end{aligned}\quad (4.28)$$

Since $\overline{\varphi}_L(x,T)$ is linearly independent then, based on Lemma 4.5, one sees that $\|(w_L-c_L)\|^2_{l_2}\to 0$. As the set $\{\phi_j\}_1^L$ is linearly independent, it follows from Lemma 4.5 that $\|(w_L-c_L)^T\varphi_L(x)\|^2_{L_2(\Omega)}\to 0$. It thus follows that, as $L\to\infty, \|V_L^{\mu^{(i)}}-V^{\mu^{(i)}}\|^2_{L_2\Omega}\to 0$.

Similarly, since $\{d\phi_j/dx\}_1^L$ is linearly independent, from Lemma 4.5 results that $\|(w_L-c_L)^T\nabla\varphi_L(x)\|^2_{L_2(\Omega)}\to 0$, from which follows R3, $\|\nabla V_L^{\mu^{(i)}}-\nabla V^{\mu^{(i)}}\|^2_{L_2(\Omega)}\to 0$.

R4 follows immediately from R3 given that

$$\begin{aligned}\|\mu_L^{(i)}(x)-\mu^{(i)}(x)\|^2_{L_2(\Omega)} &= \|-R^{-1}g^T(x)(\nabla V_L^{\mu^{(i-1)}}(x)-\nabla V_L^{\mu^{(i-1)}}(x))\|^2_{L_2(\Omega)}\\ &\le \|-R^{-1}g^T(x)\|^2_{L_2(\Omega)}\|\nabla V_L^{\mu^{(i-1)}}(x)-\nabla V^{\mu^{(i-1)}}(x)\|^2_{L_2(\Omega)}\to 0\end{aligned}$$

(4.29)

■

Based on Theorem 4.2 the following stronger result of uniform convergence can be shown.

Corollary 4.1. If the results from Theorem 4.2 hold, then

$$\sup_{x\in\Omega}|V_L^{\mu^{(i)}}(x)-V^{\mu^{(i)}}(x)|\to 0$$

$$\sup_{x\in\Omega}|\nabla V_L^{\mu^{(i)}}(x)-\nabla V^{\mu^{(i)}}(x)|\to 0$$

$$\sup_{x\in\Omega}|\mu_L^{\mu^{(i)}}(x)-\mu^{(i)}(x)|\to 0$$

Proof: See Abu-Khalaf and Lewis (2005). ■

The IRL policy iteration algorithm with VFA is given by Algorithm 4.3. The next result shows that, given an initial admissible control policy, $\mu^{(0)}(x)$, the control policy $\mu_L^{(i)}(x)$ at each step of this algorithm is admissible provided that the number of the basis functions L in the approximation structure is sufficiently large.

Corollary 4.2. (Admissibility of ${\mu_L}^{(i)}(x)$)$\exists L_0$ such that $\forall L>L_0, {\mu_L}^{(i)}\in\Psi(\Omega)$.

Proof: See appendix.

Corollary 4.3. $\sup_{x\in\Omega}|\mu_L^{(i)}(x)-\mu^{(i)}(x)|\to 0\Rightarrow \sup_{x\in\Omega}|V_L^{(i)}(x)-V^{(i)}(x)|\to 0$

4.3.3 Convergence of approximate IRL policy iteration to solution of the HJB equation

The IRL policy iteration algorithm with value function approximation is given as Algorithm 4.3. This section shows that this algorithm converges to the solution of the HJB equation (4.7).

Theorem 4.3. Under the assumptions of Theorem 4.2 the following is satisfied $\forall i \geq 0$

(i) $\sup_{x\in\Omega} |V_L^{(i)}(x) - V^{(i)}(x)| \to 0$

(ii) $\sup_{x\in\Omega} |\mu_L^{(i+1)}(x) - \mu^{(i+1)}(x)| \to 0$

(iii) $\exists L_0 : \forall L \geq L_0 \; \mu_L^{(i)}(x) \in \Psi(\Omega)$

Proof: See Abu-Khalaf and Lewis (2005). ■

Theorem 4.4.

$$\forall \varepsilon \geq 0, \exists i_0, L_0 : \forall i \geq i_0, L \geq L_0$$

(i) $\sup_{x\in\Omega} |V_L^{(i)}(x) - V^*(x)| < \varepsilon$

(ii) $\sup_{x\in\Omega} |\mu_L^{(i+1)}(x) - \mu^*(x)| < \varepsilon$

(iii) $\mu_L^{(i)}(x) \in \Psi(\Omega)$

Proof: The proof follows directly from Theorems 4.1 and 4.3. ■

4.4 Online IRL actor–critic algorithm for optimal adaptive control

This section discusses the online implementation of the IRL policy iteration algorithm on the actor–critic structure. The main features of the online adaptive IRL structure are presented while noting similarities with learning mechanisms in the mammal brain. Section 3.2 about the implementation and structure of the IRL optimal adaptive controller for the LQR is also germane.

4.4.1 Actor–critic structure for online implementation of adaptive optimal control algorithm

IRL policy iteration is given by Algorithm 4.2. This can be implemented in practice using IRL policy iterations with value function approximation given as Algorithm 4.3. To implement this, one measures the data set ($x(t)$, $x(t+T)$) along the system trajectory at the sample times t, $t+T$, $t+2T$, ... The integral reinforcement in

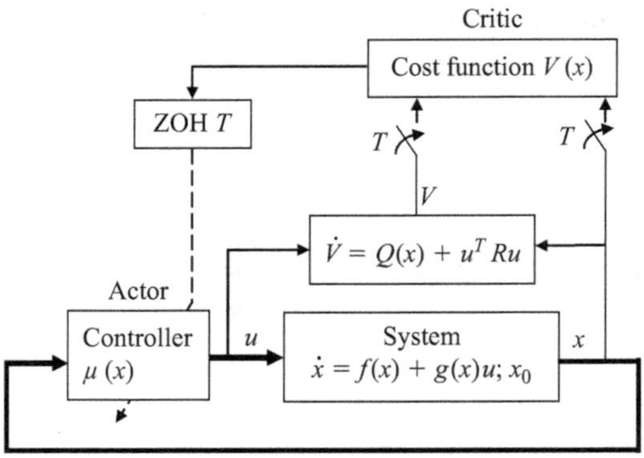

Figure 4.1 Hybrid actor–critic structure of IRL optimal adaptive controller

(4.9) and (4.16) can be evaluated by introducing a controller state

$$\dot{V} = r(x, u) = Q(x) + u^T Ru$$

This provides a dynamic memory that enables one to extract the information regarding the value associated with the given policy. If one resets $V(t)$ to zero at the beginning of each sample interval $[t, t+T)$, then the measurement $V(t+T)$ gives the integral reinforcement over time interval $[t, t+T)$ required to implement the policy evaluation step in (4.9), that is $V(t+T)$ gives the integral reinforcement term in (4.9). Thus, the IRL adaptive controller result is a dynamic controller whose memory is exactly the value $V(t)$ of using the current policy.

The structure of the IRL adaptive controller is presented in Figure 4.1. The policy iteration technique in this chapter has led us to adaptive control system that converges in real time to an optimal control solution by measuring data along the system trajectories and without knowing the drift dynamics of the system. We term this optimal adaptive control. The IRL optimal adaptive controller has an actor–critic structure, where the critic solves (4.9) for the value and the actor computes (4.11) for the new policy at each iteration. The controller dynamics is the value state equation $\dot{V} = r(x, u) = Q(x) + u^T Ru$. This structure is not a standard one in the control systems literature. It is a hybrid continuous-time/discrete-time adaptive control structure that has continuous-time controller dynamics and a discrete-time sampled data portion for policy evaluation and policy updates.

In another non-standard feature, the integration period T is not a sample period in the standard sense. The integration period T can change and need not be constant. According to Lemma 4.3, T is related to the non-singularity of $\mathbf{\Phi}$ in (4.21) and hence to the persistence of excitation of the regression vector $[\varphi_L(x(t+T)) - \varphi_L(x(t))]$ in (4.17). See the discussion on PE in Section 3.2.

One can consider the two approximation structures as neural networks. Then, the critic NN is (4.15) and the actor control NN is (4.22). These are implemented

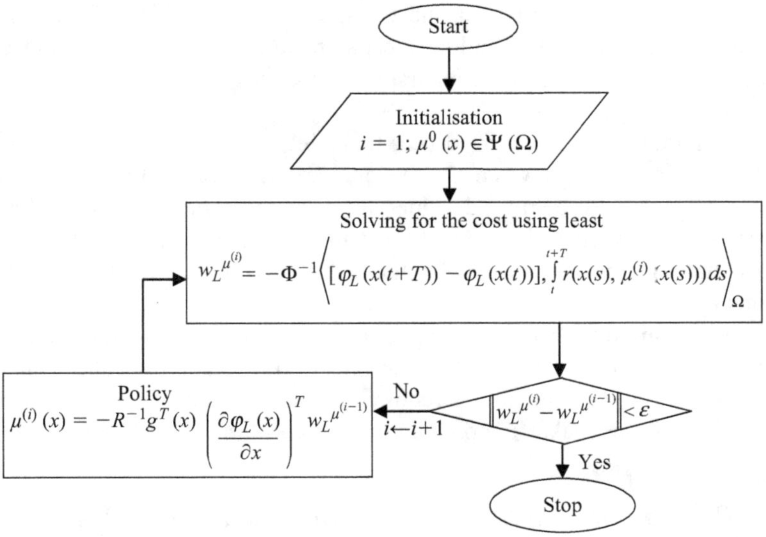

Figure 4.2 Flowchart of the online IRL policy iteration algorithm

with the basis or activation functions $\varphi_L(x)$ and $\nabla\varphi_L(x)$, respectively. The weights are updated by solving (4.16) at discrete instants of time, yet the actor NN operates in continuous time and generates a continuous control signal. See the figures and discussion in Section 3.2.

The flowchart of the online IRL algorithm is presented in Figure 4.2. All the calculations involved are performed at a supervisory level that operates based on discrete-time data $(x(t), x(t+T), V(t+T))$ measured in real time along the system trajectories. This high-level intelligent control structure implements the IRL policy iteration algorithm and uses the critic neural network to solve (4.16) to parameterize the performance of the continuous-time control system associated with a certain control policy. The high-level supervisory structure makes the decisions relative to the discrete-time moments at which both the actor and the critic parameters will be updated. The actor neural network (4.22) is part of the control system structure and performs continuous-time control, while its constant gain is updated at discrete moments in time. The algorithm converges to the solution of the continuous-time optimal control problem, as proved in Section 4.3, since the critic's update is based on the observations of the continuous-time cost over a finite sample interval. The net result is a *continuous-time controller* incorporated in a *continuous-time/discrete-time adaptive structure*, which includes the continuous-time dynamics of the cost function and operates based on sampled data, to perform the policy evaluation and policy update steps at discrete moments in time.

The cost function solution, given by (4.21), can be obtained in real time after a sufficient number of data points are collected along state trajectories in the region of interest Ω. In practice, the matrix inversion in (4.21) is not performed, the solution of the equation being obtained using algorithms that involve techniques such as Gaussian elimination, back substitution and Householder reflections. Also, the least-squares method for finding the parameters of the cost function can be replaced with any other suitable, recursive or not recursive, method of parameter identification.

The iterations will be stopped (i.e. the critic will stop updating the control policy) when the error between the system performance evaluated at two consecutive steps will cross below a designer-specified threshold. Also, when this error becomes bigger than the above-mentioned threshold, indicating a change in the system dynamics, the critic will take again the decision to start tuning the actor parameters.

We note again that there is no required knowledge about the system dynamics $f(x)$ for the evaluation of the cost or the update of the control policy. However, knowledge on the $g(x)$ function is required for the update of the control policy, using (4.22), and this makes the online tuning algorithm only partially model free.

4.4.2 Relation of adaptive IRL control structure to learning mechanisms in the mammal brain

It is interesting to note the rough similarity between the IRL optimal adaptive controller structure and learning mechanisms in the mammal brain. The critic structure learns, in an episodic manner and based on samples of the reward signal from the environment, the parameters of a function that describes the actor performance. Once a performance evaluation episode was completed, the critic passes this information to the actor structure, which uses it to adapt for improved performance. At all times the actor must perform continuous-time control for the system (the environment in which optimal behavior is sought). This description of the way in which the actor–critic structure works while searching for continuous-time optimal control policies points out the existence of two time scales for the mechanisms involved:

- a fast time scale that characterizes the continuous-time control process, and
- a slower time scale that characterizes the learning processes at the levels of the critic and the actor.

Thus, the actor and critic structures perform tasks at different operation frequencies in relation with the nature of the task to be performed (i.e. learning or control). Evidence regarding the oscillatory behavior naturally characterizing biological neural systems is presented in a comprehensive manner in Levine *et al.* (2000). Different oscillation frequencies are connected with the way in which different areas of the brain perform their functions of processing the information received from the sensors. Low-level control structures must quickly react to new information received from the environment while higher level structures slowly evaluate the results associated with the present behavior policy. Low-level control in the human nervous system corresponds to the muscle motor control system, which operates at a sample period of about 200 Hz. On the other hand, higher level control functions may operate in the hippocampus and thalamus, where they might be characterized by theta rhythms of 4–10 Hz.

Section 4.3 showed that though it has only limited information ($x(t)$, $x(t+T)$, $V(t+T)$) measured at discrete times about the system states and the controller state, that is $V(t)$, the critic is able to evaluate the infinite-horizon continuous-time performance of the system associated with a given control policy described in terms of the actor parameters. The critic learns the cost function associated with a certain control behavior by solving (4.16) based on a computed temporal difference (TD) error signal, given by $V(t+T) - V(t)$.

It is interesting to mention here that in a number of reports (for example Sandberg (1998), Schultz *et al.* (1997)) it is argued that the temporal difference error between the received and the expected rewards is physically encoded in the dopamine signal produced by basal ganglia structures in the mammal brain. At the same time, it is known that the dopamine signal encoding the temporal error difference favors the learning process by increasing the synaptic plasticity of certain groups of neurons.

4.5 Simulation results

In this section, the adaptive optimal control algorithm is tested in simulation considering two non-linear systems for which the optimal cost function and optimal controller are known. These examples were constructed using the converse HJB approach (Nevistic and Primbs, 1996), which allows one to construct systems for which the solution to the HJB equation is known. This allows verification that the IRL adaptive controller converges to the correct value function and optimal control solution.

4.5.1 Non-linear system example 1

The first non-linear system is given by

$$\begin{cases} \dot{x}_1 = -x_1 + x_2 \\ \dot{x}_2 = f(x) + g(x)u \end{cases} \tag{4.30}$$

with $f(x) = -\frac{1}{2}(x_1 + x_2) + \frac{1}{2}x_2 \sin^2(x_1)$, $g(x) = \sin(x_1)$. The cost function is $V^u(x(t)) = \int_t^\infty (Q(x) + u^2)\, d\tau$ with $Q(x) = x_1^2 + x_2^2$. Then, the optimal value for this system is $V^*(x) = \frac{1}{2}x_1^2 + x_2^2$ and the optimal controller is $u^*(x) = -\sin(x_1)x_2$.

The simulation was conducted using data obtained from the system at every 0.1 s. We note here that the value of this sample time is not relevant for the cost function identification procedure. In fact data does not have to be measured with a fixed sample time, as long as it is suitable for learning (i.e. caries new information on the cost function to be identified). In this sense, as long as the measured signals did not reach steady-state values, meaning that the measured data is not redundant, the sampling could be executed as fast as the hardware permits it. At the same time, as we used a batch method for identifying the cost function parameters in the least-squares sense a larger number of samples will lead to better approximation of the cost function. However, this batch procedure can be replaced with a recursive one, or a recursive procedure on time windows, such that the parameters of the cost function will be adapted over time as more data is acquired.

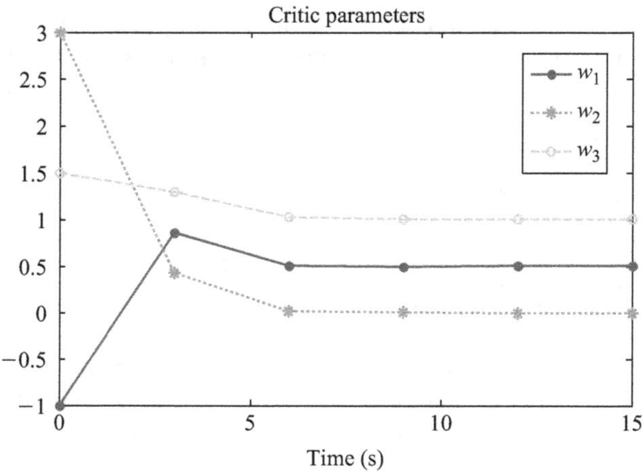

Figure 4.3 Convergence of the critic parameters

For the purpose of demonstrating the algorithm the initial state of the system is taken to be different than zero. For each iteration we considered data measured along five trajectories defined by five different initial conditions chosen randomly in $\Omega = \{-1 \le x_i \le 1; i = 1, 2\}$. The initial stabilizing controller was taken as $\mu^{(0)}(x) = -\frac{3}{2}\sin(x_1)(x_1 + x_2)$. The cost function $V^{\mu^{(i)}}(x)$ was approximated by the following smooth function, for $x \in \Omega$, $V_L^{\mu^{(i)}}(x) = (w_L^{\mu^{(i)}})^T \varphi_L(x)$ with $L = 3, w_3^{\mu^{(i)}} = \begin{bmatrix} w_1^{\mu^{(i)}} & w_2^{\mu^{(i)}} & w_3^{\mu^{(i)}} \end{bmatrix}^T$ and $\varphi_3(x) = \begin{bmatrix} x_1^2 & x_1 x_2 & x_2^2 \end{bmatrix}^T$.

To solve online for the parameters $w_3^{\mu^{(i)}}$ of the cost function, at each iteration step we setup a least-squares problem with the solution given by (4.21). At each iteration step we solved for $w_3^{\mu^{(i)}}$ using 30 data points consisting of the measured cost function associated with a given control policy over 30 time intervals $T = 0.1$ s, the initial state and the system state at the end of each time interval, 6 points measured over each of the 5 trajectories in the state space. In this way, every 3 s, the cost function was solved for and a policy update was performed. The result of applying the algorithm is presented in Figure 4.3.

One can see from the figure that the parameters of the critic converged to the coefficients of the optimal cost function: $V^*(x) = \frac{1}{2}x_1^2 + x_2^2$, that is $w_3^{\mu^*} = \begin{bmatrix} 0.5 & 0 & 1 \end{bmatrix}^T$.

4.5.2 Non-linear system example 2

In this example, we present the results obtained for a system that has stronger non-linearities and quartic cost. We consider the non-linear system given by

$$\begin{cases} \dot{x}_1 = -x_1 + x_2 + 2x_2^3 \\ \dot{x}_2 = f(x) + g(x)u \end{cases} \tag{4.31}$$

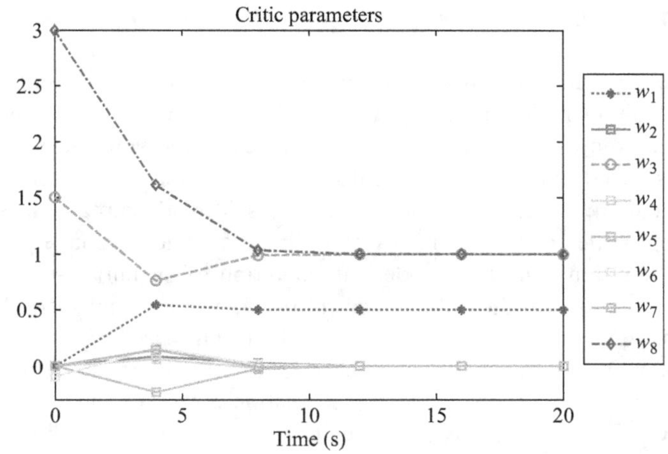

Figure 4.4 Convergence of the critic parameters

with $f(x) = -\frac{1}{2}(x_1 + x_2) + \frac{1}{2}x_2(1 + 2x_2^2)\sin^2(x_1), g(x) = \sin(x_1)$. Define $Q(x) = x_1^2 + x_2^2 + 2x_2^4$ and the cost function $V^u(x(t)) = \int_t^\infty (Q(x) + u^2)\, d\tau$. Then the optimal value for this system is $V^*(x) = \frac{1}{2}x_1^2 + x_2^2 + x_2^4$ and the optimal controller is $u^*(x) = -\sin(x_1)(x_2 + 2x_2^3)$.

The simulation was conducted using data obtained from the system at every 0.1 s. For each iteration we considered data measured along five trajectories defined by five different initial conditions chosen randomly in $\Omega = \{-1 \le x_i \le 1; i = 1, 2\}$. The initial stabilizing controller was taken as $\mu^{(0)}(x) = -\frac{1}{2}\sin(x_1)(3x_2 - 0.2x_1^2 x_2 + 12x_2^3)$. The cost function $V^{\mu^{(i)}}(x)$ was approximated on Ω as $V_L^{\mu^{(i)}}(x) = (w_L^{\mu^{(i)}})^T \varphi_L(x)$ with $L = 8$, $w_8^{\mu^{(i)}} = \begin{bmatrix} \mathbf{w}_1^{\mu^{(i)}} & \ldots & \mathbf{w}_8^{\mu^{(i)}} \end{bmatrix}^T$ and $\varphi_8(x) = \begin{bmatrix} x_1^2 & x_1x_2 & x_2^2 & x_1^4 & x_1^3x_2 & x_1^2x_2^2 & x_1x_2^3 & x_2^4 \end{bmatrix}^T$.

At each iteration step we solved for $w_8^{\mu^{(i)}}$ using 40 data points, that is 8 points measured on each of the 5 trajectories in Ω. Each data point consists of the measured cost function associated with the present control policy, over a time interval $T = 0.1$ s, and the system state at both ends of this interval. In this way, at every 4 s, the cost function was solved for and a policy update was performed. One notes that each data point set measured on each trajectory is sufficient to identify the parameters of the cost function corresponding to that given trajectory. However, it is often not the case that cost function parameters associated with one trajectory are equal to the cost function parameters associated with another trajectory. The result of applying the algorithm is presented in Figure 4.4.

The figure clearly shows that the parameters of the critic neural network converged to the coefficients of the optimal cost function $V^*(x) = \frac{1}{2}x_1^2 + x_2^2 + x_2^4$, that is $w_8^{u^*} = [0.5 \quad 0 \quad 1 \quad 0 \quad 0 \quad 0 \quad 0 \quad 1]$. One observes that after three iteration steps the parameters of the controller, obtained based on the update equation (4.22), are very close to the parameters of the optimal controller $u^*(x) = -\sin(x_1)(x_2 + 2x_2^3)$.

4.6 Conclusion

This chapter presents a continuous-time adaptive controller, based on policy iteration, which adapts online to learn the continuous-time optimal control policy without using knowledge about the drift dynamics of the non-linear system. The controller is based on the integral reinforcement learning (IRL) form of the CT Bellman equation, which does not contain the system dynamics and shares the beneficial properties of the discrete-time Bellman equations. Convergence of the IRL policy iteration algorithm, under the condition of an initial stabilizing controller, to the solution of the optimal control problem was established. A practical method was given for implementing the algorithm based on value function approximation (VFA). Proof of convergence for the IRL policy iteration algorithm using VFA, taking into account the approximation error, was also provided. The resulting IRL optimal adaptive controller is a hybrid continuous/discrete control structure with an actor–critic structure that converges online to the optimal control solution. Knowledge of the drift dynamics is not needed. The simulation results support the effectiveness of the online adaptive optimal controller.

Chapter 5
Generalized policy iteration for continuous-time systems

As discussed in Chapter 2, reinforcement learning methods have been applied for many years in the optimal feedback control of discrete-time (DT) systems. Applications of policy iteration and value iteration to continuous-time (CT) dynamical systems $\dot{x} = f(x) + g(x)u$ have lagged due to the fact that the CT Bellman equation $0 = r(x, u) + (\nabla V_x)^T(f(x) + g(x)u)$ explicitly contains the system dynamics. (In this equation, the cost function integrand is $r(x, u)$ and ∇V_x is the gradient vector.) The DT Bellman equation, on the other hand, is $V(x_k) = r(x_k, u_k) + \gamma V(x_{k+1})$, which does not contain the system dynamics.

Chapter 4 gives a policy iteration algorithm for non-linear CT systems that does not rely on knowing the full system dynamics. This algorithm is based on the integral reinforcement learning (IRL) form of the CT Bellman equation (Vrabie *et al.*, 2008, 2009; Vrabie, 2009; Vrabie and Lewis, 2009) that is $V(x(t)) = \int_t^{t+T} r(x(\tau), \mu(x(\tau)))\, d\tau + V(x(t + T))$ and does not contain the system dynamics. In fact, the IRL form of the CT Bellman equation has the same structure as the DT Bellman equation. Algorithm 4.1 is the standard form of policy iterations for CT systems, while Algorithm 4.2 is the IRL form and Algorithm 4.3 shows how to implement IRL policy iterations practically by using value function approximation.

A further difficulty with the CT Bellman equation $0 = r(x, u) + (\nabla Vx)^T(f(x) + g(x)u)$ is that it has only one occurrence of the value function. Therefore, it is not easy to imagine extending generalized policy iterations (GPIs) and value iterations as described in Chapter 2 to CT systems. These algorithms rely on the special form of the DT Bellman equation $V(x_k) = r(x_k, u_k) + \gamma V(x_{k+1})$, which has two occurrences of the value function evaluated at two times k and $k+1$.

Chapter 2 discusses generalized policy iteration (GPI) for DT systems. The chapter gives a class of algorithms for GPI for CT systems (Vrabie *et al.*, 2008, 2009; Vrabie, 2009; Vrabie and Lewis, 2009). These algorithms rely on the structure of the IRL Bellman equation $V(x(t)) = \int_t^{t+T} r(x(\tau), \mu(x(\tau)))\, d\tau + V(x(t + T))$. The practical usefulness of GPI is that it avoids the explicit solution of the IRL Bellman equation in IRL Algorithm 4.2 by developing a set of simplified iterations that converges to its solution. It is shown in this chapter that GPI provides a spectrum of iterative algorithms that includes at one end the PI Algorithm 4.2. At the other end of the spectrum GPI becomes the value iteration algorithm for CT systems presented in Chapter 6, where it is shown how to extend value iterations to CT systems using IRL.

The development of GPI for CT systems in this chapter is based on the IRL Bellman equation. The analysis is based on contraction maps and the Banach fixed-point theorem. Section 5.1 reviews the standard policy iteration (PI) approach to the solution of the infinite-horizon optimal control problem for non-linear systems, that is the optimal control problem discussed in Section 4.1, and IRL is reviewed. Section 5.2 presents the main result: a new formulation for the PI algorithm, with convergence proof, followed by the extension of GPI to CT systems. Section 5.3 briefly discusses the implementation aspects of the GPI algorithms using function approximators in an actor–critic structure. Section 5.4 presents simulation results obtained, first for a linear quadratic regulator (LQR) problem and second for the case of a non-linear system. It is shown that GPI solves the Riccati equation, or the Hamilton–Jacobi–Bellman (HJB) equation for the non-linear case, online in real time without knowing the full system dynamics.

5.1 Policy iteration algorithm for optimal control

The setup in this chapter is the non-linear CT system dynamics and cost function given in Section 4.1. Thus, consider the time-invariant affine-in-the-input dynamical system given by

$$\dot{x}(t) = f(x(t)) + g(x(t))u(x(t)); \quad x(0) = x_0$$

with state $x(t) \in \mathbb{R}^n$, $f(x(t)) \in \mathbb{R}^n$, $g(x(t)) \in \mathbb{R}^{n \times m}$ and control input $u(t) \in U \subset \mathbb{R}^m$. Assumptions on this system are as given in Section 4.1.

The cost function or *value function* associated with any admissible control policy $u(t) = \mu(x(t))$ is

$$V^{\mu}(x(t)) = \int_t^{\infty} r(x(\tau), \mu(x(\tau)))\, d\tau$$

Take here the cost integrand as $r(x, u) = Q(x) + u^T R u$ with $Q(x)$ positive definite (i.e. $\forall x \neq 0, Q(x) > 0$ and $x = 0 \Rightarrow Q(x) = 0$) and $R \in \mathbb{R}^{m \times m}$ a symmetric positive definite matrix. The value $V^{\mu}(x)$ is C^1.

A control policy $u(t) = \mu(x)$ is defined as admissible with respect to this system on a set $\Omega \subset \mathbb{R}^n$, denoted by $\mu \in \Psi(\Omega)$, if $\mu(x)$ is continuous on Ω, $\mu(0) = 0$, $\mu(x)$ stabilizes (4.1) on Ω, and $V(x_0)$ is finite $\forall x_0 \in \Omega$.

In the optimal control problem it is desired to select an admissible control policy that minimizes the value.

5.1.1 Policy iteration for continuous-time systems

Using Leibniz's formula, the infinitesimal version of the infinite integral value function is found to be the following.

CT Bellman equation

$$0 = r(x, \mu(x)) + (\nabla V_x^{\mu})^T (f(x) + g(x)\mu(x)), \quad V^{\mu}(0) = 0 \tag{5.1}$$

Here, ∇V_x^μ (a column vector) denotes the gradient of the cost function V^μ with respect to x. The CT Hamiltonian function is defined as

$$H(x, u, \nabla V_x) = r(x(t), u(t)) + (\nabla V_x)^T (f(x(t)) + g(x(t))u(t))$$

Based on Bellman equation and Hamiltonian, the standard policy iteration algorithm for CT systems is described as follows.

Algorithm 5.1. Continuous-time policy iteration

1. Select $u_0 \in \Psi(\Omega)$.
2. (Policy evaluation step.) Solve for V^i using the Bellman equation

$$H(x,\ u_{i-1}, \nabla V_x^i) = r(x(t), u_{i-1}(t)) + (\nabla V_x^i)^T (f(x(t)) + g(x(t))u_{i-1}(t)) = 0 \tag{5.2}$$

3. (Policy improvement step.) Find u_i that satisfies

$$u_i = \arg\min_{v \in \Psi(\Omega)} [H(x, v, \nabla V_x^i)] \tag{5.3}$$

which is explicitly

$$u_i(x) = -\frac{1}{2} R^{-1} g^T(x)\, \nabla V_x^i \tag{5.4}$$

■

The convergence of this algorithm has been proven in Leake and Liu (1967) for the case that (5.2) is exactly solved. This is an offline algorithm. It is further deficient in that the full system dynamics are needed to solve the Bellman equation (5.2). It is not clear how to provide online learning mechanisms to solve (5.2) in real time.

Conditioned by a suitable initialization, that is an admissible initial policy, PI provides the solution of the optimal control problem based on recursively solving (5.2) and (5.3) as the index $i \to \infty$. The solution V^i of (5.2) represents the value function associated with using the control policy u_{i-1}.

5.1.2 Integral reinforcement learning for continuous-time systems

To avoid the requirement in Algorithm 5.1 for knowing the drift dynamics of the system, $f(x)$, and to allow online implementation, Chapter 4 developed an equivalent formulation of CT policy iteration based on IRL.

Write the value function as the following equivalent form.

Integral reinforcement form of value function: IRL Bellman equation

$$V^\mu(x(t)) = \int_t^{t+T} r(x(\tau), \mu(x(\tau)))\, d\tau + V^\mu(x(t+T))$$

Note that this form of the Bellman equation does not contain the system dynamics ($f(\cdot)$, $g(\cdot)$). Lemma 4.1 showed that this equation is equivalent to the CT Bellman equation (5.1) in the sense that both equations have the same solution. The integral

$$\rho(x(t), t, t+T) = \int_t^{t+T} r(x(\tau), \mu(x(\tau)))\, d\tau$$

is known as the *integral reinforcement* on the time interval $[t, t+T]$.

Based on the IRL form of the Bellman equation one can write the following algorithm.

Algorithm 5.2. CT policy iteration using integral reinforcement learning

1. Select $u_0 \in \Psi(\Omega)$.
2. (Policy evaluation step.) Solve for V^i using IRL Bellman equation

$$\int_t^{t+T_0} r(x, u_{i-1})\, d\tau + V^i(x_{t+T_0}) - V^i(x_t) = 0, \quad V^i(0) = 0 \tag{5.5}$$

3. (Policy improvement step.) Find u_i that satisfies

$$u_i = \arg\min_{v \in \Psi(\Omega)} [H(x, v, \nabla V_x^i)] \tag{5.6}$$

■

In (5.5), x_t and x_{t+T_0} are shorthand notations for $x(t)$ and $x(t+T_0)$.

It was proven in Chapter 4 that (5.2) and (5.5), corresponding to the policy evaluation step, have the same solution. The advantage of using (5.5) is that this equation can be solved online based on real-time measurements, without any requirement for knowing the system drift dynamics $f(x)$. Online approaches to policy iterations were given in Section 4.3 using value function approximators to solve (5.5) and (5.6). Implementation is further discussed in Section 5.3 and is similar to the structure discussed in Section 4.4.

5.2 Generalized policy iteration for continuous-time systems

Methods for solving the Bellman equation (5.5) were given in Section 4.3. Nevertheless, (5.5) is a system of equations that must be solved either for the value (as in Algorithm 4.2) or (as in Algorithm 4.3) for the coefficients in a basis function expansion that approximates the value function. GPI was introduced for DT systems in Chapter 2 as a simplified method for solving the Bellman equation that uses iterations instead of requiring solution of a set of equations. Development of GPI methods for CT systems has been hindered by the inconvenient structure of the CT Bellman equation (5.1).

This section formulates the GPI algorithms for CT systems (Vrabie *et al.*, 2008, 2009; Vrabie, 2009; Vrabie and Lewis, 2009). These algorithms are based on the IRL Bellman equation. The analysis is based on contraction maps and the Banach fixed-point theorem. Section 5.1.1 introduces the mathematical tools that provide the mechanisms for the PI and GPI algorithm formulations given in Sections 5.2.2 and 5.2.3.

5.2.1 Preliminaries: Mathematical operators for policy iteration

Let X denote the space of bounded functionals $V(\cdot):\Omega\to\mathbb{R}$ with $V(x_t)>0, \forall x_t\in\Omega, V(0)=0$. X is a Banach space with the norm $\|V\|=\sup_{x\in\Omega}|V(x)|$. Define the dynamic programming operator $T_\mu : X\to X$

$$T_\mu V(x_t)=\int_t^{t+T_0} r(x,\mu)\,d\tau+V(x_{t+T_0}) \tag{5.7}$$

where x_{t+T_0} is the value of $x(\tau)$ at $\tau=t+T_0$, with $x(\tau)$ the solution of (3.1) for initial condition $x(t)$ (denoted x_t) and input $\{\mu(\tau);\tau\geq t\}$. Also, define the operator $T: X\to X$

$$TV(x_t)=\min_{u\in\Psi(\Omega)}\left\{\int_t^{t+T_0} r(x,u)\,d\tau+V(x_{t+T_0})\right\} \tag{5.8}$$

The first operator, $T_\mu : X\to X$, maps the cost functional $V(\cdot)\in X$ into the cost functional denoted $T_\mu V(\cdot)\in X$, while using the control policy $\mu\in\Psi(\Omega)$ over the time interval $[t, t+T_0]$. The sample period T_0 must be chosen such that $\forall x_t\in\Omega$ the solution of (3.1) at time $t+T_0$, using the control policy μ, satisfies $x_{t+T_0}\in\Omega_1\subseteq\Omega$. Note that if $\mu\in\Psi(\Omega)$ there exists a lower bound T_l such that $\forall T_0\geq T_l$ and $\forall x_t\in\Omega$ then $x_{t+T_0}\in\Omega$.

The formulation of the operator T_μ when the state-feedback optimal control problem for linear systems with quadratic performance index, that is LQR, is considered is given now. Using the parametric description of the value function $V(x)=x^TPx, \forall x\in\mathbb{R}^n$ and $T_\mu V(x)=x^TP_1^\mu x, \forall x\in\mathbb{R}^n$ and the control policy $\mu(x)=-K^\mu x$, this operator can be written as

$$x_t^TP_1^\mu x_t=\int_t^{t+T_0} x^T(\tau)[Q+(K^\mu)^TRK^\mu]x(\tau)\,d\tau+x_t^T\,e^{(A-BK^\mu)^TT_0}Pe^{(A-BK^\mu)T_0}x_t \tag{5.9}$$

It is interesting to note that this has a structure similar to the DT Lyapunov equation. We shall show these DT-like Lyapunov recursions converge to solutions to (5.5), which is equivalent in the LQR case to a CT Lyapunov equation.

Let X^P denote the set of all positive definite matrices that can serve as parametric representations of quadratic value functions. Denoting by $A_d^{T_0}\triangleq e^{(A-BK^\mu)T_0}$ the discrete version of the CT dynamics of the linear system, when a sampling period of T_0 is used, and writing the integral term as

$$x_t^TM^\mu x_t\equiv\int_t^{t+T_0} x^T(\tau)[Q+(K^\mu)^TRK^\mu]x(\tau)\,d\tau$$

with $M^{\mu} > 0$, introduce the operator $T'_{\mu} : X^P \to X^P$ defined as

$$T'_{\mu}P = P_1^{\mu} = M^{\mu} + (A_d^{T_0})^T P A_d^{T_0} \tag{5.10}$$

where P_k^{μ} denotes the composition of k copies of T'_{μ} applied on the parametric representation of the quadratic cost function $V(x) = x^T Px, x \in \mathbb{R}^n$, that is matrix P.

The operator defined by (5.8), $T : X \to X$, maps the cost functional $V(\cdot) \in X$ into the cost functional denoted $TV(\cdot) \in X$, while using over the time interval $[t, t + T_0]$ the control policy u that is solution to the finite-horizon optimal control problem defined by the right-hand side of (5.8).

It is important to see that the control solution of (5.8) is not the same as $u = \arg\min_{v \in \Psi(\Omega)} [H(x, v, \nabla V_x)]$; in the first case a time-varying control policy is obtained while in the latter case the resulting policy is time-invariant. For example, in a linear system case with quadratic cost function the control solution given by (5.8) is $u(\tau, x) = -R^{-1}B^T P(\tau)x$, where $P(\tau)$ is the solution of the differential Riccati equation over the time interval $[t, t + T_0]$ with final condition $P_{t+T_0} = P$. On the other hand, the solution obtained using $u = \arg\min_{v \in \Psi(\Omega)} [H(x, v, \nabla V_x)]$; is $u(x) = -R^{-1}B^T Px$. Thus, one can see that using (5.8) as basis for the policy improvement step would lead to a significant modification of the policy iteration algorithm.

Using the introduced operators we can also write

$$TV(x_t) = \min_{u \in \Psi(\Omega)} \{T_u V(x_t)\} \tag{5.11}$$

Also, Bellman's optimality principle can be formulated as

$$TV^*(x_t) = \min_{u \in \Psi(\Omega)} \{T_u V^*(x_t)\} = V^*(x_t) \tag{5.12}$$

In the following T^k and T_u^k will denote the composition of k copies of T and T_u.

5.2.2 Contraction maps for policy iteration

Two equivalent formulations of continuous-time PI were given as Algorithm 5.1, equations (5.2), (5.3), and Algorithm 5.2, equations (5.5), (5.6). This section presents results that allow a third formulation of the policy iteration algorithm based on the functional mapping operators introduced in the previous section. The analysis is based on contraction maps and the Banach fixed-point theorem. The following results are required.

Lemma 5.1. Let $\mu \in \Psi(\Omega)$. Then $V^{\mu} \in X$ is a fixed point of the mapping $T_{\mu} : X \to X$.

Proof: Let V^{μ} denote the cost associated with the policy μ. Then, using the definition in (5.7), we have

$$T_{\mu}V^{\mu}(x_t) = \int_t^{t+T_0} r(x, \mu)\, d\tau + V^{\mu}(x_{t+T_0}) \tag{5.13}$$

which is

$$T_\mu V^\mu(x_t) = V^\mu(x_t) \tag{5.14}$$

thus $V^\mu(x_t)$ is a fixed point of the mapping T_μ.

Note that the equation

$$T_\mu V(x_t) = \int_t^{t+T_0} r(x,\mu)\, d\tau + V(x_{t+T_0}) = V(x_t) \tag{5.15}$$

has a unique solution denoted by $V^\mu(x_t)$, which is also the solution of $H(x, \mu, \nabla V_x) = 0$. ■

Lemma 5.2. Let $\mu \in \Psi(\Omega)$. Then $T_\mu : X \to X$ is a contraction mapping on X.

Proof: Let $V,\ W \in X$, then

$$T_\mu V(x_t) = \int_t^{t+T_0} r(x,\mu)\, d\tau + V(x_{t+T_0}) \tag{5.16}$$

$$T_\mu W(x_t) = \int_t^{t+T_0} r(x,\mu)\, d\tau + W(x_{t+T_0}) \tag{5.17}$$

Subtracting the two equations one gets

$$T_\mu V(x_t) - T_\mu W(x_t) = V(x_{t+T_0}) - W(x_{t+T_0}) \tag{5.18}$$

T_0 is chosen such that $\forall x_t \in \Omega$ and $x_{t+T_0} \in \Omega_1 \subseteq \Omega$.

Then

$$\sup_{\Omega}(|V - W|)(x) \le \sup_{\Omega}(|V - W|)(x) \tag{5.19}$$

which together with (5.16) gives

$$\sup_{\Omega_1}(|T_\mu V - T_\mu W|)(x) \le \sup_{\Omega}(|V - W|)(x)$$

This is

$$\|T_\mu V - T_\mu W\| \le \|V - W\| \tag{5.20}$$

This proves the lemma. ■

Subtracting (5.13) from (5.16), and making use of (5.14), one obtains

$$T_\mu V(x_t) - V^\mu(x_t) = V(x_{t+T_0}) - V^\mu(x_{t+T_0})$$

resulting in

$$\|T_\mu V - V^\mu\| \leq \|V - V^\mu\|, \quad \forall V \in X \tag{5.21}$$

The next corollary of Lemma 5.2 considers the formulation of the infinite-horizon optimal control problem for linear systems with quadratic performance index, that is the LQR problem. In this case it is known that the value function associated with a given admissible state-feedback policy can be exactly represented by the parametric description $V(x) = x^T Px, \forall x \in \mathbb{R}^n$. The operator T'_μ defined by (5.10) is now used. X^P, equipped with the spectral radius matrix norm defined as $\rho(A) \equiv \|A\|_\rho \triangleq \max_i(|\lambda_i|)$, where λ_i, the eigenvalue of A, is a Banach space.

Corollary 5.1. T'_μ: $X^P \to X^P$ is a contraction map on X^P.

Proof: See appendix.

Lemma 5.3. The mapping $T_\mu : X \to X$ has a unique fixed point on X that can be obtained using

$$V^\mu(x_t) = \lim_{k\to\infty} T_\mu^k V(x_t), \quad \forall V(x_t) \in X \tag{5.22}$$

Proof: Using the Banach fixed-point theorem, since $T_\mu : X \to X$ is a contraction on X (Lemma 5.2) then its fixed point $V^\mu \in X$ (Lemma 5.1) is the unique fixed point. Moreover, the unique fixed point can be determined as the limit of the iterative sequence defined by $V_k^\mu(\cdot) = T_\mu V_{k-1}^\mu(\cdot) = T_\mu^k V_0^\mu(\cdot)$, $k \geq 1$, $V_0^\mu(\cdot) = V(\cdot) \in X$, that is $V^\mu(x_t) = \lim_{k\to\infty} T_\mu^k V(x_t), \forall V(x_t) \in X$. ■

5.2.3 A new formulation of continuous-time policy iteration: Generalized policy iteration

Using the result in Lemma 5.3 is now given a third formulation for the policy iteration algorithm that makes use of the operator defined by (5.7). This algorithm is known as iterative or GPI and replaces solution of the IRL Bellman equation by an iterative procedure that converges to its solution. This simplifies the computations required.

Algorithm 5.3. CT generalized policy iteration using IRL

1. Select $u_0 \in \Psi(\Omega)$
2. (Policy evaluation step.) Solve for $V^{u_{i-1}}(x_t)$ (denoted with $V^i(x_t)$) using the iteration

$$V^i(x_t) = \lim_{k\to\infty} T_{u_{i-1}}^k V(x_t) \tag{5.23}$$

starting with any $V(x_t) \in X$, and T_μ defined by (5.7).

3. (Policy improvement step.) Find u_i that satisfies

$$u_i = \arg\min_{v \in \Psi(\Omega)} [H(x, v, \nabla V_x^i)] \tag{5.24}$$

■

A variant of the above policy iteration algorithm is obtained when one starts the policy evaluation step for $V^{u_{i-1}}(x_t) = V^i(x_t)$ with the cost functional obtained at the previous step

$V^{u_{i-2}}(x_t) = V^{i-1}(x_t)$, that is (5.23) becomes

$$V^i(x_t) = \lim_{k \to \infty} T_{u_{i-1}}^k V^{i-1}(x_t) \tag{5.25}$$

5.2.4 *Continuous-time generalized policy iteration*

The formulation of the GPI algorithm for CT systems with continuous state and action space is now given.

Algorithm 5.4. Generalized policy iteration

1. Select $u_0 \in \Psi(\Omega)$.
2. (Approximate policy evaluation step.) Approximately solve the Bellman equation for $V^i(x_t)$ using the iteration

$$V^i(x_t) \triangleq V_k^i(x_t) = T_{u_{i-1}}^k V(x_t) \tag{5.26}$$

starting with any $V(x_t) \in X$, for some $k \geq 1$.

Note that this step can also be replaced by

$$V^i(x_t) \triangleq V_k^i(x_t) = T_{u_{i-1}}^k V^{i-1}(x_t) \tag{5.27}$$

3. (Policy improvement step.) Find u_i that satisfies

$$u_i = \arg\min_{v \in \Psi(\Omega)} [H(x, v, \nabla V_x^i)] \tag{5.28}$$

■

One clearly sees that when $k \to \infty$ in (5.22) we encounter the regular PI algorithm with the policy evaluation step given by (5.23). For the case $\infty > k \geq 1$, one obtains the so-called *optimistic policy iteration* (Sutton and Barto, 1998), with the policy evaluation step given by

$$V_0^i(x_t) = T_{u_{i-1}}^k V(x_t) \tag{5.29}$$

In this 'optimistic' case the policy update step is executed prior to the convergence to the true value associated with the current control policy. In (5.29), the notation $V_0^i(\cdot)$ was used to make the point that the value resulting from (5.26) is not the true value associated with the current control policy $u_{i-1}(x)$, that is $V^i(\cdot)$.

When $k = 1$, the GPI becomes the next algorithm.

Algorithm 5.5. CT value iteration variant with initial stabilizing policy

1. Select $u_0 \in \Psi(\Omega)$.
2. (Value function update step.) Solve for $V^i(x_t)$ using only one value update step

$$V^i(x_t) = T_{u_{i-1}} V^{i-1}(x_t) \tag{5.30}$$

3. (Policy update step.) Find u_i that satisfies

$$u_i = \arg\min_{v \in \Psi(\Omega)} [H(x, v, \nabla V_x^i)] \tag{5.31}$$

■

This algorithm is similar in form to the value iteration algorithm given for DT systems in Chapter 2. A key difference between Algorithm 5.5 and the DT value iteration algorithm discussed in Chapter 2 is that, in the case of true value iteration, the requirement for a stabilizing initial policy $u_0 \in \Psi(\Omega)$ is removed. This is a significant improvement, since for non-linear systems it may be difficult to find stabilizing controls. Chapter 6 presents a VI algorithm for CT systems that does not require a stabilizing initial policy.

The flowchart of the GPI Algorithm 5.4 is presented in Figure 5.1.

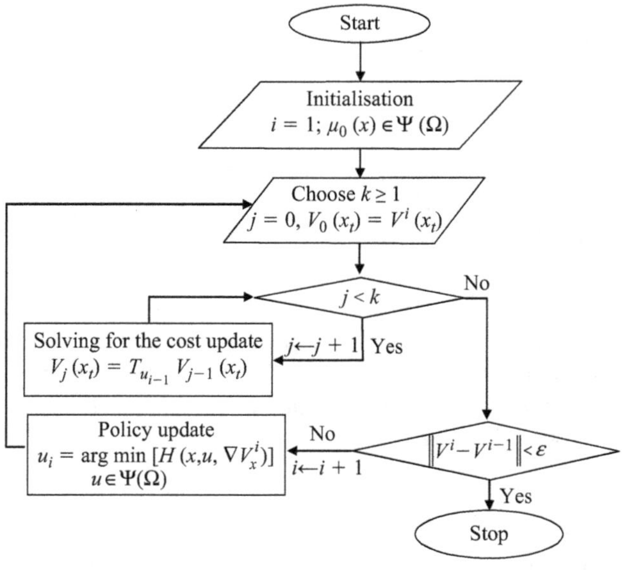

Figure 5.1 Flow chart of the generalized policy iteration (GPI) Algorithm 5.3

5.3 Implementation of generalized policy iteration algorithm

The online learning GPI Algorithm 5.4 is implemented on an actor–critic structure. In a general case the two structures can be neural networks (NNs) that are universal approximators (Hornik *et al.*, 1990). See Section 4.4 about implementation of IRL policy iterations for CT systems. Those remarks are germane here.

For value function approximation the cost $V^i(x) \in X$ will be represented as

$$V^i(x) = \sum_{j=1}^{L} w_j^i \phi_j(x) = (w_L^i)^T \varphi_L(x) \tag{5.32}$$

This can be interpreted as a neural network with L neurons in the hidden layer and activation functions $\phi_j(x) \in C^1(\Omega)$, $\phi_j(0) = 0$. $\varphi_L(x)$ denotes the vector of activation functions and w_L^i denotes the vector of the parameters of the neural network, with w_j^i the weights of the neural network. The activation functions should be selected to provide a complete basis for the space of value functions over Ω.

To solve for the cost function $V^i(x)$ in (5.27), the jth step of the value function update, which is

$$V_j^i(x_t) = T_{u_{i-1}} V_{j-1}^{i-1}(x_t), \quad V_0^i = V^{i-1}, 1 \le j \le k \tag{5.33}$$

can be written as

$$(w_L^j)^T \varphi_L(x_t) = \int_t^{t+T_0} r(x, u_i(x))\, d\tau + (w_L^{j-1})^T \varphi_L(x_{t+T}) \tag{5.34}$$

with $V^i(x_t) = (w_L^k)^T \varphi_L(x_t)$. The parameters of the value function approximation will be tuned, at each iterative step (5.33) of (5.27), to minimize, in the least-squares sense, the objective

$$S = \int_{\Omega^{ui}_{\{x_0\}_n}} \delta_L^j(x) \delta_L^j(x)\, dx \tag{5.35}$$

In (5.35), $\Omega^{u_i}_{\{x_0\}_n}$ denotes a set of trajectories generated by the policy u_i, from the initial conditions $\{x_0\}_n \subset \Omega$, and

$$\delta_L^j(x_t) = \int_t^{t+T_0} r(x, u_i(x))\, d\tau + (w_L^{j-1})^T \varphi_L(x_{t+T}) - (w_L^j)^T \varphi_L(x_t)$$

The quantity $\delta_L^j(x_t)$ can be viewed as the *temporal difference* residual error.

Using the inner product notation for the Lebesgue integral, the least-squares solution of (5.34) is

$$w_L^j = -\Phi^{-1} \left\langle \varphi_L(x_t), \int_t^{t+T_0} r(x, u_i(x)\, d\tau + (w_L^{j-1})^T \varphi_L(x_{t+T})) \right\rangle_{\Omega^{u_i}_{\{x_0\}_n}} \tag{5.36}$$

where $\Phi = \left\langle \varphi_L(x_t), \varphi_L(x_t)^T \right\rangle_{\Omega^{u_i}_{\{x_0\}_n}}$.

After updating the value function to solve equation (5.34) k times, that is once the approximate solution of (5.27) had been obtained, the policy update step, given by (5.28), is

$$u_{i+1}(x) = -R^{-1}g^T(x)\left(\frac{\partial \varphi_L(x)}{\partial x}\right)^T w_L^i \tag{5.37}$$

Note that in this implementation, after the policy update step is executed, the parameters of the two approximation structures, namely, actor and critic, are the same.

Finally, it is noted that all approximate dynamic programming (ADP) algorithms are developed on the assumption that correct estimation of the value function is possible. This means that, in order to successfully apply the online learning algorithm, enough excitation must be present in the system to guarantee correct estimation of the value function at each value update step. In relation with this, one must also note that the greedy policies obtained at the policy update steps are admissible policies and thus not excitatory. This is the well known exploration/exploitation dilemma (Sutton and Barto, 1998), which characterizes adaptive controllers that have simultaneous conflicting goals such as optimal control and fast and effective adaptation/learning.

5.4 Simulation results

This example shows how GPI Algorithm 5.3 can be used to solve the optimal control problem online in real time without solving the Riccati equation or, in the non-linear case, the HJB equation. Practically, the GPI algorithm solves these equations online by measuring data along the system trajectories and without knowing the system drift dynamics A, or in the non-linear case $f(x)$.

5.4.1 Example 1: Linear system

This section presents comparative simulation results using the GPI approach to solving the LQR problem considering the linear model of the F16 short period dynamics given in Sontag and Sussmann (1995).

The description of the linear system is given by matrices

$$A = \begin{bmatrix} -1.01887 & 0.90506 & -0.00215 \\ 0.82225 & -1.07741 & -0.1755 \\ 0 & 0 & -20.2 \end{bmatrix}, \qquad B = \begin{bmatrix} 0 \\ 0 \\ 20.2 \end{bmatrix}$$

The infinite-horizon quadratic cost function to be minimized is characterized by the identity matrices Q and R of appropriate dimensions. The optimal value function obtained by solving the algebraic Riccati equation (ARE) is

$$P = \begin{bmatrix} 1.4116 & 1.1539 & -0.0072 \\ 1.1539 & 1.4191 & -0.0087 \\ -0.0072 & -0.0087 & 0.0206 \end{bmatrix}$$

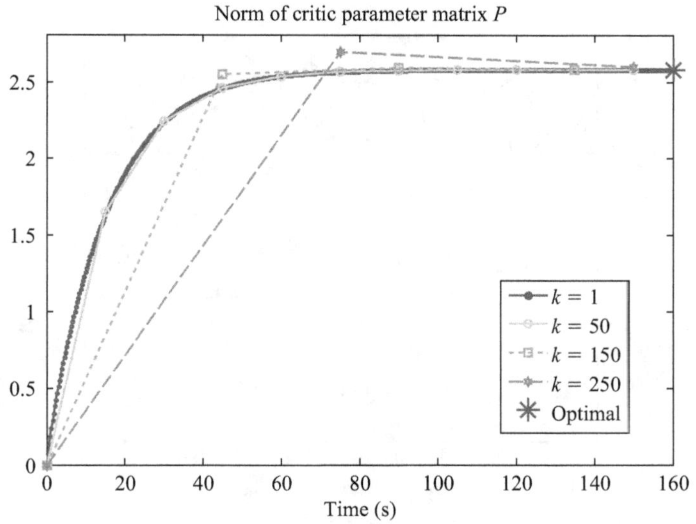

Figure 5.2 Comparative view of the results obtained while using the GPI algorithm for different values of the parameter k in terms of the norm of the critic parameters given by matrix P; the relevant values are indicated by the marker points while the connecting lines are only intended to provide ease of visualization

Figure 5.2 presents a comparative view of the results obtained with the various GPI algorithms (for different values of the parameter k) in terms of the norm of the cost function matrix P, considering the F-16 system.

As the system is stable, the GPI algorithms (applied for different values of k) were initialized using the state-feedback controller $K_0 = 0$. The simulation was conducted using data obtained from the system at every 0.05 s. In this way, at each 0.3 s, enough data is collected from the system to solve for the value of the matrix P and perform a value function update, as there are six independent elements in the symmetric matrix P, which parameterizes the value function associated with any admissible state-feedback controller. After a number of k updates the solution given by (5.27) is used to perform a policy update step.

One can see from Figure 5.2 that the number of iterative steps (i.e. value function and policy update steps) required for the GPI algorithm to converge is inversely proportional to the number of iterative updates used to solve the value function update step (i.e. parameter k of the GPI algorithm).

5.4.2 Example 2: Non-linear system

We now consider the non-linear system described by the equation

$$\begin{cases} \dot{x}_1 = -x_1 + x_2 \\ \dot{x}_1 = -\dfrac{1}{2}x_1 - \dfrac{1}{2}x_2((1-(\cos(2x_1)+2)^2)) + (\cos(2x_1)+2)u \end{cases} \tag{5.38}$$

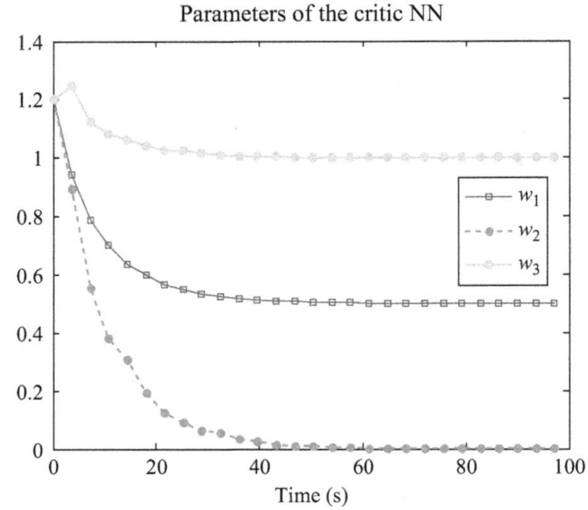

Figure 5.3 Convergence of the critic parameters to the optimal values using sequential GPI with K = 1

This system was designed using the converse HJB approach (Nevistic and Primbs, 1996), such that, when the cost function to be minimized is described by $r(x, u) = x^T Qx + u^T Ru$, with Q, R identity matrices of appropriate dimensions, the optimal cost function is $V^*(x) = \frac{1}{2}x_1^2 + x_2^2$. The critic is given by the equation

$$V_{W_1}(x) = [w_1 \quad w_2 \quad w_3][x_1^2 \quad x_1x_2 \quad x_2^2]^T + \varepsilon(x) \tag{5.39}$$

Figures 5.3–5.5 show the convergence of the parameters of the critic when the sequential GPI algorithm was used. The number of iterative steps to solve for the value function using iteration (5.27) were $K = 1$ (i.e. heuristic dynamic programming (HDP) algorithm), $K = 5$ and $K = 50$.

All the results were obtained while measuring data from the system using a sampling time interval $T_0 = 0.1$ s. At each iterative step, (5.34) was solved in the least-squares sense using 36 DT measurements from the system along six different, randomly chosen, trajectories in $\Omega = \{(x_1, x_2); -1 \leq x_1 \leq 1, \ -1 \leq x_2 \leq 1\}$.

The number of points and of trajectories used to solve (5.34) are a matter of fine tuning the algorithm and depends on the experience of the engineer relative to the system dynamics, in a similar manner with the choice of the sample time T_0.

Comparing the results in Figures 5.3–5.5, one observes that increasing the number of steps in the iterative algorithm used at the critic training phase leads to an increase in the time until the critic converges to the optimal critic. Nonetheless, in all instances the sequential GPI algorithm leads to convergence to the optimal cost and optimal control policy.

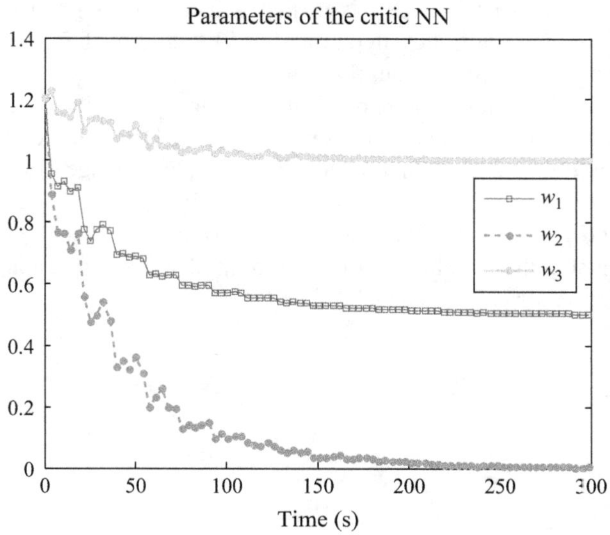

Figure 5.4 Convergence of the critic parameters to the optimal values using sequential GPI with K = 5

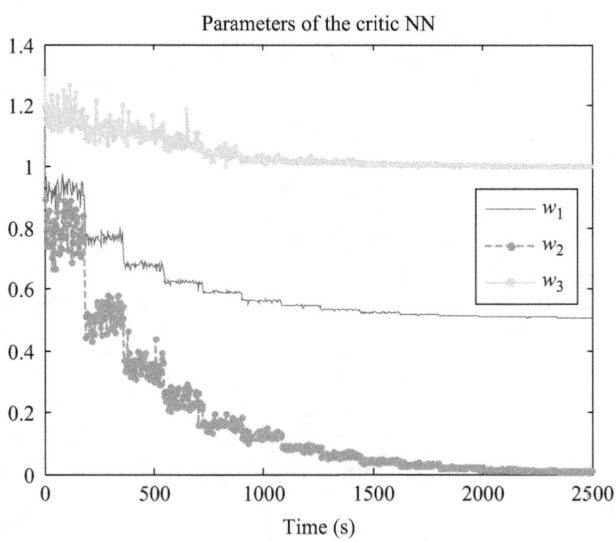

Figure 5.5 Convergence of the critic parameters to the optimal values using sequential GPI with K = 50

5.5 Conclusion

This chapter gives the formulation of GPI algorithms in a CT framework. This is based on the IRL form of the Bellman equation, which allows the extension of GPI to CT systems. It is seen that GPI is in fact a spectrum of iterative algorithms, which has at one end the policy iteration algorithm and at the other a variant of the value iteration algorithm. The algorithms can be implemented in a partially model-free setup using function approximators on an actor–critic structure. That is, GPI solves the Riccati equation in the LQR case, or the HJB equation for non-linear optimal control, online in real time without requiring knowledge of the system drift dynamics $f(x)$.

The new GPI algorithm could lead to a CT formulation of the Q function. This is an important step that would result in a set of model-free direct adaptive optimal control algorithms for CT systems.

Chapter 6
Value iteration for continuous-time systems

The idea of value iteration has been applied to the online learning of optimal controllers for discrete-time (DT) systems for many years (see Chapter 2). In the work of Werbos (1974, 1989, 1991, 1992, 2009) a family of DT learning control algorithms based on value iteration ideas has been developed. These techniques are known as approximate dynamic programming or adaptive dynamic programming (ADP). ADP includes heuristic dynamic programming (HDP) (which is value iteration), dual heuristic programming and action-based variants of those algorithms, which are equivalent to Q learning for DT dynamical system $x_{k+1} = f(x_k) + g(x_k)u_k$. Value iteration algorithms rely on the special form of the DT Bellman equation $V(x_k) = r(x_k, u_k) + \gamma V(x_{k+1})$, with $r(x_k, u_k)$ the utility or stage cost of the value function. This equation has two occurrences of the value function evaluated at two times k and $k+1$ and does not depend on the system dynamics $f(x_k), g(x_k)$.

The extension of value iteration techniques to continuous-time (CT) systems of the form $\dot{x} = f(x) + g(x)u$ has been hindered by the form of the CT Bellman equation $0 = r(x, u) + (\nabla V_x)^T (f(x) + g(x)u)$. This equation has only one occurrence of the value function. Based on this equation, it is not easy to imagine extending generalized policy iterations (GPIs) or value iterations, as described in Chapter 2, to CT systems. A further deficiency of the CT Bellman equation is that the dynamics $f(x) + g(x)$ appears, and so must be known for its solution.

In Chapter 5 was given a class of algorithms for generalized policy iteration (GPI) for continuous-time systems. These algorithms were introduced in Vrabie *et al.* (2008, 2009), Vrabie (2009), Vrabie and Lewis (2009) and rely on the structure of the integral reinforcement learning (IRL) Bellman equation $V(x(t)) = \int_t^{t+T_0} r(x(\tau), \mu(x(\tau)))\, d\tau + V(x(t+T))$, which shares the beneficial properties of the DT Bellman equation. It was shown in Chapter 5 that CT GPI provides a family of algorithms that contains policy iteration at one end and a variant of value iteration at the other.

Based on the IRL Bellman equation, in this chapter is formulated the value iteration or HDP algorithm for continuous-time systems (Vrabie *et al.*, 2008, 2009; Vrabie, 2009; Vrabie and Lewis, 2009). The continuous-time linear quadratic regulation (LQR) problem is considered, where the dynamics have the form $\dot{x} = Ax + Bu$. Focusing on the LQR problem allows easy comparison between the policy iteration algorithm (i.e. Newton's method) and the HDP (value iteration) algorithm, while providing some insight relative to the convergence of the HDP algorithm. The chapter is short, but this material is given the form of a separate chapter due to the importance of the extension of value iteration to CT systems.

In this chapter, bringing together concepts from ADP and control systems theory, and based on the formulation of the GPI algorithm introduced in Chapter 5, we formulate and analyze an ADP technique that offers solution, obtained online in a forward-in-time fashion, to the continuous-time infinite-horizon optimal control problem for linear systems. The online learning method makes use only of partial knowledge on the system dynamics (i.e. the drift dynamics, specified by the system matrix A need not be known).

This technique is a continuous-time approach to HDP (CT-HDP) for linear systems. CT-HDP solves online for the continuous-time optimal value function – an approach also known as *V-learning*. It will be shown that the proposed iterative ADP algorithm is in fact a quasi-Newton method. It solves the algebraic Riccati equation (ARE) in real time without knowing the system A matrix. Unlike the case of the policy iteration algorithm, for CT-HDP an initial stabilizing control policy is not required. CT-HDP yields an online direct adaptive control algorithm that determines the optimal control solution without knowing the system A matrix. We call this optimal adaptive control.

The CT-HDP algorithm is presented in Section 6.1. A mathematical formulation of the algorithm is given in Section 6.2 that proves the equivalence of the CT-HDP with a certain Quasi-Newton method. The algorithm is then tested in simulation in Section 6.3. There, the optimal controller for a linear power system model is obtained. The ARE is effectively solved online in real time without making use of any knowledge regarding the system matrix A.

6.1 Continuous-time heuristic dynamic programming for the LQR problem

Consider the LQR problem described in Section 3.1. Thus, consider the continuous linear time-invariant dynamical system described by

$$\dot{x}(t) = Ax(t) + Bu(t)$$

with state $x(t) \in \mathbb{R}^n$, control input $u(t) \in \mathbb{R}^m$ and (A, B) stabilizable. To this system associate the infinite-horizon quadratic cost or value function

$$V(x(t_0), t_0) = \int_{t_0}^{\infty} (x^T(\tau)Qx(\tau) + u^T(\tau)Ru(\tau))\, d\tau$$

with $Q \geq 0$, $R > 0$ such that $(Q^{1/2}, A)$ is detectable.

The LQR optimal control problem requires finding the control policy that minimizes the cost. The solution of this optimal control problem is given by the state feedback $u(t) = -Kx(t)$ with

$$K = R^{-1}B^TP$$

where the matrix P is the unique positive definite solution of the algebraic Riccati equation (ARE)

$$A^T P + PA - PBR^{-1}B^T P + Q = 0$$

Under the detectability condition for $(Q^{1/2}, A)$ the unique positive semidefinite solution of the ARE determines a stabilizing closed-loop controller.

To solve ARE, complete knowledge of the model of the system is needed, that is both the system matrix A and control input matrix B must be known.

An offline CT policy iteration iterative algorithm to solve the ARE from Chapter 3 is as follows. Given K_0 such that $A_0 = A - BK_0$ is Hurwitz, a sequence $\{P_i\}_{i \geq 1}$ is determined by solving successively the Lyapunov equation

$$0 = A_i^T P_{i+1} + P_{i+1} A_i + Q + P_i BR^{-1} B^T P_i \tag{6.1}$$

where $A_i = A - BR^{-1}B^T P_i$. Kleinman (1968) showed that the sequence $\{P_i\}_{i \geq 1}$ is monotonically decreasing and lower bounded by the unique positive definite solution of the ARE. This is an offline Policy Iteration (PI) algorithm that requires knowledge of the dynamics A, B and a stabilizing initial gain.

As seen in Chapter 3, (6.1) can be expressed equivalently in a Newton's method-like setting as

$$P_{i+1} = P_i - (Ric'_{P_i})^{-1} Ric(P_i) \tag{6.2}$$

where

$$Ric(P_i) = A^T P_i + P_i A + Q - P_i BR^{-1} B^T P_i \tag{6.3}$$

and Ric'_{P_i} denotes the Frechet derivative of $Ric(P_i)$ taken with respect to P_i. The matrix function Ric'_{P_i} evaluated at a given matrix M will thus be $Ric'_{P_i}(M) = A_i{}^T M + MA_i$.

6.1.1 Continuous-time HDP formulation using integral reinforcement learning

As discussed in Chapter 3, the LQR problem can be solved in real time using online policy iteration without knowing the A matrix. This is an online version of the underlying Newton method. In Chapter 3 was given a method based on Integral Reinforcement Learning (IRL) for implementing the CT policy iteration algorithm online in real time. It is based on the following.

IRL form of the CT Bellman equation

$$V(x(t)) = \int_t^{t+T_0} r(x(\tau), \mu(x(\tau)))\, d\tau + V(x(t+T))$$

This form does not contain the system dynamics. The variable T_0 is known as the integral reward time period. It is not like a typical sampling period, and can vary with time t. The IRL policy iteration Algorithm 3.1 solves the ARE online without knowing the system A matrix by measuring data along the system trajectories. It is an optimal adaptive control algorithm that converges to the optimal LQR solution. A drawback is that it requires initialization using a stabilizing feedback gain, which cannot be determined in complete ignorance of the A matrix.

Based on the IRL formulation, a generalized policy iterations algorithm was given in Chapter 5 as Algorithm 5.4. This algorithm also required a stabilizing initial gain. We now formulate value iterations, or the heuristic dynamic programming (HDP) algorithm of ADP, for continuous-time systems. CT-HDP provides an optimal adaptive control algorithm that converges to the optimal LQR solution without knowing the A matrix and without knowing an initial stabilizing gain.

Let $T_\mu : \mathrm{X} \to \mathrm{X}$ be the dynamic programming operator defined as

$$T_\mu V(x_t) = \int_t^{t+T_0} r(x, \mu)\, d\tau + V(x_{t+T_0}) \tag{6.4}$$

where x_{t+T_0} denotes the value of $x(\tau)$ at $\tau = t + T_0$, with $x(\tau)$ the solution of (3.1) for initial condition $x(t)(\textit{denoted } x_t)$ and input $\{\mu(\tau); \tau \geq t\}$.

Algorithm 6.1. Heuristic dynamic programming: Value iteration for CT systems

1. Select $V^0(x_t) : \Omega \to \mathbb{R}; V^0(\cdot) \subset \mathrm{X}, i = 0$.
2. (Policy update step.) Find u_i, which satisfies

$$u_i = \arg\min_{v \in \Psi(\Omega)} [H(x, v, \nabla V_x^i)] \tag{6.5}$$

3. (Value function update step.) Solve for $V^i(x_t)$ using

$$V^{i+1}(x_t) = T_{u_i} V^i(x_t) \tag{6.6}$$

■

The key difference between Algorithm 5.5, namely, GPI with $k = 1$, and the value iteration Algorithm 6.1 is that in the latter case the requirement of $u_0 \in \Psi(\Omega)$ is removed, that is the initial policy need not be stabilizing.

For the case of the LQR problem the value iteration scheme is the following.

Algorithm 6.2. Heuristic dynamic programming: Value iteration for CT LQR

1. Select $P_0 \geq 0$ such that $V_0(x_t) : \mathbb{R}^n \to \mathbb{R}; V_0(x_t) = x_t^T P_0 x_t \geq 0, i = 0$.
2. (Policy update step.) Find u_i, which satisfies

$$u_i = -R^{-1} B^T P_i x = K_i x \tag{6.7}$$

3. (Value function update step.) Solve for $V_{i+1}(x_t)$ using

$$V_{i+1}(x(t)) = \int_t^{t+T_0} (x^T Q x + u_i^T R u_i)\, d\tau + V_i(x(t + T_0)) \tag{6.8}$$

with the V-function parameterized as $V_i(x) = x^T P_i x$, then set $i \to i + 1$. ■

Equation (6.8) can be explicitly written in parametric form as

$$x^T(t)P_{i+1}x(t) = \int_t^{t+T_0} (x^TQx + u_i^TRu_i)\, d\tau + x^T(t+T_0)P_ix(t+T_0) \tag{6.9}$$

Then, Algorithm 6.2 amounts to the update of the kernel matrix P_i. A restriction on the initial matrix P_0 such that the corresponding K_0 be a stabilizing controller is not required. In fact the VI algorithm can simply be initialized with $P_0 = 0$.

6.1.2 Online tuning value iteration algorithm for partially unknown systems

The continuous-time VI algorithms can be implemented online without having any knowledge about the plant drift dynamics, that is the A matrix need not be known (matrix B is required), and without starting with an initial stabilizing policy. The information about the A matrix of the system is embedded in the states $x(t)$ and $x(t+T_0)$, which are observed online.

To find the parameters of V_{i+1} in (6.8), the left-hand side of (6.9) is written as

$$V_{i+1}(\overline{x}(t), \overline{p}_{i+1}) = x^T(t)P_{i+1}x(t) = \overline{p}_{i+1}^T\overline{x}(t) \tag{6.10}$$

where $\overline{x}(t)$ denotes the Kronecker product quadratic polynomial basis vector with the elements $\{x_i(t)x_j(t)\}_{i=1,n;\, j=i,n}$ and $\overline{p} = v(P)$, where $v(\cdot)$ is a vector-valued matrix function that acts on $n \times n$ matrices and gives a column vector by stacking the column elements of the symmetric matrix into a vector with the off-diagonal elements summed as $P_{ij} + P_{ji}$. Then, $\overline{p} = v(P)$ contains the $n(n+1)/2$ independent elements of the kernel matrix, ordered by columns and with redundant symmetric elements removed. The right-hand side of (6.8), using (6.7), is

$$d(x(t), P_i) = \int_t^{t+T_0} x(\tau)^T(Q + P_iBR^{-1}B^TP_i)x(\tau)\, d\tau + \overline{p}_i^T\overline{x}(t+T_0) \tag{6.11}$$

Denote the *integral reinforcement* at time t by

$$r(t) = r(x(t), P_i, T_0) = \int_t^{t+T_0} x^T(\tau)(Q + P_iBR^{-1}B^TP_i)x(\tau)\, d\tau \tag{6.12}$$

which is the observed reward over the integral reward time interval $[t, t+T_0]$. Equating (6.10) and (6.11), the iterations (6.7) and (6.8) can be written as

$$\overline{p}_{i+1}^T\overline{x}(t) = r(x(t), P_i, T_0) + \overline{p}_i^T\overline{x}(t+T_0) \tag{6.13}$$

This is a scalar equation for the $n(n+1)/2$ unknown parameters $\overline{p}_{i+1}$. This is the type of equation often confronted in adaptive control (Ljung, 1999). Quadratic vector $\overline{x}(t)$ is known as the regression vector, and is a known function of observed state data.

At each iteration step, after a sufficient number of state-trajectory points are collected using the same control policy K_i, a least-squares method is employed to solve for the V-function parameters, $\bar{p}_{i+1}$, which will then yield the kernel matrix P_{i+1}. The parameter vector $\bar{p}_{i+1}$ is found by minimizing, in the least-squares sense, the error between the target function given by (6.11) and the parameterized relation (6.10) over a compact set $\Omega \subset R^n$. Evaluating (6.13) at $N \geq n(n+1)/2$ points $\bar{x}^i$ in the data space, the least-squares solution is obtained as

$$\bar{p}_{i+1} = (XX^T)^{-1}XY \tag{6.14}$$

where $X = \begin{bmatrix} \bar{x}^1 & \bar{x}^2 & \dots & \bar{x}^N \end{bmatrix}$ and $Y = \begin{bmatrix} d(\bar{x}^1, P_i) & d(\bar{x}^2, P_i) & \dots & d(\bar{x}^N, P_i) \end{bmatrix}^T$.

To obtain a solution for the least-squares problem (6.14) one requires at least $N = n(n+1)/2$ points, which is the number of independent elements in the matrix P.

Equation (6.13) can be solved in real time for the unknown parameters $\bar{p}_{i+1}$ by observing the data sets $(x(t), x(t+T_0), r(t))$ along the system trajectories – that is the current state, the next state and the integral reinforcement for times t. At The least-squares problem can be solved in real time after a sufficient number of data points are collected along a single state trajectory. The solution of (6.13) can alternatively be obtained using the recursive least-squares algorithm (RLS) (Ljung, 1999). In either case, a persistence of excitation (PE) condition is required. Specifically, the quadratic basis vector $\bar{x}(t)$ must be PE, that is there exist $\beta_0, \beta_1, T > 0$ such that

$$\beta_0 I < \int_t^{t+T} \bar{x}(\tau)\bar{x}^T(\tau)\, d\tau < \beta_1 I, \quad \forall t > 0$$

The selection of the integral reward time interval T_0 is based on this PE requirement.

This online solution procedure requires only measurements of the states at discrete moments in time, t and $t + T_0$, as well as knowledge of the observed reward over the sample time interval $[t, t+T_0]$. Therefore, there is no required knowledge about the system A matrix for the update of the critic or the action. However, the B matrix is required for the update of the control policy (actor), using (6.7), and this makes the tuning algorithm only partially model-free.

It should be observed that the update of both the actor and the critic is performed at discrete moments in time. However, the control action (6.7) is a continuous-time control, with gain updated at the sample points. Moreover, the critic update is based on the observations of the continuous-time cost over a finite integral reward time interval. As a result, the algorithm converges to the solution of the continuous-time optimal control problem.

Further discussion of the actor–critic structure for implementing CT-HDP online in real time is given in Chapters 3–5.

6.2 Mathematical formulation of the HDP algorithm

In this section, the HDP Algorithm 6.2 is analyzed and placed in relation with known results and form optimal control theory.

The first result shows that CT-HDP is equivalent to an underlying quasi-Newton method.

Lemma 6.1. The ADP iteration between (6.7) and (6.8) is equivalent to the quasi-Newton method

$$P_{i+1} = P_i - (Ric'_{P_i})^{-1}(Ric(P_i) - (e^{A_iT_0})^T Ric(P_i)e^{A_iT_0}) \tag{6.15}$$

Proof:
Differentiating (6.8) with respect to time one obtains

$$\begin{aligned}\dot{V}_{i+1}(x(t)) &= -x^T(t)Qx(t) - u_i^T(t)Ru_i(t) + x^T(t+T_0)Qx(t+T_0)\\ &\quad + u_i^T(t+T_0)Ru_i(t+T_0) + \dot{V}_i(x(t+T_0))\end{aligned} \tag{6.16}$$

which can be written as

$$\begin{aligned}&(A+BK_i)^T P_{i+1} + P_{i+1}(A+BK_i) + K_i^T RK_i + Q\\ &\quad = (e^{(A+BK_i)T_0})^T (A^T P_i + P_iA - P_iBR^{-1}B^T P_i + Q)e^{(A+BK_i)T_0}\end{aligned} \tag{6.17}$$

Adding and subtracting $A_i^T P_i + P_iA_i$ and making use of (6.3), (6.17) becomes

$$A_i^T(P_{i+1} - P_i) + (P_{i+1} - P_i)A_i = -Ric(P_i) + (e^{A_iT_0})^T Ric(P_i)e^{A_iT_0} \tag{6.18}$$

and can be written in a Quasi-Newton formulation as

$$P_{i+1} = P_i - (Ric'_{P_i})^{-1}(Ric(P_i) - (e^{A_iT_0})^T Ric(P_i)e^{A_iT_0})$$

■

Remark 6.1. If $A_i = A + BK_i$ is stable and $T_0 \to \infty$ one may recover from (6.15) the standard Newton method, (6.2), to solve ARE, for which Kleinman (1968), proved convergence conditioned by an initial stabilizing control gain K_0. This Newton's method is given in Lemma 3.3. It seems that the last term, appearing in the formulation of the new ADP algorithm, compensates for the need of an initial stabilizing gain.

Equations (6.7) and (6.8) can be written as

$$P_{i+1} = \int_0^{T_0} (e^{A_it})^T (Q + P_iBR^{-1}B^T P_i)e^{A_it}\, dt + (e^{A_iT_0})^T P_i e^{A_iT_0} \tag{6.19}$$

so that we obtain the next result.

Lemma 6.2. Iteration (6.19) is equivalent to

$$P_{i+1} = P_i + \int_0^{T_0} (e^{A_i t})^T Ric(P_i) e^{A_i t} \, dt \tag{6.20}$$

Proof: From matrix calculus one may write

$$P_i - (e^{A_i T_0})^T P_i e^{A_i T_0} = -\int_0^{T_0} (e^{A_i t})^T (A_i{}^T P_i + P_i A_i) e^{A_i t} \, dt \tag{6.21}$$

From (6.19) and (6.21) follow (6.20).
Note that (6.20) is just a different way of writing (6.19). ■

Remark 6.2. As $T_0 \to 0$, (6.20) becomes

$$\dot{P} = A^T P + PA - PBR^{-1}B^T P + Q \tag{6.22}$$

$$P(0) = P_0$$

which is a *forward-in-time* computation of the ARE solution, with the terminal boundary condition considered at the starting time, $P_{t_f} = P_0$.

Remark 6.3. The term $e^{A_i T_0}$ is the discrete-time version, obtained for the sample time T_0, of the closed-loop system matrix A_i. Therefore, (6.19) is the expression of a hybrid discrete-time/continuous-time Riccati equation recursion.

Note that CT-HDP Algorithm 6.2, does not require the knowledge of the system matrix A, yet has just been shown equivalent to several underlying iterations given in terms of A and B. The next result shows that, if VI converges, it converges to the solution of the CT optimal control problem.

Lemma 6.3. Let the ADP algorithm converge so that $P_i \to P^*$. Then P^* satisfies $Ric(P^*) = 0$, that is P^* is the solution of the continuous-time ARE.

Proof:

If $\{P_i\}$ converges, then taking the limit in (6.20)

$$\lim_{P_i \to P^*} (P_{i+1} - P_i) = \lim_{P_i \to P^*} \int_0^{T_0} e^{A_i t T} Ric(P_i) e^{A_i t} \, dt \tag{6.23}$$

This implies

$$\int_0^{T_0} (e^{A^*t})^T Ric(P^*) e^{A^*t} \, dt = 0 \tag{6.24}$$

with $A^* = A - BR^{-1}B^TP^*$, and thus $Ric(P^*) = 0$. ■

Algorithm 6.2 is an online algorithm that can be implemented online without knowing the A matrix or requiring an initial stabilizing gain. Lemma 6.3 shows that it effectively solves the CT ARE online without knowing A.

6.3 Simulation results for online CT-HDP design

In this section is given a simulation of CT value iteration Algorithm 6.2 for load-frequency control of an electric power system. It is shown that the algorithm solves the ARE online in real time without knowing the system matrix A.

6.3.1 System model and motivation

In this section, the continuous-time V-learning ADP Algorithm 6.2 is used to determine an optimal controller for the power system introduced in Section 3.3. As discussed there, the non-linearity in the dynamics of such systems is determined by the load value, which under normal operation is constant. At the same time the values of the parameters defining the linear model of the actual plant are not precisely known. In view of these facts the presented HDP design technique is a good candidate for the design of the desired LQR controller, for a given operating point of the system.

The model of the system (Wang *et al.*, 1993) is

$$\dot{x} = Ax(t) + Bu(t) \tag{6.25}$$

where

$$x(t) = [\Delta f(t) \quad \Delta P_g(t) \quad \Delta X_g(t) \quad \Delta E(t)]^T$$

$$A = \begin{bmatrix} -1/T_p & K_p/T_{Pp} & 0 & 0 \\ 0 & -1/T_T & 1/T_T & 0 \\ -1/RT_G & 0 & -1/T_G & -1/T_G \\ K_E & 0 & 0 & 0 \end{bmatrix}$$

$$B^T = [0 \quad 0 \quad 1/T_G \quad 0]$$

The system states are $\Delta f(t)$ – incremental frequency deviation (Hz), $\Delta P_g(t)$ – incremental change in generator output (p.u. MW), $\Delta X_g(t)$ – incremental change in governor position (p.u. MW) and $\Delta E(t)$ – incremental change in integral control. The system parameters are T_G – the governor time constant, T_T – turbine time constant, T_p – plant model time constant, K_p – plant model gain, R – speed regulation due to governor action and KE – integral control gain.

6.3.2 Simulation setup and results

In the simulation, only the time constant T_G of the governor, which appears in the B matrix, is considered to be known, while the values for all the other parameters appearing in the system A matrix are not known.

The system parameters, necessary for simulating the system behavior are picked randomly before the simulation is started in some realistic ranges, as specified in Wang *et al.* (1993), such that

$$
\begin{aligned}
&1/T_p \in [0.033, 0.1] \\
&K_p/T_p \in [4, 12] \\
&1/T_T \in [2.564, 4.762] \\
&1/T_G \in [9.615, 17.857] \\
&1/RT_G \in [3.081, 10.639]
\end{aligned}
$$

Note that, even if the values of the parameters are known to be in the above-mentioned ranges, the algorithm does not make use of any of this knowledge, only the exact value of T_G being necessary. Also, although the values of the controller parameters K_E and R are known, as they are specified by the engineer, this information is not used by the CT-HDP algorithm to determine the optimal controller.

The simulation results that are presented next were obtained considering a randomly picked set of values (in the above-mentioned ranges) for the systems unknown parameters, that is matrix A. In all the simulations the matrix $B = [0 \quad 0 \quad 13.7355 \quad 0]$ and it is considered to be known. For the purpose of demonstrating the CT-HDP algorithm the initial state of the system is taken different than zero, $x_0 = [0 \quad 0.1 \quad 0 \quad 0]$, and the matrix $P_0 = 0$.

The online implementation requires the setup of a least-squares problem of the kind presented in Section 6.1 to solve for the values of the critic parameters, the matrix P_i, at each iteration step i. In the simulations the matrix P_i is determined after collecting 12 points for each least-squares problem. Each such point is calculated after observing the value of the reward over an integral reward time interval of $T_0 = 0.1$ s. Therefore, a least-squares problem is solved and the critic is updated at each 1.2 s. The simulations were performed over a time interval of 50 s. As such, a number of 41 iterations were performed during each simulation experiment.

For the simulation the unknown values of the system parameters were randomly picked in the specified ranges and the system matrix was

$$
A = \begin{bmatrix} -0.0596 & 5.0811 & 0 & 0 \\ 0 & -3.0938 & 3.0938 & 0 \\ -10.0912 & 0 & -13.7355 & -13.7355 \\ 0.6 & 0 & 0 & 0 \end{bmatrix}
$$

The convergence of few of the critic parameters $P(1,1)$, $P(1,3)$, $P(2,4)$ and $P(4,4)$ is presented in Figure 6.1.

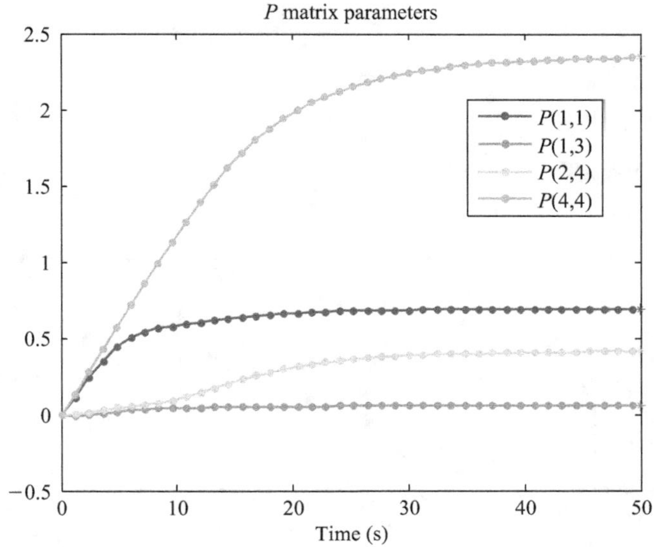

Figure 6.1 Convergence of P matrix parameters in online CT-HDP

The solution of the ARE for this given matrix A and $Q = I_4$, $R = 1$ is

$$P = \begin{bmatrix} 0.6920 & 0.5388 & 0.0551 & 0.6398 \\ 0.5388 & 0.7361 & 0.1009 & 0.4173 \\ 0.0551 & 0.1009 & 0.0451 & 0.0302 \\ 0.6398 & 0.4173 & 0.0302 & 2.3550 \end{bmatrix}$$

After 41 iteration steps the critic is characterized by

$$P_{41} = \begin{bmatrix} 0.6914 & 0.5381 & 0.0551 & 0.6371 \\ 0.5381 & 0.8922 & 0.1008 & 0.4144 \\ 0.0551 & 0.1008 & 0.0451 & 0.0299 \\ 0.6371 & 0.4144 & 0.0299 & 2.3442 \end{bmatrix}$$

Comparing the values of the two matrices it can be noted that after 41 iteration steps the error between their parameters is of order 10^{-3}, that is the algorithm converged to the solution of the ARE. It has solved ARE in real time without knowing system matrix A.

Figure 6.2, presents the evolution of the states of the system during the simulation. Figure 6.3 shows the system states in detail during the first 6 s, that is the first 5 iteration steps of the simulation. The control signal that was applied to the system during the CT-HDP tuning is presented in Figure 6.4.

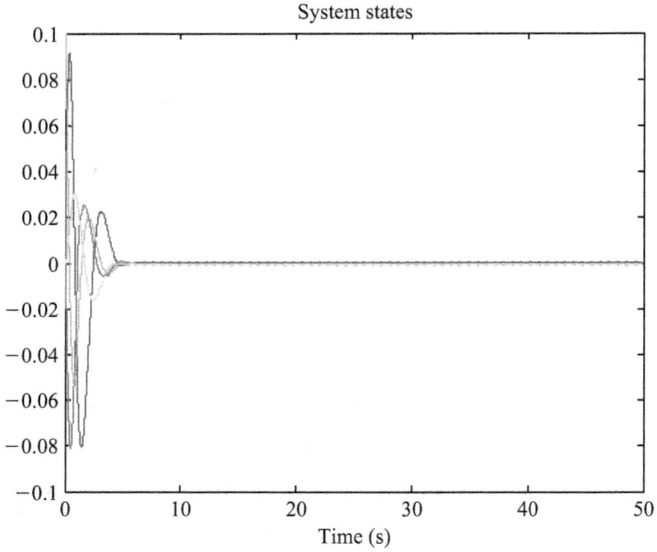

Figure 6.2 System states during the simulation

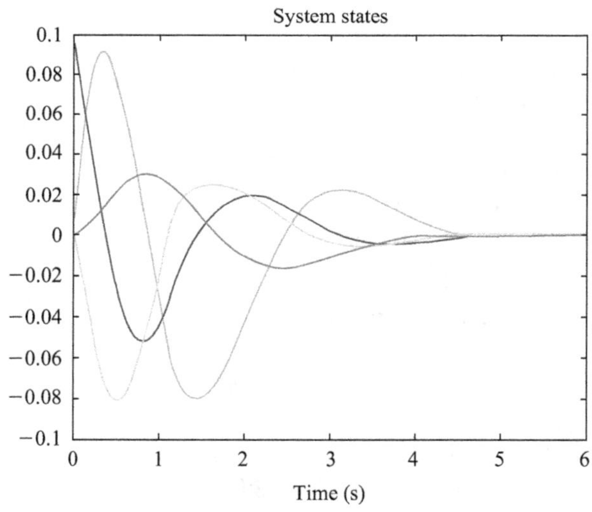

Figure 6.3 System states during the first five iteration steps

6.3.3 Comments on the convergence of CT-HDP algorithm

The relation between the integral reward time period T_0 over which the value function is observed at each step and the algorithm convergence is investigated in the following.

Figure 6.5 shows the convergence of the critic parameters, in the case of the second simulation setup, when the time period is taken $T_0 = 0.2$ s. Over the 50 s duration of

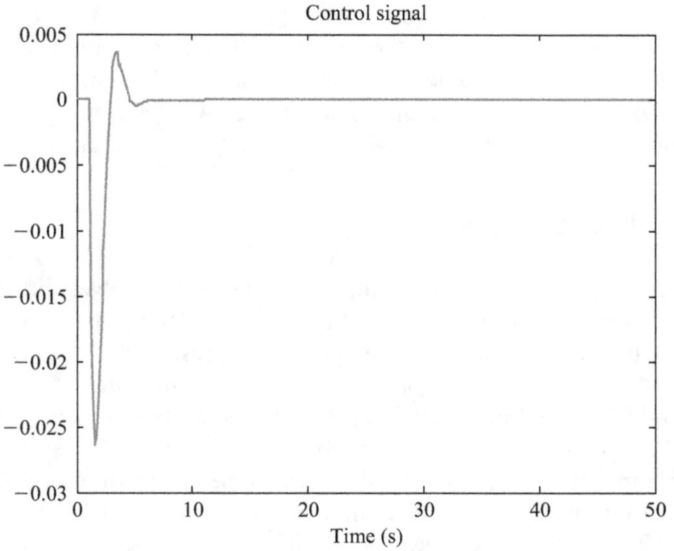

Figure 6.4 Control signal for simulation of online CT-HDP

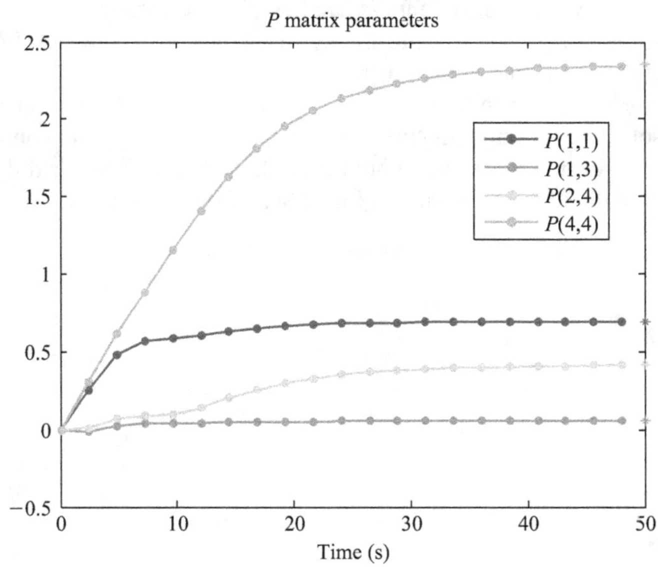

Figure 6.5 Convergence of P matrix parameters in online CT-HDP for T = 0.2 s

the simulation only 20 iterations are performed, the necessary data (12 points) for solving each least-squares problem being collected over an interval of 2.4 s.

By comparing the results plotted in Figure 6.1 with the ones presented in Figure 6.5 it becomes clear that the amount of time necessary for convergence is not dependent on the sample period that is used for observation. However, the number of iteration steps that are required for convergence is reduced when a large sample period is considered. The reason is that, in case a larger observation sample is used, an increased amount of information regarding the system is carried in the data points collected for the critic update. As such, at each step of the iteration the critic improvement is larger when the time period is increased.

6.4 Conclusion

This chapter presents a continuous-time ADP scheme that solves the continuous-time infinite-horizon optimal control problem. The heuristic dynamic programming (HDP) algorithm, also known as value iteration, is presented.

CT-HDP is implemented on an actor–critic structure as discussed in Chapters 3–5. The control signal is applied to the system in a continuous-time fashion. The actor's continuous-time performance is measured over selected integral reward time intervals along the trajectories of the system in real time. Based on this acquired information data, the critic reevaluates the infinite-horizon cost and updates the actor's parameters (i.e. the continuous-time system controller), in the sense of improving the overall system performance (i.e. to minimize the infinite-horizon continuous-time cost). As such, the system performance informational loop, which involves the critic entity, handles discrete information regarding the continuous-time performance while the system control loop, which involves the actor, operates entirely in continuous time.

The CT-HDP algorithm was shown to be equivalent to a quasi-Newton method. It solves the continuous-time ARE and obtains the optimal controller in an online, forward-in-time iteration without using knowledge of the drift dynamics of the plant and without starting with an initial stabilizing policy.

Part II
Adaptive control structures based on reinforcement learning

This book shows how to use reinforcement learning (RL) methods to design adaptive controllers that learn optimal control solutions online for continuous-time systems. We call these optimal adaptive controllers. They stand in contrast to standard adaptive control systems in the control systems literature, which do not normally converge to optimal solutions in terms of solving a Hamilton–Jacobi–Bellman equation. In Part I of the book we laid the foundation for RL applications in continuous-time systems based on a form of the continuous-time Bellman equation known as the integral reinforcement learning (IRL) Bellman equation, originally presented in Vrabie *et al.* (2008, 2009), Vrabie (2009), Vrabie and Lewis (2009). It was shown how to use IRL to formulate policy iterations and value iterations for continuous-time systems.

The controllers of Part I are of the form normally expected in RL and the convergence proofs are carried out using methods employed in RL. The actor and critic loops are tuned sequentially, that is the control policy is held fixed while the critic evaluates its performance. This strategy seems a bit odd for the adaptive feedback control practitioner. The RL two-loop sequential learning structures are not like standard adaptive control systems currently used in feedback control. They are hybrid controllers with a continuous inner action control loop and an outer learning critic loop that operates on a discrete-time scale.

In Part II of the book, we adopt a philosophy more akin to that of adaptive controllers in the feedback control literature (Ioannou and Fidan, 2006; Astrom and Wittenmark, 1995; Tao, 2003). These ideas were originally developed in Vamvoudakis and Lewis (2010a), Vamvoudakis and Lewis (2010b), Vamvoudakis and Lewis (2011), Vamvoudakis (2011). Precepts of RL are used to design a novel form of adaptive controller that has multiple learning loops. These controllers of Part II operate as normally expected in adaptive control in that the control loops are not tuned sequentially as in Part I, but the parameter tuning in all control loops is performed simultaneously through time. Since the critic and actor networks learn and update their parameters simultaneously, we call these structures *synchronous* optimal adaptive controllers. Moreover, the convergence proofs are carried out using methods employed in adaptive control, namely, Lyapunov energy-based techniques.

Nevertheless, in order to learn optimal control solutions, the structures of these adaptive controllers are of novel forms motivated by RL algorithms that solve

Bellman equations online. To obtain learning structures that convergence to optimal control solutions, policy iteration algorithms are first developed for continuous-time dynamical systems, and the structure of the policy iteration algorithms is used to develop novel adaptive control structures. Then, adaptive control type Lyapunov proofs are used to develop parameter tuning rules and verify the performance of these adaptive controllers. The resulting controllers have novel structures in that they are multi-loop learning networks, yet they are continuous-time controllers that are more in keeping with the philosophy of standard adaptive feedback controllers in that the critic and actor networks are tuned simultaneously in time. The proofs of performance are based on methods that are in general use from the perspective of adaptive control, namely Lyapunov methods.

In Part III of the book we develop adaptive controllers that learn optimal solutions in real time for several differential game theory problems, including zero-sum and multiplayer non–zero-sum games. The design procedure is to first formulate policy iteration algorithms for these problems, then use the structure of policy iteration to motivate the structure of multi-loop optimal adaptive controller structures. Then, tuning laws for these novel adaptive controllers are determined in Chapters 9 and 10 by adaptive control Lyapunov techniques or, in Chapter 11, by RL techniques. The result is a family of adaptive controllers that converge to optimal game theoretic solution online in real time. Methods are given for learning the solution to game Hamilton–Jacobi–Isaacs equations in real time without knowing the system drift dynamics.

Chapter 7
Optimal adaptive control using synchronous online learning

This chapter is concerned with developing an online approximate solution method, based on policy iteration (PI), for the infinite-horizon optimal control problem for continuous-time (CT) non-linear systems with known dynamics. We present an online adaptive algorithm developed in Vamvoudakis and Lewis (2010a) whose structure is based on the actor–critic structure of PI in reinforcement learning (RL). However, in PI, the critic and actor networks are tuned sequentially, that is the actor network parameters are held constant while the critic network is tuned. By contrast, the algorithm presented in this chapter involves *simultaneous tuning* of both actor and critic neural networks (i.e. the parameters in both networks are tuned at the same time). We term this algorithm *synchronous* PI or synchronous optimal adaptive control. This approach results in a CT controller that operates more along the lines of adaptive controllers standard in the literature, yet converges to an optimal control solution. It can be viewed as an extremal, or simultaneous tuning, version of the generalized PI introduced for discrete-time systems in Sutton and Barto (1998) and discussed in Chapter 2, and extended to CT systems in Chapter 5. In this chapter, the dynamics of the system are assumed known. This is relaxed in Chapter 8, where the drift dynamics are not needed.

This approach to PI is motivated by work in adaptive control (Ioannou and Fidan, 2006; Astrom and Wittenmark, 1995; Tao, 2003). Adaptive control is a powerful tool that uses online tuning of parameters to provide effective controllers for non-linear or linear systems with modeling uncertainties and disturbances. Closed-loop stability while learning the parameters is guaranteed, often by using Lyapunov design techniques. Parameter convergence, however, often requires that the measured signals carry sufficient information about the unknown parameters (persistence of excitation (PE) condition).

The design procedure in this chapter is based on Vamvoudakis and Lewis (2010a) and is as follows. First, an RL (PI) algorithm is formulated that solves the optimal control problem by online learning. This gives the structure of the critic and actor networks required to obtain an online optimal control solution using sequential tuning of the actor and critic networks. In sequential tuning, the standard learning method of RL, while one network is being tuned, the parameters of the other network are held fixed. However, it is desired in this chapter to design optimal adaptive controllers wherein the critic and actor networks are tuned simultaneously through time. This gives a learning system that operates more in line with accepted adaptive control methods. To accomplish this, value function approximation (VFA) is used to develop a critic structure that approximately solves the Bellman equation used in PI, and an actor structure that updates the control action. After the two-loop structure of the adaptive controller has been established using PI, Lyapunov proofs are used to develop tuning algorithms for the critic and actor networks that operate simultaneously and continuously in time.

By using RL methods, namely PI, to establish the two-loop structure of the controller, and adaptive control methods, namely Lyapunov proofs, to derive simultaneous tuning laws for the two loops, a novel structure of adaptive controller results that learns the optimal control solution online by solving the Bellman equation in real time using data measured along the system trajectories.

In this approach, the performance of the synchronous optimal adaptive controller does not rely on proofs of convergence of the PI algorithm. Instead, PI is used to establish the structure of the optimal adaptive controller, and adaptive control Lyapunov proofs are used to develop tuning laws that both stabilize the system and learn the optimal control solution in real time. These proofs do not rely on convergence of the underlying PI algorithm. In fact, the design method works even when no proofs have been provided for the PI algorithm, as we shall see in Part III of the book.

PI algorithms must be initialized with a stabilizing controller. Stabilizing controllers can be difficult to find for non-linear systems. By contrast, the synchronous optimal adaptive controller does not require an initial stabilizing controller.

The chapter is organized as follows. Section 7.1 reviews the optimal control problem for non-linear systems from Chapter 4 and its solution using PIs. In Section 7.2, PI is used to develop an adaptive control structure that has a critic network and an actor network. VFA is used in the critic network to provide a practical means for solving the Bellman equation online as required in PI. In Section 7.3, a special adaptive tuning algorithm for the critic network parameters is developed and shown to converge to the approximate Bellman equation solution. Adaptive control Lyapunov proof techniques are used. In Section 7.4, the structure of the actor network is developed based on the policy improvement step of PI. Tuning algorithms are then given to update the parameters of the critic network and actor network simultaneously in time. This is called synchronous tuning of the actor and critic, or synchronous PI. It is proven that this novel two-loop adaptive controller converges to the solution to the Hamilton–Jacobi–Bellman (HJB) equation in real time and so learns the optimal control solution. While learning, the controller also keeps the plant stable. Lyapunov techniques are used in the proof. The full plant dynamics must be known to implement this adaptive controller. Section 7.5 compares the novel structure of the synchronous optimal adaptive controller to other adaptive control structures appearing in the literature. Finally, Section 7.6 presents simulation examples that show the effectiveness of the online synchronous optimal adaptive controller in learning the optimal value and control for both linear systems and non-linear systems. The synchronous optimal adaptive controller effectively solves the Riccati equation in the linear case and the HJB equation in the non-linear case online using data measured along the system trajectories.

7.1 Optimal control and policy iteration

The optimal adaptive controller in this chapter has a structure that comes from CT policy iterations (PI). Therefore, first we review the optimal control problem described in Section 4.1 and its PI solution.

Consider the time-invariant affine-in-the-input non-linear dynamical system given by

$$\dot{x}(t) = f(x(t)) + g(x(t))u(x(t)), \quad x(0) = x_0$$

with state $x(t) \in \mathbb{R}^n$, $f(x(t)) \in \mathbb{R}^n$, $g(x(t)) \in \mathbb{R}^{n\times m}$ and control input $u(t) \in U \subset \mathbb{R}^m$. It is assumed that $f(0) = 0$, that $f(x) + g(x)u$ is Lipschitz continuous on a set $\Omega \subseteq \mathbb{R}^n$ that contains the origin, and that the dynamical system is stabilizable on Ω, that is there exists a continuous control function $u(t) \in U$ such that the closed-loop system is asymptotically stable on Ω.

The infinite-horizon cost function or *value* associated with any admissible control policy $u(t) = \mu(x(t)) \in \Psi(\Omega)$ is

$$V(x(t)) = \int_t^\infty r(x(\tau), u(\tau))\, d\tau = \int_t^\infty (Q(x) + u^T Ru)\, d\tau$$

where $V^\mu(x)$ is C^1. Recall from Chapter 4 that a control policy $u(t) = \mu(x)$ is admissible on Ω, denoted by $\mu \in \Psi(\Omega)$, if $\mu(x)$ is continuous on Ω, $\mu(0) = 0$, $\mu(x)$ stabilizes the dynamics on Ω, and the value is finite $\forall x_0 \in \Omega$.

Using Leibniz's formula, the infinitesimal version of the value integral is found to be the following.

CT Bellman equation

$$0 = r(x, \mu(x)) + (\nabla V_x^\mu)^T (f(x) + g(x)\mu(x)), \quad V^\mu(0) = 0$$

Here, ∇V_x^μ (a column vector) denotes the gradient of the cost function V^μ with respect to x. That is, $\nabla V_x^\mu = \partial V^\mu / \partial x$. This is a Bellman equation for non-linear CT systems that, given the control policy $\mu(x) \in \Psi(\Omega)$, can be solved for the value $V^\mu(x)$ associated with it.

The optimal control problem is formulated as follows. Given the CT dynamical system, the set $u \in \Psi(\Omega)$ of admissible control policies, and the value function, find an admissible control policy such that the value function is minimized.

Defining the Hamiltonian

$$H(x, u, V_x) = r(x(t), u(t)) + (\nabla V_x)^T (f(x(t)) + g(x(t))u(t))$$

the optimal cost function $V^*(x)$ satisfies the HJB equation

$$0 = \min_{u \in \Psi(\Omega)} [H(x, u, \nabla V_x^*)]$$

Assuming that the minimum on the right-hand side of this equation exists and is unique, then the optimal control function is

$$u^*(x) = -\frac{1}{2}R^{-1}g^T(x)\,\nabla V_x^*$$

Inserting this optimal control policy in the Hamiltonian we obtain the HJB equation in the form

$$0 = Q(x) + (\nabla V_x^*)^T f(x) - \frac{1}{4}(\nabla V_x^*)^T g(x)R^{-1}g^T(x)\,\nabla Vx_x^*, \quad V^*(0) = 0$$

This is a necessary condition for the optimal cost function. For the linear system case with a quadratic cost functional (linear quadratic regulator (LQR)), the equivalent of this HJB equation is the algebraic Riccati equation (ARE) discussed in Chapter 3.

To find the optimal control solution for the problem one can solve the HJB equation for the value function and then compute the optimal control $u^*(x)$. However, solving the HJB equation is generally difficult, and analytic solutions may not exist. Therefore, based on the CT Bellman equation, in Chapter 4, we wrote a PI Algorithm 4.1 to solve the HJB iteratively, reproduced here as follows. Index i is the iteration step number.

Algorithm 7.1. Policy iteration algorithm for continuous-time systems

Initialize. Select $\mu^{(0)}(x(t)) \in \Psi(\Omega)$ as an admissible policy.

1. (*Policy evaluation* step) Solve for the value $V^{\mu^{(i)}}(x(t))$ using the CT Bellman equation

$$0 = r(x, \mu^{(i)}(x)) + (\nabla V_x^{\mu^{(i)}})^T (f(x) + g(x)\mu^{(i)}(x)) \quad \text{with} \quad V^{\mu^{(i)}}(0) = 0$$

2. (*Policy improvement* step) Update the control policy using

$$\mu^{(i+1)} = \arg\min_{u \in \Psi(\Omega)} \left[H(x, u, \nabla V_x^{\mu^{(i)}}) \right]$$

which explicitly is

$$\mu^{(i+1)}(x) = -\frac{1}{2}R^{-1}g^T(x)\,\nabla V_x^{\mu^{(i)}}$$

■

The PI Algorithm 7.1 solves the non-linear HJB equation by iterations on Bellman equations linear in the value function gradient. Its importance is that the linear Bellman equations are easier to solve than the HJB, which is non-linear. The algorithm was proven to converge in Sutton and Barto (1998), Abu-Khalaf and Lewis (2005) and Leake and Liu (1967). In the case of the LQR, the HJB is the Riccati equation, Bellman's equation is a Lyapunov equation and PI is Kleinman's Algorithm (Kleinman, 1968) (see Sections 3.1 and 4.2).

Note that the full system dynamics $f(x), g(x)$ must be known to implement this algorithm since they appear in the Bellman equation. Moreover, it is required that a stabilizing control policy be known to initialize the algorithm. This algorithm is used in the next sections to motivate a novel adaptive control structure that solves the optimal control problem online by measuring data along the system trajectories.

7.2 Value function approximation and critic neural network

It was seen in Chapter 4 that PI Algorithm 7.1, as other RL algorithms, can be implemented on an actor–critic structure. This consists of two network structures that use value function approximation (VFA) to approximate the solutions of the two equations in steps 1 and 2 at each iteration. In Part I of the book several methods were given for developing adaptive learning structures that converge online in real time to optimal control solutions by measuring data along the system trajectories. These optimal adaptive controllers were of the sort expected in the RL literature, with the actor and critic loops being updated sequentially in time. That is, while the critic loop learns the value function update in Step 1 of Algorithm 7.1, the actor control policy is not changed. These systems are hybrid controllers where the action loop is implemented in continuous time but the outer critic loop operates on a discrete-time scale. This sort of learning system seems odd from the point of view of standard adaptive control in the feedback control systems literature.

In this chapter, we desire to formulate a multi-loop adaptive controller that converges to the optimal control solution by updating the parameters in both control loops simultaneously through time. This is not a hybrid actor–critic control structure with a continuous inner loop and an outer loop operating on a discrete-time scale as developed in Chapter 4. Rather, it is a CT controller where both loops learn simultaneously, or synchronously in time. We call this synchronous actor–critic learning or synchronous PI. This is more in keeping with the philosophy of tuning used in standard adaptive feedback control systems (Ioannou and Fidan, 2006; Astrom and Wittenmark, 1995; Tao, 2003).

Nevertheless, the structure of the synchronous actor–critic controller comes from PI. This allows it to converge online to optimal control solutions. The synchronous optimal adaptive controller has two loops based on the structure provided in PI Algorithm 7.1 In this section, we develop the outer critic control loop based on VFA. In the next section, we show that the critic converges to the solution of the Bellman equation for admissible bounded controls. Then, we develop the structure of the inner actor loop and present the synchronous critic and actor tuning algorithms for the optimal adaptive controller. In this chapter, we use proof techniques standard in adaptive control, in contrast to Part I of the book where we used proof techniques standard in RL.

The critic neural network (NN) is based on VFA. According to the form of the Bellman equation, the gradient of the value function appears. Therefore, one desires to approximate the value $V(x)$ as well as its gradient. This requires approximation in the Sobolev norm (Adams and Fournier, 2003). Therefore, some discussion is given next that relates standard NN approximation usage (Hornik *et al.*, 1990) to the Weierstrass higher-order approximation theorem (Abu-Khalaf and Lewis, 2005; Finlayson, 1990).

The solutions to the CT Bellman equation may not be smooth for general nonlinear systems, except in a generalized sense (Sontag and Sussmann, 1995). However, in keeping with other work in the literature we make the following assumptions.

Assumption 7.1. The solution to the Bellman equation is smooth, i.e. $V(x) \in C^1(\Omega)$.

Assumption 7.2. The solution to the Bellman equation is positive definite. This is guaranteed for stabilizable dynamics if the performance functional satisfies zero-state observability, which is guaranteed by the positive definite condition, that is $Q(x) > 0,\ x \in \Omega - \{0\};\ Q(0) = 0$.

Assumption 7.1 allows us to formally bring in the Weierstrass higher-order approximation theorem (Abu-Khalaf and Lewis, 2005; Finlayson, 1990) and the results of Hornik *et al.* (1990), which state that then there exists a complete independent basis set $\{\varphi_i(x)\}$ such that the solution $V(x)$ to the Bellman equation and its gradient are uniformly approximated, that is there exist coefficients c_i such that

$$V(x) = \sum_{i=1}^{\infty} c_i\varphi_i(x) = \sum_{i-1}^{N} c_i\varphi_i(x) + \sum_{i=N+1}^{\infty} c_i\varphi_i(x)$$

$$V(x) \equiv C_1^T\phi_1(x) + \sum_{i-N+1}^{\infty} c_i\varphi_i(x) \tag{7.1}$$

$$\frac{\partial V(x)}{\partial(x)} = \sum_{i-1}^{\infty} c_i\frac{\partial\varphi_i(x)}{\partial x} = \sum_{i-1}^{N} c_i\frac{\partial\varphi_i(x)}{\partial x} + \sum_{i-N+1}^{\infty} c_i\frac{\partial\varphi_i(x)}{\partial x} \tag{7.2}$$

where $\phi_1(x) = [\varphi_1(x) \quad \varphi_2(x) \quad \dots \quad \varphi_N(x)]^T : \mathbb{R}^n \to \mathbb{R}^N$ and the last terms in these equations converge uniformly to zero as the number of retained terms $N \to \infty$. (Specifically, the basis set is dense in the Sobolev norm $W^{1,\infty}$ (Finlayson, 1990; Adams and Fournier, 2003).) Standard usage of the Weierstrass higher-order approximation theorem uses polynomial approximation. However, non-polynomial basis sets have been considered in the literature (e.g. Hornik *et al.* (1990), Sandberg (1998).

Thus, it is justified to assume there exists weights W_1 of a neural network (NN) such that the value function $V(x)$ is approximated as

$$V(x) = W_1^T\phi_1(x) + \varepsilon(x) \tag{7.3}$$

Then $\phi_1(x) : \mathbb{R}^n \to \mathbb{R}^N$ is called the NN activation function vector, N the number of neurons in the hidden layer, and $\varepsilon(x)$ the NN approximation error. As per the above, the NN activation functions $\{\varphi_i(x) : i = 1, N\}$ are selected so that

$\{\varphi_i(x) : i = 1, \infty\}$ provides a complete independent basis set such that $V(x)$ and its derivative

$$\frac{\partial V}{\partial x} = \left(\frac{\partial \phi_1(x)}{\partial x}\right)^T W_1 + \frac{\partial \varepsilon}{\partial x} = \nabla \phi_1^T W_1 + \nabla \varepsilon \tag{7.4}$$

are uniformly approximated. Then, as the number of hidden-layer neurons $N \to \infty$, the approximation errors $\varepsilon \to 0, \nabla\varepsilon \to 0$ uniformly (Finlayson, 1990). In addition, for fixed N, the NN approximation errors $\varepsilon(x)$, and $\nabla\varepsilon$ are bounded by constants on a compact set (Hornik *et al.*, 1990).

Using the NN VFA, considering a fixed control policy $u(t)$, the non-linear Bellman equation becomes

$$H(x, u, W_1) = W_1^T \nabla\phi_1(f + gu) + Q(x) + u^T Ru = \varepsilon_H \tag{7.5}$$

where the residual error due to the function approximation error is

$$\begin{aligned} \varepsilon_H &= -(\nabla s)^T (f + gu) \\ &= -(C_1 - W_1)^T \nabla\phi_1(f + gu) - \sum_{i-N+1}^{\infty} c_i \nabla\varphi_i(x)(f + gu) \end{aligned} \tag{7.6}$$

Under the Lipschitz assumption on the dynamics, this residual error is bounded on a compact set.

Define $|v|$ as the magnitude of a scalar v, $\|x\|$ as the vector norm of a vector x and $\| \ \|_2$ as the induced matrix two-norm.

Definition 7.1. (Uniform convergence.) A sequence of functions $\{p_n\}$ converges uniformly to p on a set Ω if $\forall\varepsilon > 0, \exists N(\varepsilon) : \text{sub}_{x\in\Omega} \|p_n(x) - p(x)\| < \varepsilon$, $n > N(\varepsilon)\oplus$.

The following Lemma has been shown in Abu-Khalaf and Lewis (2005).

Lemma 7.1. For any admissible policy $u(t)$, the least-squares solution to (7.5) exists and is unique for each N. Denote this solution as W_1 and define

$$V_1(x) = W_1^T \phi_1(x) \tag{7.7}$$

Then, as $N \to \infty$:

a. $\sup_{x\in\Omega} |\varepsilon_H| \to 0$
b. $\|W_1 - C_1\|_2 \to 0$
c. $\sup_{x\in\Omega} |V_1 - V| \to 0$
d. $\sup_{x\in\Omega} \|\nabla V_1 - \nabla V\| \to 0$ ■

This result shows that $V_1(x)$ converges uniformly in Sobolev norm $W^{1,\infty}$ (Finlayson, 1990; Adams and Fournier, 2003) to the exact solution $V(x)$ to the Bellman equation as $N \to \infty$, and the weights W_1 converge to the first N of the weights, C_1, which exactly solves the Bellman equation.

Since the object of interest in this chapter is finding the solution of the HJB using the above-introduced function approximator, it is interesting now to look at the effect of the approximation error on the HJB equation. One has

$$W_1^T \nabla\varphi_1 f - \frac{1}{4} W_1^T \nabla\varphi_1 g R^{-1} g^T \nabla\varphi_1^T W_1 + Q(x) = \varepsilon_{HJB} \tag{7.8}$$

where the residual error due to the function approximation error is

$$\varepsilon_{HJB} = -\nabla\varepsilon^T f + \frac{1}{2} W_1^T \nabla\varphi_1 g R^{-1} g^T \nabla\varepsilon + \frac{1}{4} \nabla\varepsilon^T g R^{-1} g^T \nabla\varepsilon \tag{7.9}$$

It was also shown in Abu-Khalaf and Lewis (2005) that this error converges uniformly to zero as the number of hidden-layer units N increases. That is, $\forall\varepsilon > 0,\ \exists N(\varepsilon) : \sup_{x\in\Omega} \|\varepsilon_{HJB}\| < \varepsilon$. This is required in the proofs to come.

7.3 Tuning and convergence of critic NN

In this section, we address the issue of tuning and convergence of the critic NN weights when a fixed admissible and bounded control policy is prescribed. Therefore, the focus is on the Bellman equation for a fixed, bounded control policy u. The requirement for bounded controls is removed in Section 7.4.

In fact, this amounts to the *design of an observer for the value function* known as 'cost function' in the optimal control literature. Therefore, this algorithm is consistent with adaptive control approaches that first design an observer for the system state and unknown dynamics, and then use this observer in the design of a feedback control.

The weights of the critic NN, W_1 that provide the best approximate solution for (7.5) are unknown. Therefore, the output of the critic neural network is

$$\hat{V}(x) = \hat{W}_1^T \phi_1(x) \tag{7.10}$$

where $\hat{W}_1$ are the current estimated values of the ideal critic NN weights W_1. Recall that $\phi_1(x) : \mathbb{R}^n \to \mathbb{R}^N$ is the vector of activation functions, with N the number of neurons in the hidden layer. The approximate Bellman equation is then

$$H(x, \hat{W}_1, u) = \hat{W}_1^T \nabla\phi_1(f + gu) + Q(x) + u^T R u = e_1 \tag{7.11}$$

In view of Lemma 7.1, define the critic weight estimation error

$$\tilde{W}_1 = W_1 - \hat{W}_1$$

Then the residual error in (7.11) is

$$e_1 = -\tilde{W}_1^T \nabla\phi_1(f + gu) + \varepsilon_H$$

Given any admissible control policy u, it is desired to select $\hat{W}_1$ to minimize the squared residual error

$$E_1 = \frac{1}{2} e_1^T e_1$$

Then $\hat{W}_1(t) \to W_1$ and $e_1 \to \varepsilon_H$. We select the tuning law for the critic weights as the normalized gradient descent algorithm

$$\dot{\hat{W}}_1 = -a_1 \frac{\partial E_1}{\partial \hat{W}_1} = -a_1 \frac{\sigma_1}{(\sigma_1^T \sigma_1 + 1)^2} \left[\sigma_1^T \hat{W}_1 + Q(x) + u^T R u\right] \tag{7.12}$$

where $\sigma_1 = \nabla\phi_1(f + gu)$. This is a modified gradient descent algorithm, where $(\sigma_1^T\sigma_1 + 1)^2$ is used for normalization instead of $(\sigma_1^T\sigma_1 + 1)$. This is required in the proofs, where one needs both appearances of $\sigma_1/(1 + \sigma_1^T\sigma_1)$ in (7.12) to be bounded (Ioannou and Fidan, 2006; Tao, 2003).

Note that, from (7.5)

$$Q(x) + u^T R u = -W_1^T \nabla\varphi_1(f + gu) + \varepsilon_H \tag{7.13}$$

Substituting (7.13) in (7.12) and, with the notation

$$\overline{\sigma}_1 = \sigma_1/(\sigma_1^T\sigma_1 + 1), \quad m_s = 1 + \sigma_1^T\sigma_1 \tag{7.14}$$

we obtain the dynamics of the critic weight estimation error as

$$\dot{\tilde{W}}_1 = -a_1 \overline{\sigma}_1 \overline{\sigma}_1^T \tilde{W}_1 + a_1 \overline{\sigma}_1 \frac{\varepsilon_H}{m_s} \tag{7.15}$$

Though it is traditional to use critic tuning algorithms of the form (7.12), it is not generally understood when convergence of the critic weights can be guaranteed. We now address this issue in a formal manner. This development is motivated by adaptive control techniques that appear in Ioannou and Fidan (2006) and Tao (2003).

To guarantee convergence of $\hat{W}_1$ to W_1, the next PE assumption and associated technical lemmas are required.

Persistence of excitation assumption. Let the signal $\overline{\sigma}_1$ be persistently exciting over the interval $[t, t+T]$, that is there exist constants $\beta_1 > 0, \beta_2 > 0, T > 0$ such that, for all t

$$\beta_1 I \le S_0 \equiv \int_t^{t+T} \overline{\sigma}_1(\tau)\overline{\sigma}_1^T(\tau)\, d\tau \le \beta_2 I \tag{7.16}$$

The PE assumption is needed in adaptive control if one desires to perform system identification using, for example, RLS (Ioannou and Fidan, 2006; Tao, 2003). It is needed here because one effectively desires to perform *value junction identification*, namely, to identify the critic parameters to approximate $V(x)$.

Technical Lemma 7.1. Consider the error dynamics system with output defined as

$$\dot{\tilde{W}}_1 = -a_1 \overline{\sigma}_1 \overline{\sigma}_1^J \tilde{W}_1 + a_1 \overline{\sigma}_1 \frac{\varepsilon_H}{m_s}$$
$$y = \overline{\sigma}_1^T \dot{\tilde{W}}_1 \tag{7.17}$$

Then, the PE condition (7.16) is equivalent to the uniform complete observability (UCO) (Lewis *et al.*, 1995) of this system, that is there exist constants $\beta_3 > 0,\ \beta_4 > 0, T > 0$ such that, for all t

$$\beta_3 I \le S_1 \equiv \int_t^{t+T} \Phi^T(\tau, t)\overline{\sigma}_1(\tau)\overline{\sigma}_1^T(\tau)\Phi(\tau, t)\, d\tau \le \beta_4 I \tag{7.18}$$

with $\Phi(t_1, t_0),\ t_0 \le t_1$ the state transition matrix of (7.17).

Proof: System (7.17) and the system defined by $\dot{\tilde{W}}_1 = a_1 \overline{\sigma}_1 u,\ y = \overline{\sigma}_1^T \tilde{W}_1$ are equivalent under the output feedback $u = -y + \varepsilon_H / m_s$. Note that (7.16) is the observability gramian of this last system. ■

The importance of UCO is that bounded input and bounded output implies that the state $\tilde{W}_1(t)$ is bounded. In Theorem 7.1, we shall see that the critic tuning law (7.12) indeed guarantees boundedness of the output in (7.17). Assuming the input is bounded, this in turn implies bounded states $\tilde{W}_1(t)$.

Technical Lemma 7.2. Consider the error dynamics system (7.17). Let the signal $\overline{\sigma}_1$ be persistently exciting. Then:

a. The system (7.17) is exponentially stable. In fact, if $\varepsilon_H = 0$ then $||\tilde{W}(kT)|| \le e^{-akT}||\tilde{W}(0)||$ with

$$\alpha = -\frac{1}{T}\ln\left(\sqrt{1 - 2a_1\beta_3}\right) \tag{7.19}$$

b. Let $||\varepsilon_H|| \le \varepsilon_{\max}|$ and $|y|| \le y_{\max}$ *then* $||\tilde{W}_1||$ converges exponentially to the residual set

$$\tilde{W}_1(t) \le \frac{\sqrt{\beta_2 \mathrm{T}}}{\beta_1}\{[y_{\max} + \delta\beta_2 a_1(\varepsilon_{\max} + y_{\max})]\} \tag{7.20}$$

where δ is a positive constant of the order of 1.

Proof: See appendix. ■

The next result shows that the tuning algorithm (7.12) is effective under the PE condition, in that the weights $\hat{W}_1$ converge to the actual unknown weights W_1 that solve the Bellman equation (7.5) for a given control policy $u(t)$. That is, (7.10) converges close to the actual value function of the current control policy.

Theorem 7.1. Let $u(t)$ be any admissible bounded control policy. Let tuning for the critic NN be provided by (7.12) and assume that $\overline{\sigma}_1$ is persistently exciting. Let the residual error in (7.5) be bounded $||\varepsilon_H|| < \varepsilon_{\max}$. Then the critic parameter error converges exponentially with decay factor given by (7.19) to the residual set

$$\tilde{W}_1(t) \le \frac{\sqrt{\beta_2 \mathrm{T}}}{\beta_1}\{[1 + 2\delta\beta_2 a_1]\varepsilon_{\max}\} \tag{7.21}$$

■

Proof: Consider the following Lyapunov function candidate

$$L(t) = \frac{1}{2}\,\mathrm{tr}\left\{\tilde{W}_1^T a_1^{-1} \tilde{W}_1\right\} \tag{7.22}$$

The derivative of L is given by

$$\begin{aligned}
\dot{L} &= -\mathrm{tr}\left\{\tilde{W}_1^T \frac{\sigma_1}{m_s^2}\left[\sigma_1^T \tilde{W}_1 - \varepsilon_H\right]\right\} \\
\dot{L} &= -\mathrm{tr}\left\{\tilde{W}_1^T \frac{\sigma_1\sigma_1^T}{m_s^2}\tilde{W}_1\right\} + \mathrm{tr}\left\{\tilde{W}_1^T \frac{\sigma_1}{m_s}\frac{\varepsilon_H}{m_s}\right\} \\
\dot{L} &\le -\|\frac{\sigma_1^T}{m_s}\tilde{W}_1\|^2 + \|\frac{\sigma_1^T}{m_s}\tilde{W}_1\|\|\frac{\varepsilon_H}{m_s}\| \\
\dot{L} &\le -\|\frac{\sigma_1^T}{m_s}\tilde{W}_1\|\left[\|\frac{\sigma_1^T}{m_s}\tilde{W}_1\| - \|\frac{\varepsilon_H}{m_s}\|\right]
\end{aligned} \tag{7.23}$$

Therefore, $\dot{L} \le 0$ if

$$\|\frac{\sigma_1^T}{m_s}\tilde{W}_1\| > \varepsilon_{\max} > \|\frac{\varepsilon_H}{m_s}\| \tag{7.24}$$

since $\|m_s\| \ge 1$.

This provides an effective practical bound for $\|\overline{\sigma}_1^T \tilde{W}_1\|$, since $L(t)$ decreases if (7.24) holds.

Consider the estimation error dynamics (7.17) with the output bounded effectively by $\|y\| < \varepsilon_{\max}$, as just shown. Now, Technical Lemma 7.2 shows exponential convergence to the residual set

$$\tilde{W}_1(t) \le \frac{\sqrt{\beta_2 \mathrm{T}}}{\beta_1}\{[1 + 2a_1\delta\beta_2]\varepsilon_{\max}\} \tag{7.25}$$

This completes the proof. ■

Remark 7.1. Note that, as $N \to \infty, \varepsilon_H \to 0$ uniformly (Abu-Khalaf and Lewis, 2005). This means that $\varepsilon_{\max}$ decreases as the number of hidden-layer neurons in (7.10) increases.

Remark 7.2. This theorem requires the assumption that the control policy $u(t)$ is bounded, since $u(t)$ appears in ε_H. In the upcoming Theorem 7.2 this restriction is removed.

7.4 Action neural network and online synchronous policy iteration

We will now present an online adaptive PI algorithm variant that involves simultaneous, or synchronous, tuning of both the actor and critic neural networks. That is, the weights of both neural networks are tuned at the same time. This approach can be viewed as a version of generalized policy iteration (GPI), as introduced for discrete-time systems in Sutton and Barto (1998) and discussed in Chapter 2. In standard PI, the critic and actor NN are tuned sequentially, with one network being tuned while the weights of the other NN being held constant. By contrast, we tune both NN simultaneously in real time.

It is desired to determine a rigorously justified form for the actor NN based on the structure of PIs in RL. To this end, let us consider one step of the PI Algorithm 7.1. Suppose that the solution $V(x) \in C^1(\Omega)$ to the Bellman equation in policy evaluation step 1 for a given admissible policy $u(t)$ is given by (7.1). Then, according to (7.2) and step 2 of the PI algorithm one has for the policy update

$$u = -\frac{1}{2}R^{-1}g^T(x)\sum_{i=1}^{\infty} c_i \, \nabla\varphi_i(x) \tag{7.26}$$

for some unknown coefficients c_i. Then one has the following result.

Lemma 7.2. Let the least-squares solution to (7.5) be W_1 and define

$$u_1(x) = -\frac{1}{2}R^{-1}g^T(x)\,\nabla V_1(x) = -\frac{1}{2}R^{-1}g^T(x)\,\nabla\phi_1^T(x)W_1 \tag{7.27}$$

with V_1 defined in (7.7). Then, as $N \to \infty$:

a. $\sup_{x\in\Omega} \|u_1 - u\| \to 0$
b. There exists an N_0 such that $u_1(x)$ is admissible for $N > N_0$.

Proof: See Abu-Khalaf and Lewis (2005). ■

In light of this result, the ideal control policy update in Algorithm 7.1 is taken as (7.27), with W_1 unknown. Therefore, define the control policy in the form of an action neural network with weights $\hat{W}_2$ and compute the control input in the structured form

$$u_2(x) = -\frac{1}{2}R^{-1}g^T(x)\,\nabla\phi_1^T\hat{W}_2 \tag{7.28}$$

where $\hat{W}_2$ denotes the current estimated values of the ideal NN weights W_1. Define the actor NN weight estimation error as

$$\tilde{W}_2 = W_1 - \hat{W}_2 \tag{7.29}$$

The next definition and standard assumptions complete the machinery required for our main result.

Definition 7.2. Lewis *et al.* (1999) (Uniformly ultimately bounded) The equilibrium point $x_e = 0$ of non-linear system $\dot{x} = f(x) + g(x)u$ is said to be uniformly ultimately bounded (UUB) if there exists a compact set $S \subset R^n$ so that for all $x_0 \in S$ there exists a bound B and a time $T(B, x_0)$ such that $\|x(t) - x_e\| \leq B$ for all $t \geq t_0 + T$.

Assumptions 7.3.

a. $f(\cdot)$, is Lipschitz, and $g(\cdot)$ is bounded by a constant

$$\|f(x)\| < b_f\|x\|, \ \|g(x)\| < b_g$$

b. The NN approx error and its gradient are bounded on a compact set containing Ω so that

$$\|\varepsilon\| < b_\varepsilon$$

$$\|\nabla\varepsilon\| < b_{\varepsilon_x}$$

c. The NN activation functions and their gradients are bounded so that

$$\|\phi_1(x)\| < b_\phi$$

$$\|\nabla\phi_1(x)\| < b_{\phi_x}$$

Assumption 7.3c is satisfied, for example, by sigmoids, tanh and other standard NN activation functions.

We now present the main theorem, which provides simultanecus tuning laws for the actor and critic neural networks that guarantee convergence online to the optimal controller, while ensuring closed-loop stability.

Theorem 7.2. Synchronous online optimal adaptive controller. Let the critic NN be given by (7.10) and the control input be given by actor NN (7.28). Let tuning for the critic NN be provided by

$$\dot{\hat{W}}_1 = -a_1 \frac{\sigma_2}{(\sigma_2^T\sigma_2 + 1)^2}\left[\sigma_2^T\hat{W}_1 + Q(x) + u_2^T R u_2\right] \tag{7.30}$$

where $\sigma_2 = \nabla\phi_1(f + gu_2)$, and assume that $\overline{\sigma}_2 = \sigma_2/(\sigma_2^T\sigma_2 + 1)$ is persistently exciting. Let the actor NN be tuned as

$$\dot{\hat{W}}_2 = -\alpha_2\left\{(F_2\hat{W}_2 - F_1\overline{\sigma}_2^T\hat{W}_1) - \frac{1}{4}\overline{D}_1(x)\hat{W}_2 m^T(x)\hat{W}_1\right\} \tag{7.31}$$

where $\overline{D}_1(x) = \nabla\phi_1(x)g(x)R^{-1}g^T(x)\ \nabla\phi_1^T(x)$, $m = (\sigma_2/(\sigma_2^T\sigma_2 + 1)^2)$ and $F_1 > 0$ and $F_2 > 0$ are tuning parameters. Let Assumptions 7.1–7.3 hold, and the tuning parameters be selected as detailed in the proof. Then there exists an N_0 such that, for the number of hidden-layer units $N > N_0$ the closed-loop system state $x(t)$, the critic

NN error $\tilde{W}_1$ and the actor NN error $\tilde{W}_2$ are UUB. Moreover, Theorem 7.1 holds with $\varepsilon_{\max}$ defined in the proof below, so that exponential convergence of $\hat{W}_1$ to the approximate optimal critic value W_1 is obtained.

Proof: See appendix. ■

Remark 7.3. Note that the dynamics must be known to implement this algorithm in that $\sigma_2 = \nabla\phi_1(f + gu_2), \overline{D}_1(x)$ and (7.28) depend on $f(x), g(x)$.

Remark 7.4. Let $\varepsilon > 0$ and let N_0 be the number of hidden-layer units above which $\sup_{x\in\Omega} \|\varepsilon_{HJB}\| < \varepsilon$. In the proof it is seen that the theorem holds for $N > N_0$. Additionally, ε provides an effective bound on the critic weight residual set in Theorem 7.1. That is, $\varepsilon_{\max}$ in (7.25) is effectively replaced by ε.

Remark 7.5. The theorem shows that PE is needed for proper identification of the value function by the critic NN, and that a non-standard tuning algorithm is required for the actor NN to guarantee stability. The second term in (7.31) is a cross-product term that involves both the critic weights and the actor weights. It is needed to guarantee good behavior of the Lyapunov function, that is that the energy decreases to a bounded compact region.

Remark 7.6. The tuning parameters F_1 and F_2 in (7.31) must be selected to make the matrix M in the proof positive definite.

Remark 7.7. It is important to note that the proof of Theorem 7.2 does not rely at all on convergence of the PI Algorithm 7.1. PI, a reinforcement learning method, is used to obtain the structure of the controller, whereas the proofs of performance are carried out using adaptive control Lyapunov techniques.

Remark 7.8. The PI Algorithm 7.1 must be initialized with a stabilizing control policy. The convergence proof requires this, because the Bellman equation has a positive definite solution only if the controller is admissible. A stabilizing controller can be difficult to find for non-linear systems. By contrast, the online synchronous optimal adaptive controller does not require an initial stabilizing controller. The Lyapunov proof shows that the algorithm converges to the optimal control solution as long as the initial weight estimation errors are not large.

7.5 Structure of adaptive controllers and synchronous optimal adaptive control

The two-loop structure of the new synchronous optimal adaptive controller just developed is based on RL methods. The synchronous optimal adaptive controller should be contrasted with the optimal adaptive controllers developed in Part I of the book and displayed in Section 4.4. The controllers of Part I are in keeping with the philosophy of RL controllers where the critic and actor networks are tuned sequentially; that is, as the critic learns, the actor policy is kept constant. They are hybrid controllers with a CT inner action loop and an outer critic that learns on a discrete-time scale. By contrast, the synchronous optimal adaptive controller is in keeping with the philosophy of adaptive control (Ioannou and Fidan, 2006; Astrom and Wittenmark, 1995; Tao, 2003), in that both loops are tuned simultaneously on a CT scale.

Capabilities of the synchronous optimal adaptive controller

The synchronous controller has the following capabilities:

- Guaranteed stability of the closed-loop system dynamics. That is, the states $x(t)$ are guaranteed to be bounded near their desired values regardless of system disturbances.
- Minimization of the cost function.
- Adaptive learning of the optimal minimizing solution online in real time by measuring the states $x(t)$ and controls $u(t)$ along the system trajectories.

Due to these features we call this an *optimal adaptive controller*.

The optimal adaptive controller has the structure shown in Figure 7.1. It consists of two learning systems or networks that interact. The action network is an inner-loop feedback controller that computes the control input $u(t)$ to apply to the system. This action network has the form (7.28), where $\hat{W}_2$ are the actor NN parameters that are learned or updated online.

The second critic network learning system is not a feature of current adaptive feedback control systems. This network has the role of estimating the value function by solving the approximate Bellman equation (7.11). It has the form (7.10), where $\hat{W}_1$ are the critic NN parameters that are learned or updated online. In these equations, $\phi_1(x)$ is a set of basis functions suitably chosen depending on which specific system $\dot{x} = f(x) + g(x)u$ and which sort of value function is being considered. For linear systems with quadratic values (the LQR), these basis functions are the polynomials quadratic in the state components.

The parameters of the critic $\hat{W}_1$ and control action $\hat{W}_2$ are tuned online according to the update laws (7.30), (7.31), respectively. These are memory update laws that allow the parameters to converge to the solution to the optimal control problem that

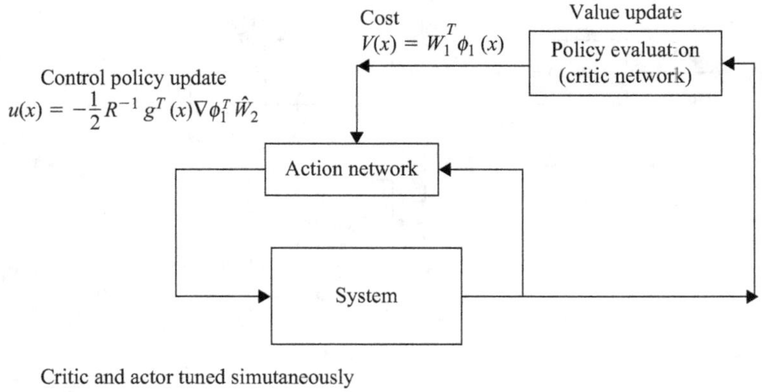

Figure 7.1 Structure of the synchronous optimal adaptive controller

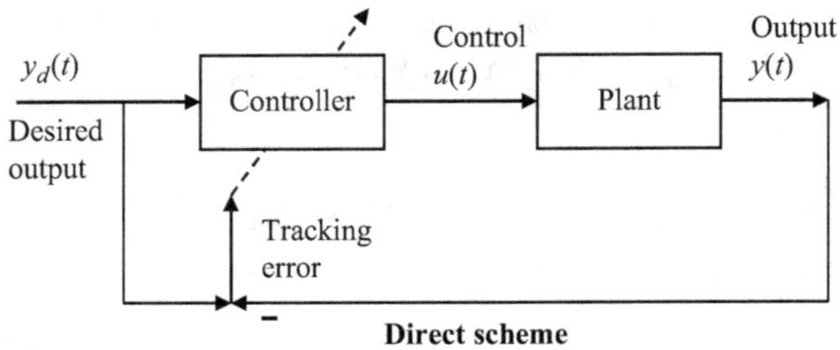

Figure 7.2 Standard form of direct adaptive controller where the controller parameters are updated in real time

minimizes the value function. Specifically, the critic network learns the minimum value, while action network learns the optimal control that produces this best value.

Contrast with standard adaptive control structures used in feedback control

The new controller has the structure in Figure 7.1. This is not like standard adaptive control system structures such as those now described next (Ioannou and Fidan, 2006; Astrom and Wittenmark, 1995; Tao, 2003).

Direct adaptive controller. Standard adaptive control systems can have one of several structures. In the direct adaptive controller of Figure 7.2, the controller is parameterized in terms of unknown parameters, and adaptive tuning laws are given for updating the controller parameters. These tuning laws depend on the tracking error, which it is desired to make small.

Indirect adaptive controller. In the indirect adaptive controller in Figure 7.3, two networks are used, one of which learns online. One network is a system identifier that has the function of identifying or learning the system dynamics model $\dot{x} = f(x) + g(x)u$ online in real time. The tuning law for the system identifier depends on the identification error, which it is desired to make small. After the system has been

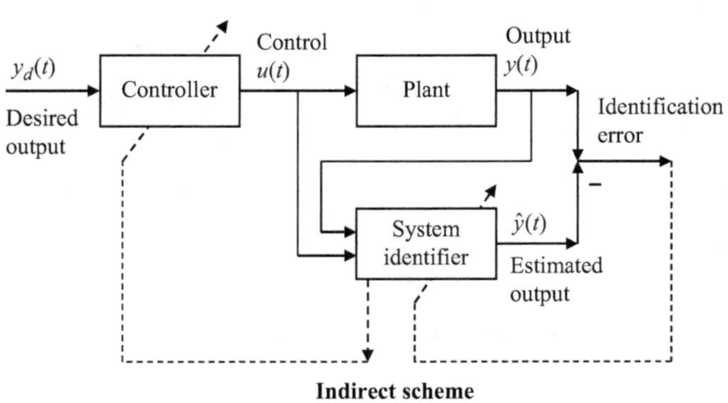

Figure 7.3 Standard form of indirect adaptive controller where the parameters of a system identifier are updated in real time

identified, the controller parameters in the controller network can be computed using a variety of methods, including for instance solution of the HJB equation.

Synchronous optimal adaptive control: Identification of the value function

The function of the direct adaptive controller is to tune the parameters of the control loop to identify a good controller. The function of the indirect adaptive controller is to tune parameters to identify the system model. Following this progression naturally, the function of the optimal adaptive controller is to tune parameters in order to identify the value function. In summary:

- Direct adaptive control identifies the controller.
- Indirect adaptive control identifies the system model.
- Optimal adaptive control identifies the value function.

Neural-network controller. The critic and action networks in Figure 7.1 have been interpreted here as neural networks. A standard NN controller (Lewis *et al.*, 1999) is shown in Figure 7.4. This has one learning network based on NN and an outer proportional-plus-derivative (PD) tracking loop.

In the NN controller, the function of the NN is to learn the unknown dynamics $f(x)$ by approximating it so that

$$\hat{f}(x) = \hat{W}^T \phi(x)$$

The parameters are updated online using the learning rule

$$\dot{\hat{W}} = F\hat{\sigma}r^T - F\hat{\sigma}'\hat{V}^T xr^T - \kappa F\|r\|\hat{W}$$

This has the effect of linearizing the dynamics $\dot{x} = f(x) + g(x)u$ so that the outer PD tracking loop performs correctly. This controller is known as a feedback linearization NN control structure.

The NN controller in Figure 7.4 cannot deliver optimal performance in terms of minimizing a value function. It does guarantee closed-loop stability of the system without knowing the system dynamics function $f(x)$.

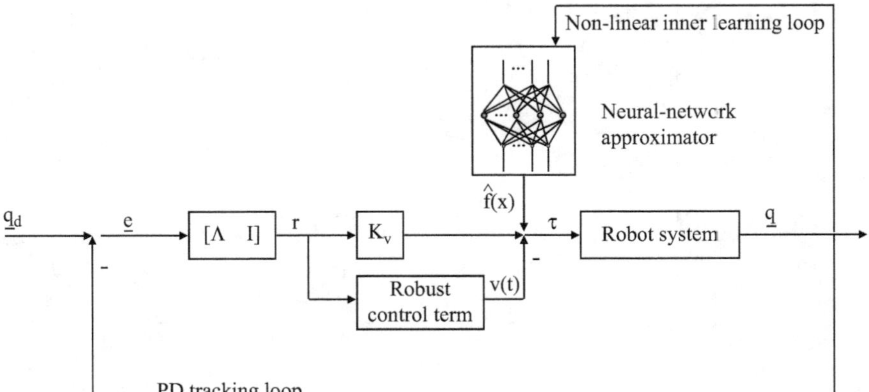

Figure 7.4 Standard neural-network controller

7.6 Simulations

7.6.1. Linear system example

Consider the CT F-16 longitudinal aircraft dynamics with quadratic cost function used in Stevens and Lewis (2003)

$$\dot{x} = \begin{bmatrix} -1.01887 & 0.90506 & -0.00215 \\ 0.82225 & -1.07741 & -0.17555 \\ 0 & 0 & -1 \end{bmatrix} x + \begin{bmatrix} 0 \\ 0 \\ 1 \end{bmatrix} u = Ax + Bu$$

where Q and R in the cost function are identity matrices of appropriate dimensions. In this linear case the solution of the HJB equation is given by the solution of the algebraic Riccati equation (ARE) $A^T P + PA + Q - PBR^{-1}B^T P = 0$. Since the value is quadratic in the LQR case, the critic NN basis set $\phi_1(x)$ was selected as the quadratic vector in the state components. Solving the ARE offline gives the parameters of the optimal critic $W_1^* = [p_{11} \quad 2p_{12} \quad 2p_{13} \quad p_{22} \quad 2p_{23} \quad p_{33}]^T$ with $P = [P_{ij}]$ the Riccati solution matrix. The solution is

$$W_1^* = [1.4245 \quad 1.1682 \quad -0.1352 \quad 1.4349 \quad -0.1501 \quad 0.4329]^T$$

The synchronous PI algorithm is implemented as in Theorem 7.2. PE was ensured by adding a small exponentially decaying probing noise to the control input. Figure 7.5 shows that the critic parameters, denoted by

$$\hat{W}_1 = [W_{c1} \quad W_{c2} \quad W_{c3} \quad W_{c4} \quad W_{c5} \quad W_{c6}]^T$$

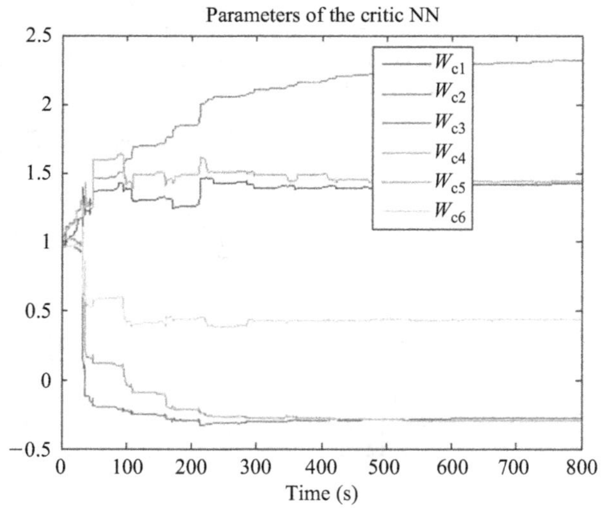

Figure 7.5 Convergence of the critic parameters to the parameters of the optimal critic

converge to the optimal values. In fact after 800 s the critic parameters converged to

$$\hat{W}_1(t_f) = [1.4279 \quad 1.1612 \quad -0.1366 \quad 1.4462 \quad -0.1480 \quad 0.4317]^T$$

The actor parameters after 800 s converge to the values of $\hat{W}_2(t_f) = \hat{W}_1(t_f)$. Then, the actor NN is given by (7.28) as

$$\hat{u}_2(x) = -\frac{1}{2}R^{-1}\begin{bmatrix}0\\0\\1\end{bmatrix}^T \begin{bmatrix} 2x_1 & 0 & 0 \\ x_2 & x_1 & 0 \\ x_3 & 0 & x_1 \\ 0 & 2x_2 & 0 \\ 0 & x_3 & x_2 \\ 0 & 0 & 2x_3 \end{bmatrix} \hat{W}_2(t_f)$$

This is approximately the correct optimal control solution $u = -R^l B^T P x$.

The evolution of the system states is presented in Figure 7.6. One can see that after 750 s convergence of the NN weights in both critic and actor has occurred. This shows that the exponentially decreasing probing noise effectively guaranteed the PE condition. After that, the states remain very close to zero, as required.

7.6.2. Non-linear system example

Consider the following affine-in-control input non-linear system

$$\dot{x} = f(x) + g(x)u, \quad x \in R^2$$

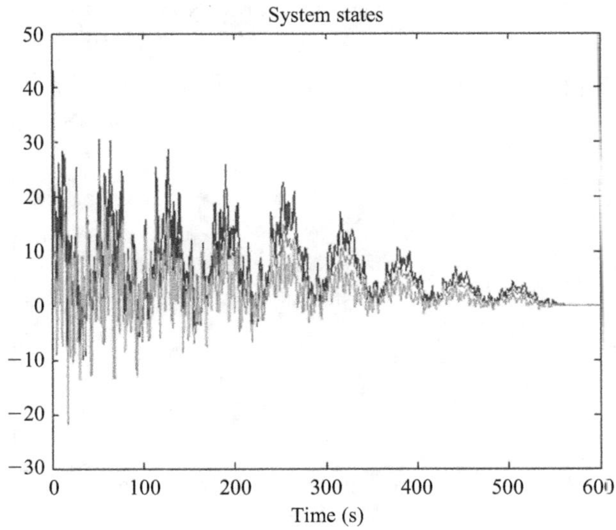

Figure 7.6 Evolution of the system states for the duration of the experiment

where

$$f(x) = \begin{bmatrix} -x_1 + x_2 \\ -0.5x_1 - 0.5x_2(1 - (\cos(2x_1) + 2)^2 \end{bmatrix}$$

$$g(x) = \begin{bmatrix} 0 \\ \cos(2x_1) + 2 \end{bmatrix}$$

One selects $Q = \begin{bmatrix} 1 & 0 \\ 0 & 1 \end{bmatrix}, R = 1$ in the value function integral.

Using the converse optimal procedure in Nevistic and Primbs (1996) the optimal value function is designed to be

$$v^*(x) = \frac{1}{2}x_1^2 + x_2^2$$

and the optimal control signal is

$$u^*(x) = -(\cos(2x_1) + 2)x_2$$

One selects the critic NN vector activation function as

$$\phi_1(x) = [x_1^2 \quad x_1x_2 \quad x_2^2]^T$$

Figure 7.7 shows the critic parameters, denoted by

$$\hat{W}_1 = [\, W_{c1} \quad W_{c2} \quad W_{c3} \,]^T$$

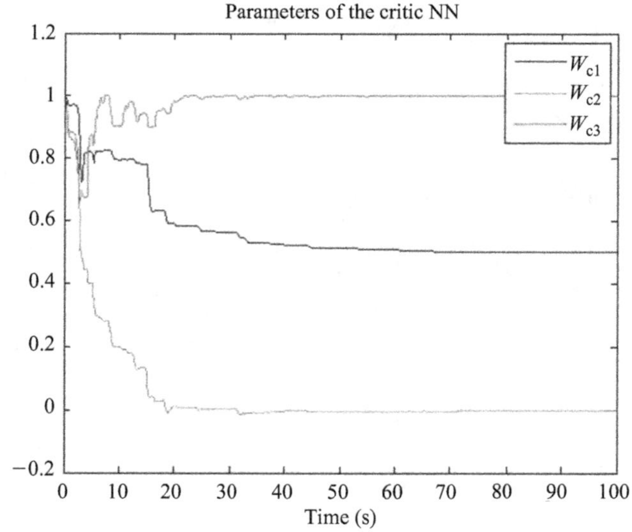

Figure 7.7 Convergence of the critic parameters

These converge after about 80 s to the approximately correct values of

$$\hat{W}_1(t_f) = [\,0.5017 \quad -0.0020 \quad 1.0008\,]^T$$

The actor parameters after 80 s converge to the correct values of

$$\hat{W}_2(t_f) = [\,0.5017 \quad -0.0020 \quad 1.0008\,]^T$$

Thus, that the actor NN (7.28) converges to the optimal control

$$\hat{u}_2(x) = -\frac{1}{2}R^{-1}\begin{bmatrix} 0 \\ \cos(2x_1)+2 \end{bmatrix}^T \begin{bmatrix} 2x_1 & 0 \\ x_2 & x_1 \\ 0 & 2x_2 \end{bmatrix}^T \begin{bmatrix} 0.5017 \\ -0.0020 \\ 1.0008 \end{bmatrix}$$

$$\approx -(\cos(2x_1)+2)x_2$$

The evolution of the system states is presented in Figure 7.8. One can see that after 80 s convergence of the NN weights in both critic and actor has occurred. The probing noise is a random excitation that was turned off at 80 s after convergence of the critic NN. This probing noise effectively guaranteed the PE condition. On convergence, the PE condition of the control signal is no longer needed, and the probing signal was turned off. After that, the states remain very close to zero, as required.

Figure 7.9 shows the optimal value function. The identified value function given by $\hat{V}_1(x) = \hat{W}_1^T\phi_1(x)$ is virtually indistinguishable. In fact, Figure 7.10 shows the 3D plot of the difference between the approximated value function, by using the online algorithm, and the optimal value. This error is close to zero. Good

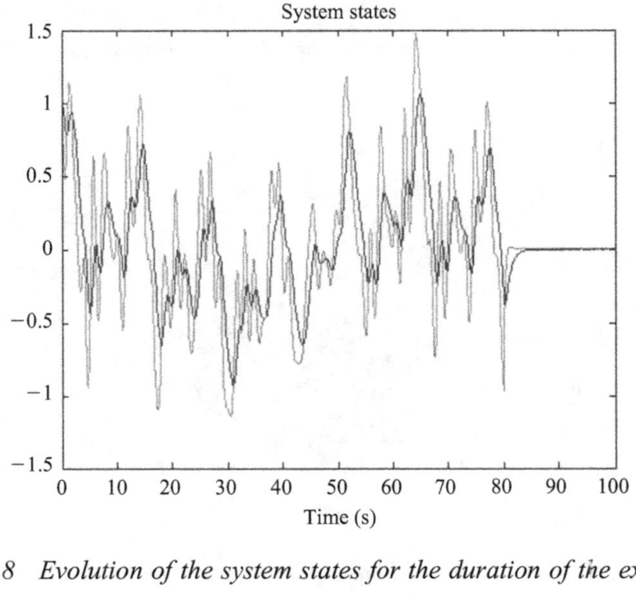

Figure 7.8 Evolution of the system states for the duration of the experiment

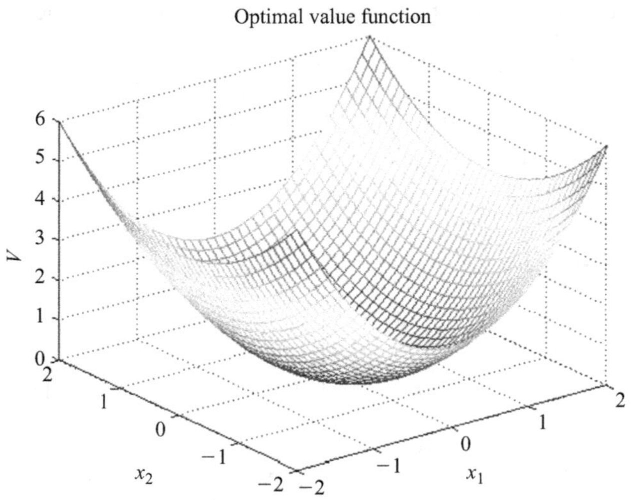

Figure 7.9 Optimal value function

approximation of the actual value function is evolved. Finally, Figure 7.11 shows the 3D plot of the difference between the approximated control, by using the online algorithm, and the optimal control. This error is close to zero.

In this example the VFA activation functions were selected as quadratic and quartic in the states. In general, the form of the value function is not known for non-linear systems. Then, one selects a polynomial basis set. If convergence is not observed, higher-order polynomials are chosen and the simulation is re-run. This is

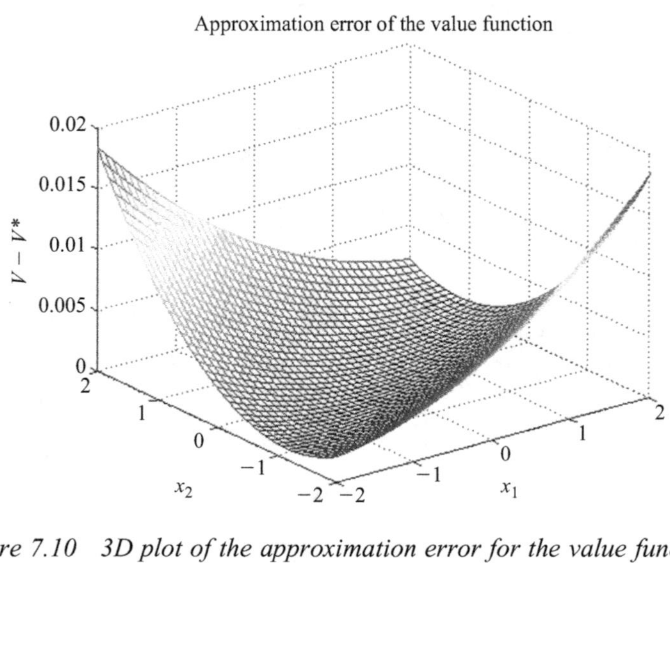

Figure 7.10 3D plot of the approximation error for the value function

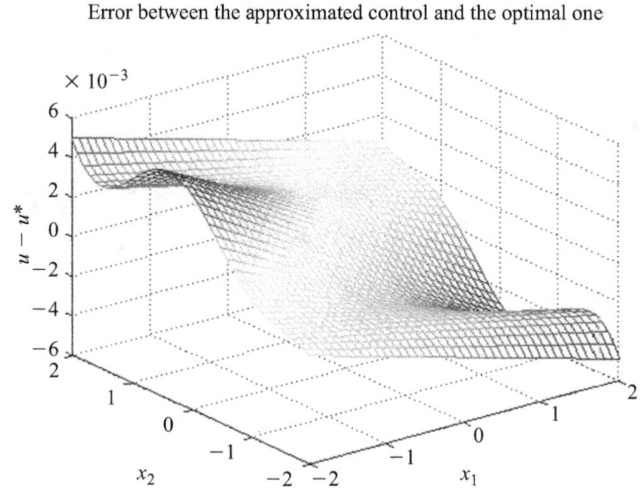

Figure 7.11 3D plot of the approximation error for the control

repeated until convergence occurs. The polynomials should be selected as even order since the value must be a Lyapunov function. Other basis sets than polynomials could also be used, such as tanh, sigmoid.

7.7 Conclusion

In this chapter, we proposed a new adaptive algorithm that solves the CT optimal control problem online in real time for affine-in-the-inputs non-linear systems. We call this algorithm synchronous online optimal adaptive control for CT systems. The structure of this controller was based on the RL PI algorithm, and the proofs were carried out using adaptive control Lyapunov function techniques. The algorithm requires complete knowledge of the system model $f(x), g(x)$. The next chapter will show an algorithm for simultaneous tuning of the neural networks in an optimal adaptive controlled without the need of the drift dynamics $f(x)$.

Chapter 8
Synchronous online learning with integral reinforcement

This chapter presents an online adaptive learning algorithm to solve the infinite-horizon optimal control problem for non-linear systems of the form $\dot{x} = f(x) + g(x)u$ that combines the advantages of the algorithms in Chapters 4 and 7. These include *simultaneous tuning* of both actor and critic NNs (i.e. both neural networks are tuned at the same time) and no need for knowledge of the drift term $f(x)$ in the dynamics. This algorithm is based on integral reinforcement learning (IRL) and solves the Hamilton-Jacobi-Bellman (HJB) equation online in real time by measuring data along the system trajectories, without knowing $f(x)$. In the linear quadratic case $\dot{x} = Ax + Bu$ it solves the algebraic Riccati equation (ARE) online without knowing the system matrix A.

IRL policy iteration, as presented in Chapter 4, is an actor–critic reinforcement learning technique, and as such it tunes the actor and critic networks sequentially. That is, while the critic network is learning the Bellman equation solution, the actor network does not learn. This sequential tuning procedure is not usual in standard adaptive control systems (Ioannou and Fidan, 2006; Astrom and Wittenmark, 1995; Tao, 2003). Therefore, in this chapter the actor–critic structure of IRL policy iteration is used to motivate a two-loop structure for an optimal adaptive controller, and then tuning laws are developed to tune the parameters of both the critic loop and the actor loop simultaneously. This is called synchronous IRL optimal adaptive control. Due to the use of IRL, the drift dynamics $f(x)$ are not needed in these tuning laws. These results were first presented in Vamvoudakis (2011).

The chapter is organized as follows. Section 8.1 reviews the optimal control problem from Chapter 4 and presents the IRL policy iteration algorithm that solves it. Section 8.2 uses value function approximation (VFA) to solve the IRL Bellman equation. In Section 8.3, the IRL policy iteration algorithm is used to motivate the form of the actor network. Two theorems are given that show how to tune the actor and critic networks simultaneously so that the optimal control solution is learned in real time while system stability is guaranteed throughout the learning process. The proofs are based on Lyapunov stability theory. Section 8.4 presents simulation examples that show the effectiveness of the synchronous IRL optimal adaptive control algorithm. In a linear design example, the IRL optimal adaptive controller learns the ARE solution online without knowing the system matrix A. In a non-linear example, the optimal adaptive controller learns the HJB solution without knowing the drift dynamics $f(x)$.

The IRL policy iteration algorithm is used to develop a novel two-loop adaptive control structure that learns optimal control solutions online. Yet, the proofs of convergence of the optimal adaptive controller do not depend on proving convergence of the policy iteration algorithm. They are based on Lyapunov techniques. Moreover, the policy iteration algorithm must be initialized with a stabilizing controller. Stabilizing controllers can be difficult to find for non-linear systems. By contrast, the IRL optimal adaptive controller does not require an initial stabilizing controller.

8.1 Optimal control and policy iteration using integral reinforcement learning

The optimal adaptive controller in this chapter has a two-loop structure that comes from continuous-time (CT) policy iterations using integral reinforcement learning (IRL), which was covered in Chapter 4. This means we shall be able to develop an adaptive controller that converges to the optimal control solution without knowing the system drift dynamics $f(x)$. Therefore, first we review the optimal control problem described in Section 4.1 and its IRL policy iteration solution. This development parallels that at the beginning of Chapter 7, where the full system dynamics was required to implement the online optimal adaptive controller.

Consider the time-invariant affine-in-the-input non-linear dynamical system given by

$$\dot{x}(t) = f(x(t)) + g(x(t))u(x(t)), \quad x(0) = x_0$$

with state $x(t) \in \mathbb{R}^n$, $f(x(t)) \in \mathbb{R}^n$, $g(x(t)) \in \mathbb{R}^{n\times m}$ and control input $u(t) \in U \subset \mathbb{R}^m$. It is assumed that $f(0) = 0$, that $f(x) + g(x)u$ is Lipschitz continuous on a set $\Omega \subseteq \mathbb{R}^n$ that contains the origin, and that the dynamical system is stabilizable on Ω, that is there exists a continuous control function $u(t) \in U$ such that the closed-loop system is asymptotically stable on Ω.

The infinite-horizon cost function or *value function* associated with any admissible control policy $u(t) = \mu(x(t)) \in \Psi(\Omega)$ is

$$V\mu(x(t)) = \int_t^\infty r(x(\tau), \mu(\tau))\, d\tau = \int_t^\infty (Q(x) + \mu^T Ru)\, d\tau$$

where $V^\mu(x)$ is C^1. Recall from Chapter 4 that a control policy $u(t) = \mu(x)$ is admissible on Ω, denoted by $\mu \in \Psi(\Omega)$, if $\mu(x)$ is continuous on Ω, $\mu(0) = 0$, $\mu(x)$ stabilizes the dynamics on Ω, and the value is finite $\forall x_0 \in \Omega$.

Using Leibniz's formula, the infinitesimal version of the value integral is found to be the following.

CT Bellman equation

$$0 = r(x, \mu(x)) + (\nabla V_x^\mu)^T (f(x) + g(x)\mu(x)), \quad V^\mu(0) = 0$$

Here, ∇V_x^μ (a column vector) denotes the gradient of the cost function V^μ with respect to x. That is, $\nabla V_x^\mu = \partial V^\mu / \partial x$. This is a Bellman equation for non-linear CT systems, which, given the control policy $\mu(x) \in \Psi(\Omega)$, can be solved for the value $V^\mu(x)$ associated with it.

The optimal control problem is formulated as follows. Given the continuous-time dynamical system, the set $u \in \Psi(\Omega)$ of admissible control policies, and the value function, find an admissible control policy such that the value function is minimized.

Defining the Hamiltonian

$$H(x, u, V_x) = r(x(t), u(t)) + (\nabla V_x)^T (f(x(t)) + g(x(t))u(t))$$

the optimal cost function $V^*(x)$ satisfies the Hamilton–Jacobi–Bellman (HJB) equation

$$0 = \min_{u \in \Psi(\Omega)} [H(x, u, \nabla V_x^*)]$$

Assuming that the minimum on the right-hand side of this equation exists and is unique, then the optimal control function is

$$u^*(x) = \frac{1}{2} R^- 1 g^T(x) \, \nabla V_x^*$$

Inserting this optimal control policy in the Hamiltonian we obtain the HJB equation in the form

$$0 = Q(x) + (\nabla V_x^*)^T f(x) - \frac{1}{4} (\nabla V_x^*)^T g(x) R^{-1} g^T(x) \, \nabla V_x^*, \quad V^*(0) = 0$$

This is a necessary condition for the optimal cost function. For the linear system case with a quadratic cost functional (linear quadratic regulator), the equivalent of this HJB equation is the algebraic Riccati equation (ARE) discussed in Chapter 3 (see Section 3.1).

To find the optimal control solution for the problem one can solve the HJB equation for the optimal value function $V^*(x)$ and then compute the optimal control $u^*(x)$. However, solving the HJB equation is generally difficult, and analytic solutions may not exist. Therefore, in Chapter 4 we wrote policy iteration Algorithm 4.1 to solve the HJB by iteratively solving the CT Bellman equation.

Algorithm 4.1, and therefore all of Chapter 7, which is based upon it, relies on complete knowledge of the system dynamics $f(x), g(x)$. To design an optimal adaptive controller that does not require knowledge of the drift dynamics $f(x)$, we can use the version of PI based on integral reinforcement learning (IRL) given as Algorithm 4.2. Given an admissible policy and an integration time interval $T > 0$ write the value function as the following equivalent form Vrabie *et al.* (2008, 2009), Vrabie (2009), Vrabie and Lewis (2009).

Integral reinforcement form of value function: IRL Bellman equation

$$V^{\mu}(x(t)) = \int_{t}^{t+T} r(x(\tau), \mu(x(\tau)))\, d\tau + V^{\mu}(x(t+T))$$

Note that this form does not contain the system dynamics $(f(\cdot), g(\cdot))$. In Lemma 4.1, it is shown that this equation is equivalent to the CT Bellman equation in the sense that both equations have the same positive definite solution. Therefore, using the IRL Bellman equation allows the formulation of continuous-time PI algorithms that share the beneficial features of discrete-time PI algorithms. The integrand

$$\rho(x(t), t, t+T) = \int_{t}^{t+T} r(x(\tau), \mu(x(\tau)))\, d\tau$$

is known as the *integral reinforcement* on the time interval $[t, t+T]$.

A policy iteration algorithm based on IRL was given as Algorithm 4.2. It is reproduced here as Algorithm 8.1 for convenience to the reader. Let $\mu^{(0)}(x(t)) \in \Psi(\Omega)$ be an admissible policy, select $T > 0$ such that, if $x(t) \in \Omega$ then also $x(t+T) \in \Omega$. The existence of such a time period $T > 0$ is guaranteed by the admissibility of $\mu^{(0)}(\cdot)$ on Ω. Then the following PI algorithm based on the IRL Bellman equation can be written.

Algorithm 8.1. Integral reinforcement learning policy iteration algorithm

Select $\mu^{(0)}(x(t)) \in \Psi(\Omega)$ as an admissible policy.

1. (*Policy evaluation* step.) Solve for the value $V^{\mu^{(i)}}(x(t))$ using the IRL Bellman equation

$$V^{\mu^{(i)}}(x(t)) = \int_{t}^{t+T} r(x(s), \mu^{(i)} x(s))\, ds + V^{\mu^{(i)}}(x(t+T)) \quad \text{with} \quad V^{\mu^{(i)}}(0) = 0$$

2. (*Policy improvement* step.) Update the control policy using

$$\mu^{(i+1)}(x) = -\frac{1}{2} R^{-1} g^{T}(x)\, \nabla V_{x}^{\mu^{(i)}}$$

■

It was shown in Chapter 4 that the IRL policy iteration Algorithm 8.1 solves the non-linear HJB equation by iterations on IRL Bellman equations that are linear in the value function gradient. Its importance is that the linear IRL Bellman equations are easier to solve than the HJB, which is non-linear.

Of major importance is the fact that nowhere in this IRL algorithm does the system drift dynamics $f(x)$ appear. The input-coupling dynamics $g(x)$ must be known to implement this algorithm since they appear in the policy improvement step. Moreover, it is required that a stabilizing control policy be known to initialize the algorithm. This algorithm is used in the next sections to motivate a novel adaptive control structure that solves the optimal control problem online by measuring data along the system trajectories, without knowing the drift dynamics $f(x)$.

8.2 Critic neural network and Bellman equation solution

The IRL policy iteration Algorithm 8.1 has an actor–critic structure and involves sequential updates of the critic network and the actor network. That is, while the critic learns the value by solving the IRL Bellman in step 1, the actor parameters (e.g. the control policy) are not changed. Though natural in reinforcement learning, this is a strange tuning philosophy from the point of view of standard adaptive control theory (Ioannou and Fidan, 2006; Astrom and Wittenmark, 1995; Tao, 2003). Therefore, we now use the structure of Algorithm 8.1 to develop an adaptive control structure that converges to the optimal control solution in real time by simultaneous or synchronous tuning of the critic and actor networks. Synchronous tuning of all loops yields a continuous-time controller that is more familiar from the point of view of standard ideas in adaptive control than the sequential tuning RL controllers developed in Part I of the book.

For IRL tuning it is necessary to modify the value function. Define there the value function as

$$V(x(t)) = \int_t^\infty r(x(\tau), u(\tau))\, d\tau = \int_t^\infty \left(Q(x) + u^T \frac{R}{T} u \right) d\tau$$

with $T > 0$ the IRL integration interval. Then the optimal control becomes

$$u^*(x) = -\frac{1}{2} T R^{-1} g^T(x) \nabla V_x^*$$

with $V^*(x)$ the HJB solution. The new value function means that as the integration time interval T becomes smaller, the control input is more heavily weighted, so that less control effort is used. This corresponds to penalizing fast changes in the control input.

Now, introduce a critic NN for value function approximation (VFA) as described in Section 7.1. Thus, value function $V(x)$ is approximated as

$$V(x) = W_1{}^T \phi(x) + \varepsilon(x)$$

where W_1 are the unknown weights of a neural network (NN), $\phi(x) : \mathbb{R}^n \to \mathbb{R}^N$ is the NN activation function vector, N is the number of neurons in the hidden layer and $\varepsilon(x)$ is the NN approximation error. As per Chapter 7, the NN activation functions $\{\varphi_i(x) : i = 1, N\}$ are selected so that $\{\varphi_i(x) : i = 1, \infty\}$ provides a complete independent basis set such that $V(x)$ and its gradient are both approximated.

Using VFA in the IRL Bellman equation yields

$$\int_{t-T}^{t}\left(Q(x)+u^T\frac{R}{T}u\right)d\tau+W_1^T\phi(x(t))-W_1^T\phi(x(t-T))=\varepsilon_B \tag{8.1}$$

where ε_B is a Bellman equation residual equation error. Note that we are integrating now over the interval $[t-T,t]$ instead of $[t,t+T]$. This is so that one obtains causal tuning laws for the critic and actor networks, where all required data is measured prior to the current time t.

The integral reinforcement signal is

$$p=\int_{t-T}^{t}\left(Q(x)+u^T\frac{R}{T}u\right)d\tau \tag{8.2}$$

Now (8.1) can be written as

$$\varepsilon_B-p=W_1^T\,\Delta\phi(x(t)) \tag{8.3}$$

where the first difference is $\Delta\phi(x(t))\equiv\phi(x(t))-\phi(x(t-T))$.

Under the Lipschitz assumption on the dynamics, the residual error ε_B is bounded on a compact set.

Remark 8.1. Note that, as $N\to\infty,\varepsilon_B\to 0$ uniformly (Abu-Khalaf and Lewis, 2005).

The weights of the critic NN, W_1, which solve (8.1) are unknown. Then the output of the critic neural network is

$$\hat{V}(x)=\hat{W}_1^T\phi(x) \tag{8.4}$$

where $\hat{W}_1$ are the current known values of the critic NN weights. The approximate Bellman equation error is then

$$\int_{t-T}^{t}\left((Q)(x)+u^T\frac{R}{T}u\right)d\tau+\hat{W}_1{}^T\phi(x(t))-\hat{W}_1^T\phi(x(t-T))=e_1 \tag{8.5}$$

which according to (8.2) can be written as

$$\hat{W}_1^T\,\Delta\phi(x(t))=e_1-p \tag{8.6}$$

This is a scalar equation and it is desired to solve for the unknown weight vector $\hat{W}_1$. The first difference $\Delta\phi(x(t))$ is a regression vector (Ljung, 1999) that can be measured along the system trajectories.

It is desired to select $\hat{W}_1$ to minimize the squared residual error

$$E_1=\frac{1}{2}e_1{}^Te_1 \tag{8.7}$$

Then $\hat{W}_1(t) \to \hat{W}_1$. Therefore, select the tuning law for the critic weights as the normalized gradient descent algorithm

$$\dot{\hat{W}} = -a_1 \frac{\Delta\phi(x(t))^T}{(1+\Delta\phi(x(t))^T \Delta\phi(x(t)))^2} \left[\int_{t-T}^{t} \left(Q(x) + u^T \frac{R}{T} u \right) d\tau - \Delta\phi(x(t))^T \hat{W}_1 \right] \tag{8.8}$$

Note that the data required in this tuning algorithm at each time are $(\Delta\phi(t), p(t))$. The system dynamics $f(x), g(x)$ are not needed to implement this critic network tuning algorithm, in contrast to Chapter 7, where the full dynamics appeared in the critic NN tuning algorithm.

Define the critic weight estimation error $\tilde{W}_1 = W_1 - \hat{W}_1$ and substitute (8.1) in (8.8). With the notation $\overline{\Delta\phi}(t) = \Delta\phi(t)/(\Delta\phi(t)^T \Delta\phi(t) + 1)$ and $m_s = 1 + \Delta\phi(t)^T \Delta\phi(t)$, we obtain the dynamics of the critic weight estimation error as

$$\dot{\tilde{W}}_1 = -a_1 \overline{\Delta\phi}(t) \overline{\Delta\phi}(t)^T \tilde{W}_1 + a_1 \overline{\Delta\phi}(t) \frac{\varepsilon_B}{m_s} \tag{8.9}$$

In the upcoming development it is important to note that the activation function difference can be written as

$$\Delta\phi(x(t)) = \phi(x(t)) - \phi(x(t-T)) = \int_{t-T}^{t} \nabla\phi(x)\dot{x}\,d\tau = \int_{t-T}^{t} \nabla\phi(f+gu)\,d\tau \tag{8.10}$$

Though it is traditional to use critic tuning algorithms of the form (8.8), it is not generally understood when convergence of the critic weights to the Bellman equation solution can be guaranteed. To guarantee convergence of $\hat{W}_1$ to W_1, the next persistence of excitation (PE) assumption is required.

According to (8.6) the regression vector $\Delta\phi(t)$, or equivalently $\overline{\Delta\phi}(t)$ must be persistently exciting to solve for $\hat{W}_1$ in a least-squares sense. Therefore, make the following assumption.

Persistence of excitation assumption. Let the signal $\overline{\Delta\phi}(t)$ be persistently exciting over the interval $[t-T, t]$, that is there exist constants $\beta_1 > 0$, $\beta_2 > 0$, $T > 0$ such that, for all t

$$\beta_1 I \le S_0 \equiv \int_{t-T}^{t} \overline{\Delta\phi}(\tau) \overline{\Delta\phi}^T(\tau)\,d\tau \le \beta_2 I \tag{8.11}$$

Technical Lemma 8.1. Consider the error dynamics (8.9) with output

$$y_1 = \overline{\Delta\phi}(t)^T \tilde{W}_1$$

Assume $\overline{\Delta}\phi(t)$ is PE. Let $||\varepsilon_B|| \leq \varepsilon_{max}$ and $||y_1|| \leq y_{max}$. Then $||\hat{W}_1||$ converges exponentially to the residual set

$$\hat{W}_1(t) \leq \frac{\sqrt{\beta_2 T_{PE}}}{\beta_1}\{[y_{max} + \delta\beta_2 a_1(\varepsilon_{max} + y_{max})]\}$$

where δ is a positive constant of the order of 1.

Proof: See Chapter 7 (Technical Lemma 7.2). ■

8.3 Action neural network and adaptive tuning laws

The policy improvement step in IRL policy iteration Algorithm 8.1, according to (8.4), is

$$u(x) = -\frac{1}{2}TR^{-}1g^T(x)\,\nabla\phi^T W_1 \tag{8.12}$$

with critic weights W_1 unknown. Therefore, define the control policy in the form of an action neural network that computes the control input in the structured form

$$u_2(x) = -\frac{1}{2}TR^{-1}g^T(x)\,\nabla\phi^T \hat{W}_2 \tag{8.13}$$

where $\hat{W}_2$ denotes the current known values of the actor NN weights.

Based on (8.12) and (8.1), define the approximate HJB equation

$$\int_{t-T}^{t}\left(-Q(x) - \frac{1}{4}W_1^T\overline{D}_1(x)W_1 + \varepsilon_{HJB}(x)\right)d\tau = W_1^T\,\Delta\phi(x(t)) \tag{8.14}$$

with the notation $\overline{D}_1(x) = T\,\nabla\phi(x)g(x)R^{-1}g^T(x)\,\nabla\phi^T(x)$, where W_1 denotes the ideal unknown weights of the critic and actor neural networks that solve the HJB.

We now present the main theorems, which provide the synchronous tuning laws for the actor and critic NNs that guarantee convergence to the optimal controller along with closed-loop stability. The practical notion of uniform ultimate boundedness (UUB) from Chapter 7 is used. The next definition is needed in the proof.

Definition 8.1. Khalil (1996) A continuous function $\alpha : [\alpha, 0) \to [0, \infty)$ is said to belong to class K if it is strictly increasing and $\alpha(0) = 0$. It is said to belong to class K_∞ if $a = \infty$ and $\alpha(r) = \infty$ as $r \to \infty$.

Theorem 8.1. Online synchronous IRL optimal adaptive controller. Let the tuning for the critic NN be provided by

$$\dot{\hat{W}}_1 = -a_1\frac{\Delta\phi_2(t)}{(\Delta\phi_2(t)^T\Delta\phi_2(t)+1)^2}\left(\Delta\phi_2(t)^T\hat{W}_1 + \int_{t-T}^{t}\left(Q(x) + \frac{1}{4}\hat{W}_2^T\overline{D}_1\hat{W}_2\right)d\tau\right) \tag{8.15}$$

where $\Delta\phi_2(x(t)) = \int_{t-T}^{t} \nabla\phi(f + gu_2)\, d\tau = \int_{t-T}^{t} \sigma_2\, d\tau \equiv \phi(x(t)) - \phi(x(t-T)) \equiv \Delta\phi_2(t)$ and assume that $\Delta\overline{\phi}(t)$ is persistently exciting. Let the actor NN be tuned as

$$\hat{W}_2 = -a_2\left\{(F_2\hat{W}_2 - F_1 T\, \Delta\overline{\phi}_2^T \hat{W}_1) - \frac{1}{4m_s}\overline{D}_1(x)\hat{W}_2\, \Delta\overline{\phi}_2^T \hat{W}_1\right\} \tag{8.16}$$

where $F_1 > 0, F_2 > 0$ are tuning parameters chosen as in the proof. Let Assumptions 7.1–7.3 hold. Then there exists a N_0 and a time T_0 such that, for the number of hidden-layer units $N > N_0$ and the time interval $T < T_0$, the closed-loop system state, the critic NN error $\tilde{W}_1$ and the actor NN error $\tilde{W}_2$, are UUB.

Proof: See appendix. ■

Theorem 8.2. Optimal control solution. Suppose the hypotheses of Theorem 8.1 hold. Then:

a. $H(\hat{u}, \hat{W}_1, x) \equiv \int_{t-T}^{t}(Q(x) + \hat{u}^T \frac{R}{T}\hat{u} - \varepsilon_{HJB})\, d\tau + \hat{W}_1^T(x(t)) - \hat{W}_1^T\phi(x(t-T))$ is UUB, where $\hat{u} = -\frac{1}{2}TR^{-1}g^T(x)\,\nabla\phi^T(x)\hat{W}_1$. That is, $\hat{W}_1$ converge to the approximate HJB solution.
b. $\hat{u}_2(x)$ converges to the optimal control solution, where $\hat{u}_2 = -\frac{1}{2}TR^{-1}g^T(x)\, \nabla\phi(x)\hat{W}_2$.

Proof:

a. Consider the weights $\tilde{W}_1, \tilde{W}_2$ to be UUB as proved in Theorem 8.1.

$$H(\hat{u}, \hat{W}_1, x) = \int_{t-T}^{t}\left(Q(x) + \hat{u}^T \frac{R}{T}\hat{u} - \varepsilon_{HJB}\right) d\tau - \hat{W}_1^T\phi(x(t)) - \hat{W}_1^T\phi(x(t-T))$$

After adding zero we have

$$H(\hat{u}, \hat{W}_1, x) \equiv \int_{t-T}^{t}\left(\frac{1}{4}\tilde{W}_1^T\overline{D}_1(x)\tilde{W}_1 - \frac{1}{2}\tilde{W}_1^T\overline{D}_1(x)W_1 - \varepsilon_{HJB}\right) d\tau$$
$$- \tilde{W}_1^T\phi(x(t)) + \tilde{W}_1^T\phi(x(t-T))$$

By taking norms on both sides and taking into account that $\sup_{x\in\Omega}\|\varepsilon_{HJB}\| < \varepsilon$ Facts 7.1 and letting

$$\|H(\hat{u}, \hat{W}_1, x)\| \equiv \int_{t-T}^{t}\left(\frac{1}{4}\|\tilde{W}_1\|^2\|\overline{D}_1(x)\| + \frac{1}{2}\|\tilde{W}_1\|\|W_{1\,\max}\|\overline{D}_1(x)\| + \varepsilon\right) d\tau$$
$$+ \|\tilde{W}_1\|\|\phi(x(t))\| + \|\tilde{W}_1\|\|\phi(x(t-T))\|$$

All the signals on the right-hand side of are UUB. So $H(\hat{u}, \hat{W}_1, x)$ is UUB and convergence to the approximate HJB solution is obtained.

b. According to Theorem 8.1 and equations (8.12) and (8.13), $||u_2 - u||$ is UUB because $||\hat{W}_2 - \hat{W}_1||$ is UUB.

So $\mu_2(x)$ gives the optimal solution.
This completes the proof. ∎

Remark 8.2. Note that the data required in the critic tuning algorithm (8.15) at each time are $\Delta\phi_2(t)$ and the integral reinforcement. The system dynamics $f(x)$, $g(x)$ are not needed. The input coupling dynamics $g(x)$ are needed for the actor tuning algorithm (8.16). Therefore, the IRL synchronous adaptive controller does not need to know the drift dynamics $f(x)$, yet converges to the solution of the HJB equation online.

Remark 8.3. The theorems show that PE is needed for proper identification of the value function by the critic NN, and that nonstandard tuning algorithm is required for the actor NN to guarantee stability while learning the optimal control solution.

Remark 8.4. The tuning parameters F_1, F_2 are selected appropriately to ensure stability as detailed in the proof of Theorem 8.1.

Remark 8.5. The proof reveals that the integral reinforcement time interval T cannot be selected too large nor the number of hidden-layer units N too small.

Remark 8.6. The assumption $Q(x) > 0$ is sufficient but not necessary for this result. If this condition is replaced by zero state observability, the proof still goes through, however it is tedious and does not add insight.

Remark 8.7. It is important to note that the proof of Theorem 8.1 does not rely at all on convergence of the IRL policy iteration Algorithm 8.1. IRL policy iteration, a reinforcement learning method, is used to obtain the structure of the controller, whereas the proofs of performance are carried out using adaptive control Lyapunov techniques.

Remark 8.8. The IRL policy iteration Algorithm 8.1 must be initialized with a stabilizing control policy. The convergence proof requires this, since the IRL Bellman equation has a positive definite solution only if the controller is admissible. A stabilizing controller can be difficult to find for non-linear systems. By contrast, the online synchronous IRL optimal adaptive controller does not require an initial stabilizing controller. The Lyapunov proof shows that the algorithm converges to the optimal control solution as long as the initial weight estimation errors are not large.

The implementation aspects and structure of the IRL online optimal adaptive controller are the same as in Section 7.5.

8.4 Simulations

This section presents simulation examples to show that the novel IRL adaptive control structure in Theorem 8.1 solves the optimal control problem online by measuring data along the system trajectories in real time. The system drift dynamics need not be known. In the linear quadratic case, therefore, the IRL optimal adaptive controller solves the algebraic Riccati equation without knowing the system A matrix. In the non-linear case, it solves the HJB equation without knowing the system drift term $f(x)$.

8.4.1 Linear system

Consider the continuous-time F16 longitudinal dynamics aircraft plant with quadratic cost function used in Stevens and Lewis (2003)

$$\dot{x} = \begin{bmatrix} -1.01887 & 0.90506 & -0.00215 \\ 0.82225 & -1.07741 & -0.17555 \\ 0 & 0 & -1 \end{bmatrix} x + \begin{bmatrix} 0 \\ 0 \\ 1 \end{bmatrix} u = Ax + Bu$$

where Q and R in the cost function are identity matrices of appropriate dimensions and the IRL time period was selected as $T = 0.01$. In this linear case the solution of the HJB equation is given by the solution of the algebraic Riccati equation (ARE) $A^TP + PA + Q - PBR^{-1}B^TP = 0$. Since the value is quadratic in the LQR case, the critic NN basis set $\phi(x)$ was selected as the quadratic vector in the state components. Solving the ARE gives the parameters of the optimal critic as $W_1^* = [p_{11} \quad 2p_{12} \quad 2p_{13} \quad p_{22} \quad 2p_{23} \quad p_{33}]^T$ with $P = [p_{ij}]$ the Riccati solution matrix. Numerical values for this example are $W_1^* = [1.4245 \quad 1.1682 \quad -0.1352 \quad 1.4349 \quad -0.1501 \quad 0.4329]T$.

The integral reinforcement optimal adaptive control algorithm is implemented as in Theorem 8.1. PE was ensured by adding a small exponentially decaying probing noise to the control input. Figure 8.1 shows the critic parameters, denoted by $\hat{W}_1 = [W_{c1} \quad W_{c2} \quad W_{c3} \quad W_{c4} \quad W_{c5} \quad W_{c6}]^T$ converging to the optimal values. In fact after 250 s the critic parameters converged to $\hat{W}_1(t_f) = [1.4282 \quad 1.1641 \quad -0.1372 \quad 1.4404 \quad -0.1495 \quad 0.4321]^T$.

The actor parameters after 300 s converge to the values of $\hat{W}_2(t_f) = \hat{W}_1(t_f)$.

The actor NN is given by $\hat{u}_2(x) = -\frac{1}{2}T \begin{bmatrix} 0 \\ 0 \\ 1 \end{bmatrix}^T \begin{bmatrix} 2x_1 & 0 & 0 \\ x_2 & x_1 & 0 \\ x_3 & 0 & x_1 \\ 0 & 2x_2 & 0 \\ 0 & x_3 & x_2 \\ 0 & 0 & 2x_3 \end{bmatrix}^T \hat{W}_2(t_f)$

The evolution of the system states is presented in Figure 8.2. As the probing noise decays, the states go to zero.

It is seen that after 250 s convergence of the NN weights in both critic and actor has occurred. The IRL optimal adaptive controller has determined the solution of the ARE equation in real time using measure data along the system trajectories without knowing the drift dynamics matrix A.

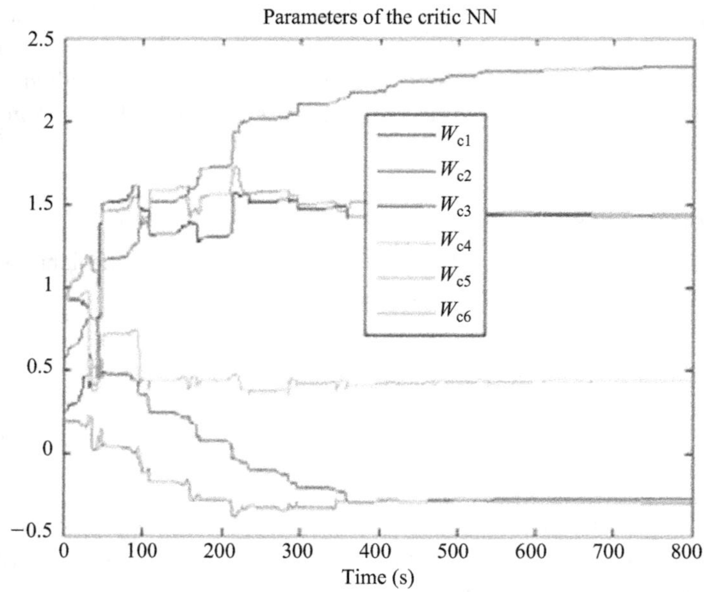

Figure 8.1 Convergence of the critic parameters to the parameters of the optimal critic

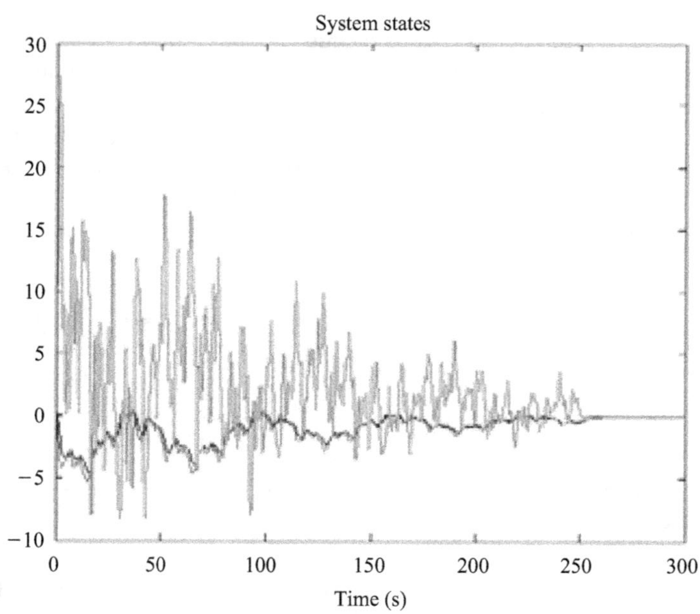

Figure 8.2 Evolution of the system states for the duration of the experiment

8.4.2 Non-linear system

Consider the following affine in control input non-linear system, with a quadratic cost constructed as in Nevistic and Primbs (1996)

$$\dot{x} = f(x) + g(x)u, \quad x \in \mathbb{R}^2$$

where

$$f(x) = \begin{bmatrix} -x_1 + x_2 \\ -x_1^3 - x_2 - \dfrac{x_1^2}{x_2} + 0.25x_2(\cos(2x_1 + x_1^3) + 2)^2 \end{bmatrix}, g(x) = \begin{bmatrix} 0 \\ \cos(2x_1 + x_1^3) + 2 \end{bmatrix}$$

One selects $Q = \begin{bmatrix} 1 & 0 \\ 0 & 1 \end{bmatrix}$, $R = 1$ and $T = 0.01$. The optimal value function constructed as in Nevistic and Primbs (1996) is $V^*(x) = \frac{1}{4}x_1^4 + \frac{1}{2}x_2^2$ and the optimal control signal is $u^*(x) = -\frac{1}{2}T(\cos(2x_1 + x_1^3) + 2)x_2$. One selects the critic NN vector activation function as $\phi(x) - \begin{bmatrix} x_1^2 & x_2^2 & x_1^4 & x_2^4 \end{bmatrix}$.

Figure 8.3 shows the critic parameters, denoted by $\hat{W}_1 = [W_{c1} \ W_{c2} \ W_{c3} \ W_{c4}]^T$. After the simulation by using the integral reinforcement learning optimal adaptive control algorithm we have

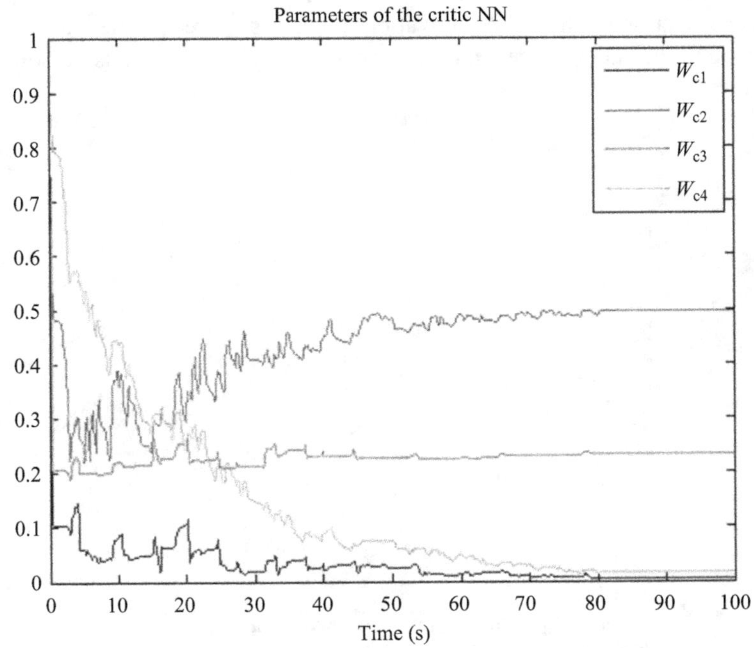

Figure 8.3 Convergence of the critic parameters to the parameters of the optimal critic

$$\hat{W}_1(t_f) = [0.0033 \quad 0.4967 \quad 0.2405 \quad 0.0153]^T$$

The actor parameters after 80 s converge to the values of

$$\hat{W}_2(t_f) = [0.0033 \quad 0.4967 \quad 0.2405 \quad 0.0153]^T$$

so that the actor NN is given by

$$\hat{u}_2(x) = -\frac{1}{2}T\begin{bmatrix} 0 \\ \cos(2x_1 + x_1^3) + 2 \end{bmatrix}^T \begin{bmatrix} 2x_1 & 0 & 4x_1^3 & 0 \\ 0 & 2x_2 & 0 & 4x_2^3 \end{bmatrix} \hat{W}_2(t_f)$$

One can see that after 80 s convergence of the NN weights in both critic and actor has occurred. The evolution of the system states is presented in Figure 8.4, which converge to zero as the probing noise decay to zero.

Figure 8.5 shows the 3D plot of the difference between the approximated value function, by using the online algorithm and the optimal one. This error is close to zero. Good approximation of the actual value function is being evolved. Figure 8.6 shows the 3D plot of the difference between the approximated control, by using the online integral reinforcement learning algorithm and the optimal one. This error is close to zero.

The adaptive controller has determined the solution of the non-linear HJB equation without knowing the drift dynamics $f(x)$.

In this example, the VFA activation functions were selected as quadratic and quartic in the states. In general, the form of the value function is not known for

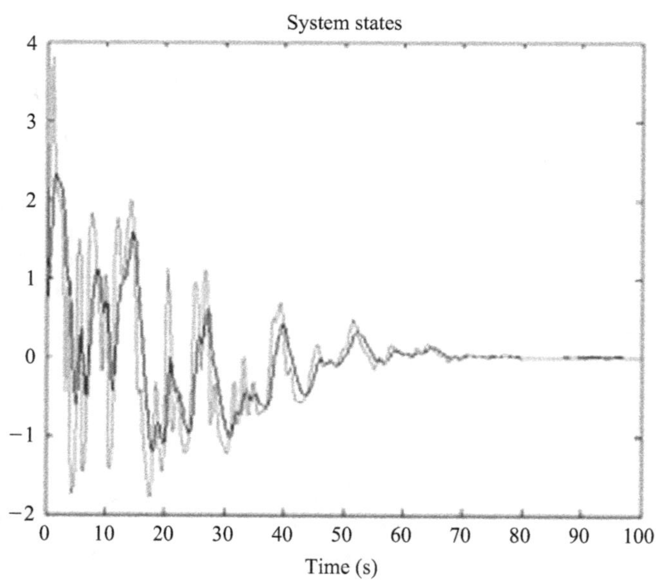

Figure 8.4 Evolution of the system states for the duration of the experiment

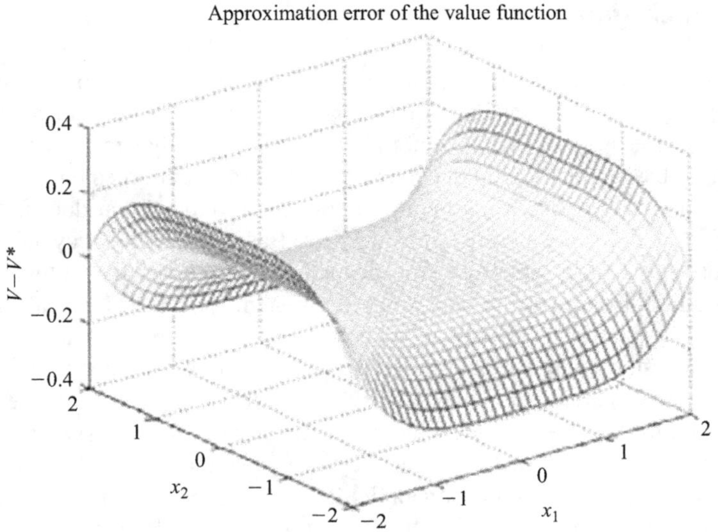

Figure 8.5 Error between the optimal and approximated value function

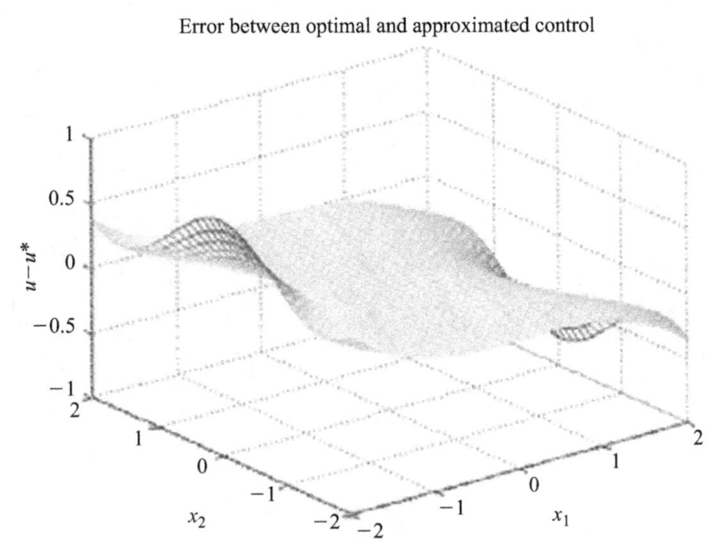

Figure 8.6 Error between the optimal and the approximated control input

non-linear systems. Then, one selects a polynomial basis set. If convergence is not observed, higher-order polynomials are chosen and the simulation is re-run. This is repeated until convergence occurs. The polynomials should be selected as even order since the value must be a Lyapunov function. Other basis sets than polynomials could also be used, such as tanh, sigmoid.

8.5 Conclusion

This chapter considered the optimal control problem for systems of the form $\dot{x} = f(x) + g(x)u$. A novel form of Bellman equation was developed for these continuous-time systems that does not involve the system dynamics. This Bellman equation is based on a technique known as integral reinforcement learning. A corresponding IRL policy iteration algorithm was developed that does not require knowledge of the system drift dynamics $f(x)$. Based on IRL policy iterations, a novel adaptive control structure was developed that solves the optimal control problem online by measuring data along the system trajectories in real time. The system drift dynamics $f(x)$ need not be known. The IRL adaptive controller has two approximator structures, one to estimate the value function and one to estimate the optimal control. Parameter tuning algorithms were given that learn the optimal control solution online while at the same time guaranteeing system stability. Simulation examples showed that, in the linear quadratic case, the IRL optimal adaptive controller solves the algebraic Riccati equation without knowing the system A matrix. In the non-linear case, it solves the HJB equation without knowing the system drift term $f(x)$.

Part III
Online differential games using reinforcement learning

This book shows how to use reinforcement learning (RL) methods to design adaptive controllers that learn optimal control solutions for continuous-time systems. We call these optimal adaptive controllers. They stand in contrast to standard adaptive control systems in the control systems literature, which do not normally converge to optimal solutions in terms of solving a Hamilton–Jacobi–Bellman equation. In Part I of the book we laid the foundations for RL applications in continuous-time systems based on a form of the continuous-time Bellman equation known as the integral reinforcement learning (IRL) Bellman equation. It was shown how to use IRL to formulate policy iterations and value iterations for continuous-time systems.

The controllers of Part I are of the form normally expected in RL and the convergence proofs are carried out using methods employed in RL. The actor and critic loops are tuned sequentially, that is, the control policy is held fixed while the critic evaluates its performance. These two-loop sequential learning structures are not like standard adaptive control systems currently used in feedback control.

In Part II of the book, we adopt a philosophy more akin to that of adaptive controllers in the feedback control literature. The controllers of Part II are multiple loop learning systems with novel structures designed using RL precepts. Yet, they operate as normally expected in adaptive control in that the control loops are not tuned sequentially as in Part I, but the parameter tuning in all control loops is performed simultaneously through time. Since the critic and actor networks learn and update their parameters simultaneously, we call these structures *synchronous* optimal adaptive controllers. The convergence proofs are carried out using methods employed in adaptive control, namely, Lyapunov energy-based techniques.

In Part III of the book, we develop adaptive controllers that learn optimal solutions in real time for several differential game problems, including zero-sum and multiplayer non–zero-sum games. The design procedure is to first formulate policy iteration algorithms for these game problems, then use the structure of policy iteration to motivate novel structures of multi-loop adaptive learning systems. Then, tuning laws for the different adaptive loops are determined by adaptive control Lyapunov techniques. The result is a family of adaptive controllers that converge to optimal game theoretic solutions online in real time using data measured along the system trajectories. An algorithm is given in Chapter 11 that solves the linear-quadratic zero-sum game problem online by learning the solution of the Hamilton-Jacobi-Isaacs equation in real time without knowing the system drift matrix A.

Chapter 9

Synchronous online learning for zero-sum two-player games and H-infinity control

Zero-sum (ZS) two-player game theory and H-infinity solutions rely on solving the Hamilton–Jacobi–Isaacs (HJI) equations, a generalized version of the Hamilton–Jacobi–Bellman equations appearing in optimal control problems. In the non-linear case the HJI equations are difficult or impossible to solve, and may not have global analytic solutions even in simple cases (e.g. scalar system, bilinear in input and state). Solution methods are generally offline and generate fixed control policies that are then implemented in online controllers in real time.

In this chapter, we provide methods for *online gaming*, that is for solution of two-player zero-sum infinite-horizon games online, through learning the saddle-point strategies in real time (Vamvoudakis and Lewis, 2010b). The dynamics may be non-linear in continuous time and are assumed known in this chapter. This requirement is lifted in Chapter 11. A novel neural-network (NN) adaptive control technique is given here that is based on reinforcement learning techniques, whereby the control and disturbance policies are tuned online using data generated in real time along the system trajectories. Also tuned is a 'critic' approximator structure whose function is to identify the value or outcome of the current control and disturbance policies. Based on this value estimate, the policies are continuously updated. All three loops are tuned simultaneously, or synchronously. This is a sort of indirect adaptive control algorithm, yet, due to the direct form dependence of the policies on the learned value, it is affected online as 'direct' optimal adaptive control.

The chapter is based on Vamvoudakis and Lewis (2010b) and is organized as follows. Section 9.1 reviews the formulation of the two-player ZS differential game for non-linear dynamical systems. The concepts of Nash equilibrium and the Isaacs condition are introduced. The ZS game Bellman equation and the HJI equation are developed. The use of ZS games in the H_∞ control problem is detailed. The special case of linear quadratic ZS games is covered. In Section 9.2, a ZS game policy iteration algorithm is given to solve the HJI equation by successive solutions on the ZS game Bellman equation. This essentially extends Kleinman's algorithm (Kleinman, 1968) to non-linear zero-sum two-player differential games. Section 9.3 uses the structure of the ZS policy iteration algorithm to obtain a suitable structure for an online adaptive controller for solving the ZS games in real time. This structure turns out to have one critic network with two actor networks, one for each player in the game. Value function approximation (Werbos,

1974, 1989, 1991, 1992) is used to solve the Bellman equation. First a suitable 'critic' approximator structure is developed for the value function and its tuning method is pinned down. Next, suitable 'actor' approximator structures are developed for the control and disturbance policies. Finally, in Section 9.4 the main results of the chapter are presented. Theorem 9.2 shows how to tune all three approximators simultaneously by using data measurements along the system trajectories in real time. Theorem 9.3 proves exponential convergence of the critic NN to the approximate Nash solution. Proofs using Lyapunov techniques guarantee convergence and closed-loop stability. Section 9.5 presents simulation examples that show the effectiveness of the online synchronous zero-sum game algorithm in learning in real time the optimal game value and the Nash equilibrium solution for both linear and non-linear systems.

9.1 Two-player differential game and H_∞ control

This section presents a background review of two-player zero-sum (ZS) differential games. The objective is to lay a foundation for the structure needed in subsequent sections for online solution of these problems in real time. The policy iteration algorithm for two-player games presented in Section 9.2 gives the structure needed for an adaptive controller that learns the ZS game solution online. It turns out to have a single critic network and two actor networks, one for each player in the game.

Consider the non-linear time-invariant affine-in-the-input dynamical system given by

$$\begin{aligned} \dot{x} &= f(x) + g(x)u(x) + k(x)d(x) \\ y &= h(x) \end{aligned} \tag{9.1}$$

where state $x(t) \in \mathbb{R}^n$, output $y(t) \in \mathbb{R}^p$, and smooth functions $f(x(t)) \in \mathbb{R}^n$, $g(x) \in \mathbb{R}^{nxm}$, $k(x) \in \mathbb{R}^{nxq}, h(x) \in \mathbb{R}^{pxn}$. This system has two inputs, the control input $u(x(t)) \in \mathbb{R}^m$ and the disturbance input $d(x(t)) \in \mathbb{R}^q$. In zero-sum games, the control and disturbance are considered as inputs that can be selected by two players to optimize their benefit as described below. Assume that $f(x)$ is locally Lipschitz and $f(0) = 0$ so that $x = 0$ is an equilibrium point of the system.

Define the performance index (Başar and Olsder, 1999; Başar and Bernard, 1995).

$$J(x(0), u, d) = \int_0^\infty \left(Q(x) + u^T Ru - \gamma^2 \|d\|^2 \right) dt \equiv \int_0^\infty r(x, u, d) dt \tag{9.2}$$

where $Q(x) = h^T(x)h(x) \geq 0$, $R = R^T > 0$, utility $r(x,u,d) = Q(x) + u^T Ru - \gamma^2 \|d\|^2$ and $\gamma \geq \gamma^* \geq 0$. Here, γ^* is the smallest γ for which the system is stabilized (Van Der Schaft, 1992). The quantity γ^* is called the H-infinity gain. For linear systems, an explicit formula can be given for γ^* (Chen *et al.*, 2004). For non-linear systems, γ^* may depend on the region within which solutions are sought.

For fixed feedback policies $u(x)$ and disturbance policies $d(x)$, define the value function or cost of the policies as

$$V(x(t), u, d) = \int_t^\infty \left(Q(x) + u^T Ru - \gamma^2 \|d\|^2\right) dt \tag{9.3}$$

When the value is finite, a differential equivalent to this is the non-linear zero-sum game Bellman equation

$$0 = r(x, u, d) + (\nabla V)^T (f(x) + g(x)u(x) + k(x)d(x)), V(0) = 0 \tag{9.4}$$

where $\nabla V = \partial V / \partial x \in R^n$ is the (transposed) gradient. The Hamiltonian of dynamics (9.1) and value function (9.3) is

$$H(x, \nabla V, u, d) = r(x, u, d) + (\nabla V)^T (f(x) + g(x)u(x) + k(x)d) \tag{9.5}$$

For stabilizing control and disturbance feedback policies yielding finite values (Başar and Bernard, 1995), a solution $V(x) \geq 0$ to the Bellman equation (9.4) is the value (9.3) for the given feedback policy $u(x)$ and disturbance policy $d(x)$.

9.1.1 Two-player zero-sum differential games and Nash equilibrium

Define the two-player ZS differential game (Başar and Olsder, 1999; Başar and Bernard, 1995)

$$\begin{aligned} V^*(x(0)) &= \min_u \max_d J(x(0), u, d) \\ &= \min_u \max_d \int_0^\infty \left(Q(x) + u^T Ru - \gamma^2 \|d\|^2\right) dt \end{aligned} \tag{9.6}$$

subject to the dynamical constraints (9.1). By the definition of this game, u is a minimizing player while d is a maximizing one. This puts the control and the disturbance inputs in opposition with one another with opposite objectives. This two-player optimal control problem has a unique solution if a game theoretic saddle point (u^*, d^*) exists, that is if the Nash condition holds

$$\min_u \max_d J(x(0), u, d) = \max_d \min_u J(x(0), u, d^*) \tag{9.7}$$

This is equivalent to the Nash equilibrium condition

$$J(x(0), u^*, d) \leq J(x(0), u^*, d^*) \leq J(x(0), u, d^*)$$

holding for all policies u, d.

A necessary condition for this is Isaacs' condition

$$\min_u \max_d H(x, \nabla V, u, d) = \max_d \min_u H(x, \nabla V, u, d) \tag{9.8}$$

or, equivalently

$$H(x, \nabla V, u^*, d) \leq H(x, \nabla V, u^*, d^*) \leq H(x, \nabla V, u, d^*) \tag{9.9}$$

for some saddle-point solution (u^*, d^*). The Nash conditions depend on the value integral and the system dynamics. On the other hand, the Isaacs condition holds pointwise in time and is easier to check. Under certain conditions detailed below, Isaacs' condition implies Nash equilibrium.

To this game is associated the Hamilton–Jacobi–Isaacs (HJI) equation

$$0 = Q(x) + \nabla V^T(x)f(x) - \frac{1}{4}\nabla V^T(x)g(x)R^{-1}g^T(x)\Delta V(x) + \frac{1}{4\gamma^2}\nabla V^T(x)kk^T\nabla V(x),\ V(0) = 0 \tag{9.10}$$

Given a solution $V^*(x) \geq 0 : \mathbb{R}^n \to \mathbb{R}$ to the HJI (9.10), denote the associated control and disturbance as

$$u^* = -\frac{1}{2}R^{-1}g^T(x)\nabla V^* \tag{9.11}$$

$$d^* = \frac{1}{2\gamma^2}k^T(x)\nabla V^* \tag{9.12}$$

These policies are called the optimal control and the worst-case disturbance. The HJI can be written in terms of the Hamiltonian as

$$0 = H(x, \nabla V, u^*, d^*) = Q(x) + \nabla V^T(x)f(x) - \frac{1}{4}\nabla V^T(x)g(x)R^{-1}g^T(x)\nabla V(x) + \frac{1}{4\gamma^2}\nabla V^T(x)kk^T\nabla V(x) \tag{9.13}$$

Lemma 9.1. Isaacs condition satisfied. Select $\gamma > \gamma^* \geq 0$. Suppose $V(x) : \mathbb{R}^n \to \mathbb{R}$ is smooth, $V(x) \geq 0$, and is a solution to the HJI equation (9.10). Then the Isaacs condition (9.8) is satisfied for control $u^*(t)$ given by (9.11) and $d^*(t)$ given by (9.12) in terms of $V(x)$. ■

Proof: Complete the squares to obtain, for any control $u(t)$ and disturbance $d(t)$

$$H(x,\nabla V,u,d) = H(x,\nabla V,u^*,d^*) - \gamma^2\|d-d^*\|^2 + (u-u^*)^T R(u-u^*) \tag{9.14}$$

with $u(t) = u^*(t)\, and\, d(t) = d^*(t)$ given, respectively, by (9.11), and (9.12). If $V(x) \geq 0$ satisfies the HJI, then $H(x, \nabla V, u^*, d^*) = 0$ and $H(x, \nabla V, u, d) = -\gamma^2\|d - d^*\|^2 + (u - u^*)^T R(u - u^*)$. This implies (9.9). ■

The next result shows that this implies Nash equilibrium (9.7) under certain conditions. System (9.1) is said to be *zero state observable* if $u(t) \equiv 0, y(t) \equiv 0 \Rightarrow x(t) \equiv 0$. The HJI equation (9.13) may have more than one nonnegative definite solution $V(x) \geq 0$. A minimal nonnegative definite solution $V_a(x) \geq 0$ is one such that there exists no other nonnegative definite solution $V(x) \geq 0$. such that $V_a(x) \geq V(x) \geq 0$.

Lemma 9.2. Solution of two-player zero-sum game. Van Der Schaft (1992). Select $\gamma > \gamma^* \geq 0$. Suppose $V_a(x) : \mathbb{R}^n \to \mathbb{R}$ is smooth, $V_a(x) \geq 0$ and is a minimal nonnegative definite solution to the HJI equation (9.13). Assume (9.1) is zero state observable. Then condition (9.7) is satisfied for control $u^*(t)$ given by (9.11) and $d^*(t)$ given by (9.12) in terms of $V_a(x)$. Then the system is in Nash equilibrium, the game has a value given by $V_a(x(0))$, and (u^*, d^*) is a saddle-point equilibrium solution among strategies in $L_2[0, \infty)$. Moreover, the closed-loop systems $(f(x) + g(x)u^*)$ and $(f(x) + g(x)u^* + k(x)d^*)$ are locally asymptotically stable. ■

Proof: One has for any smooth $V(x) : \mathbb{R}^n \to \mathbb{R}$:

$$\begin{aligned}
J_T(x(0), u, d) &\equiv \int_0^T (h^T(x)h(x) + u^T Ru - \gamma^2 \|d\|^2)\, dt \\
&= \int_0^T (h^T(x)h(x) + u^T Ru - \gamma^2 \|d\|^2)\, dt \\
&\quad + \int_0^T \dot{V}\, dt - V(x(T)) + V(x(0)) \\
&= \int_0^T H(x, \nabla V, u, d) dt - V(x(T)) + V(x(0))
\end{aligned}$$

Now, suppose $V(x)$ satisfies the HJI equation (9.13) and use (9.14) to obtain

$$\begin{aligned}
J_T(x(0), u, d) = \int_0^T \left((u - u^*)R(u - u^*)^T - \gamma^2 \|d - d^*\|^2 \right) dt - V^*(x(T)) \\
+ V^*(x(0))
\end{aligned}$$

Therefore, for finite $T, J_T(x(0), u, d)$ has a saddle point induced by that of the Hamiltonian. The remainder of the proof involves showing that, for the specific HJI solution selected, the closed-loop system $(f(x) + g(x)u^* + k(x)d^*)$ is locally asymptotically stable. Then taking the limit as $T \to \infty$ yields the result. ■

The issue in this proof is that existence of a Nash equilibrium on an infinite horizon requires that the closed-loop system $(f(x) + g(x)u^* + k(x)d^*)$ be locally asymptotically stable. It is proven in Van Der Schaft (1992) that the minimum nonnegative definite solution to the HJI is the unique nonnegative definite solution for which this is true. Linearize the system (9.1) about the origin to obtain the game algebraic Ricati equation (GARE) (9.24). Of the nonnegative definite solutions to the GARE, select the one corresponding to the stable invariant manifold of the Hamiltonian matrix (9.25). Then, the minimum nonnegative definite solution of the HJI is the one having this stabilizing GARE solution as its Hessian matrix evaluated at the origin (Van Der Schaft, 1992).

Note that global solutions to the HJI (9.13) may not exist. Moreover, if they do, they may not be smooth. For a discussion on viscosity solutions to the HJI, see Başar and Bernard (1995), Ball and Helton (1996) and Bardi and Capuzzo-Dolcetta (1997).

9.1.2 Application of Zero-sum games to H_∞ control

Consider the system (9.1) and define a performance output $z(t) = [u^T(t)y^T(t)]^T$. This setup is shown in Figure 9.1. In the bounded L_2 − gain problem (Başar and Olsder, 1999; Van Der Schaft, 1992; Zames, 1981) one desires to find a feedback control policy $u(x)$ such that, when $x(0) = 0$ and for all disturbances $d(t) \in L_2[0, \infty)$ one has

$$\frac{\int_0^T \|z(t)\|^2 dt}{\int_0^T \|d(t)\|^2\, dt} = \frac{\int_0^T (h^T h + u^T Ru) dt}{\int_0^T \|d(t)\|^2 dt} \le \gamma^2 \tag{9.15}$$

for a prescribed $\gamma > 0$ and $R = R^T > 0$ and for all $T > 0$. That is, the L_2 − gain from the disturbance to the performance output is less than or equal to γ.

The H-infinity control problem is to find, if it exists, the smallest value $\gamma^* > 0$ such that for any $\gamma > \gamma^*$ the bounded L_2-gain problem has a solution. In the linear case an explicit expression can be provided for the H-infinity gain γ^* Chen *et al.* (2004).

To solve the bounded L_2 − gain problem, one may use the machinery of two-player ZS games just developed. In the two-player ZS game, both inputs can be controlled, with the control input seeking to minimize a performance index and the disturbance input seeking to maximize it. By contrast, here $d(t)$ is a disturbance which cannot be controlled, and $u(t)$ is the control input used to offset the deleterious effects of the disturbance.

The next result provides a solution to the bounded L_2 − gain problem in terms of a solution to the HJI equation.

Lemma 9.3. Solution to the bounded L_2 gain problem. Van Der Schaft (1992). Select $\gamma > \gamma^* > 0$. Suppose $V(x) : \mathbb{R}^n \to \mathbb{R}$ is smooth, $V(x) \ge 0$, and is a solution to the HJI equation (9.13). Assume (9.1) is zero state observable. Then the control $u^*(t)$ selected as (9.11) in terms of the HJI solution renders the equilibrium point locally

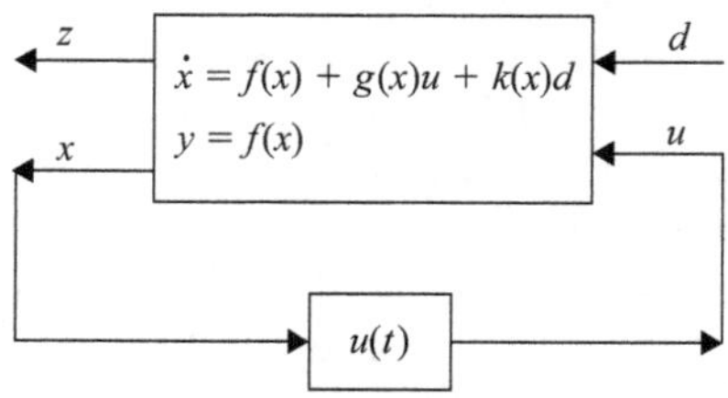

Figure 9.1 Bounded L_2-gain control

asymptotically stable (when $d(t) = 0$, that is $(f(x) + g(x)u^*)$ is locally asymptotically stable.) Moreover, system (9.1) has L_2 gain $\leq \gamma$. Finally, $u^*(t) \in L_2[0, \infty)$. ■

Note that under the hypotheses of the lemma, specifically zero state observability, if $V(x) \geq 0$ is a solution to the HJI equation (9.13), then one has in fact $V(x) > 0$ (Van Der Schaft, 1992). The next result shows when smooth solutions exist to the HJI. System (9.1) is locally detectable if its linearized version has all unobservable modes stable.

Lemma 9.4. Van Der Schaft (1992). Select $\gamma > 0$. Assume (9.1) is locally detectable from $y = h(x)$, and there exists a control policy $u(x)$ so that locally the system has L_2 gain $\leq \gamma$ and the system is asymptotically stable. Then there exists a local smooth solution $V(x) \geq 0$ to the HJI equation (9.13). Furthermore, the control given in terms of this solution by (9.11) renders the L_2 gain $\leq \gamma$ for all trajectories originating at the origin locally. ■

9.1.3 Linear quadratic zero-sum games

In the linear quadratic (LQ) ZS game one has linear dynamics

$$\dot{x} = Ax + Bu + Dd \tag{9.16}$$

with $x(t) \in R^n$. The value is the integral quadratic form

$$V(x(t),u,d) = \frac{1}{2}\int_t^\infty \left(x^T H^T Hx + u^T Ru - \gamma^2 \|d\|^2\right) d\tau \equiv \int_t^\infty r(x,u,d)\, d\tau \tag{9.17}$$

where $R = R^T > 0$ and $\gamma > 0$. Assume that the value is quadratic in the state so that

$$V(x) = \frac{1}{2} x^T Sx \tag{9.18}$$

for some matrix $S > 0$. Select state variable feedbacks for the control and disturbance so that

$$u = -Kx \tag{9.19}$$

$$d = Lx \tag{9.20}$$

Substituting this into Bellman equation (9.4) gives

$$S(A - BK + DL) + (A - BK + DL)^T S + H^T H + K^T RK - \gamma^2 L^T L = 0 \tag{9.21}$$

It has been assumed that the Bellman equation holds for all initial conditions, and the state $x(t)$ has been cancelled. This is a Lyapunov equation for S in terms of the prescribed state feedback policies K and L. If $(A - BK + DL)$ is stable, (A, H) is observable, and $\gamma > \gamma^* > 0$ then there is a positive definite solution $S \geq 0$, and (9.18) is the value (9.17) for the selected feedback policies K and L.

The stationary point control (9.11) and disturbance (9.12) are given by

$$u(x) = -R^{-1}B^T Sx = -Kx \tag{9.22}$$

$$d(x) = \frac{1}{\gamma^2} D^T Sx = Lx \tag{9.23}$$

Substituting these into (9.21) yields the game algebraic Riccati equation (GARE)

$$0 = A^T S + SA + H^T H - SBR^{-1}B^T S + \frac{1}{\gamma^2} SDD^T S \tag{9.24}$$

To solve the ZS game problem, one solves the GARE equation for the non-negative definite optimal value kernel $S \geq 0$, then the optimal control is given as a state variable feedback in terms of the ARE solution by (9.22) and the worst-case disturbance by (9.23). There exists a solution $S > 0$ if (A, B) is stabilizable and (A, H) is observable and $\gamma > \gamma^*$, the H-infinity gain.

Since we have developed a solution to the problem, it is verified that the assumption that the value has the quadratic form (9.18) holds.

There may be more than one PSD solution to the GARE. To solve the ZS game problem, one requires gains such that the poles of $(A - BK + DL)$ are in the open left-half plane. The minimum PSD solution of the GARE is the unique stabilizing solution. It is shown in Başar and Olsder (1999) and Van Der Schaft (1992) that the stabilizing solution of the GARE corresponds to the stable eigenspace of the Hamiltonian matrix

$$H = \begin{bmatrix} A & \left(\frac{1}{\gamma^2} DD^T - BR^{-1}B^T\right) \\ H^T H & -A^T \end{bmatrix} \tag{9.25}$$

See also Lewis *et al.* (2012), Section 3.3. A method for finding the minimum PSD solution is given in Abu-Khalaf *et al.* (2006).

The work of Vamvoudakis and Lewis (2010b) presented in this chapter shows how to solve the GARE online in real time using data measured along the system trajectories. This allows the performance index weights R, γ to change slowly as the game develops. Chapter 11 presents the work of Vrabie and Lewis (2010) that provided algorithms for solving the GARE equations online in real time without knowing the system matrix A. These online solution methods rely on the ZS game policy iteration in the next section.

9.2 Policy iteration solution of the HJI equation

The HJI equation (9.13) is non-linear and is usually intractable to solve directly. One can solve the HJI iteratively using one of several algorithms that are built on iterative solutions of the Bellman equation (9.4). Included are Feng *et al.* (2009) that uses an inner loop with iterations on the control, and Van Der Schaft (1992) and Abu-Khalaf *et al.* (2006) that use an inner loop with iterations on the disturbance. The complementarity of these algorithms is shown in Abu-Khalaf *et al.* (2006). These are in effect extensions of Kleinman's algorithm (Kleinman, 1968) to non-linear two-player games. Here, we shall use the latter algorithm (e.g. Van Der Schaft, 1992; Abu-Khalaf *et al.*, 2006).

The structure given in policy iteration Algorithm 9.1 will be used as the basis for approximate online solution techniques in the next Section 9.3.

Algorithm 9.1. Policy iteration (PI) Algorithm for two-player zero-Sum differential games

Initialization: Start with a stabilizing feedback control policy u_0.

1. For $j = 0, 1, \ldots$ given u_j.
2. (*Policy evaluation*) For $i = 0, 1, \ldots$ set, $d^0 = 0$, solve for $V_j^i(x(t))$ using the ZS Bellman equation

$$0 = Q(x) + \nabla V_j^{iT}(x)(f + gu_j + kd^i) + u_j^T R u_j - \gamma^2 ||d^i||^2 \tag{9.26}$$

Update the disturbance to d^{i+1} according to

$$d^{i+1} = \arg\max_d \left[H(x, \nabla V_j^i, u_j, d)\right] = \frac{1}{2\gamma^2} k^T(x) \nabla V_j^i \tag{9.27}$$

On convergence, set $V_{j+1}(x) = V_j^i(x)$.

3. Update the control policy using

$$u_{j+1} = \arg\min_u \left[H(x, \nabla V_{j+1}), u, d\right] = -\frac{1}{2} R^{-1} g^T(x) \nabla V_{j+1} \tag{9.28}$$

Go to 1. ■

Nota bene: In practice, the iterations in i and j are continued until some convergence criterion is met, for example $||v_j^{i+1} - v_j^i||$ or, respectively $||v_{j+1} - v_j||$ is small enough in some suitable norm.

The ZS game PI algorithm is a two-loop algorithm with an inner loop to update the disturbance and an outer loop to update the control policy. Given a feedback policy $u(x)$, write the Hamilton–Jacobi (HJ) equation

$$\begin{aligned} 0 &= Q(x) + \nabla V^T(x)(f(x) + g(x)u(x)) + u^T(x) R u(x) \\ &\quad + \frac{1}{4\gamma^2} \nabla V^T(x) k k^T \nabla V(x), \; V(0) = 0 \end{aligned} \tag{9.29}$$

for fixed $u(x)$. The minimal nonnegative solution $V(x)$ to this equation is the so-called available storage for the given $u(x)$ (Van Der Schaft, 1992). Note that the inner loop of this algorithm finds the available storage for u_j, where it exists.

Assuming that the available storage at each index j is smooth on a local domain of validity, the convergence of this algorithm to the minimal nonnegative solution to the HJI equation is shown in Van Der Schaft (1992) and Abu-Khalaf *et al.* (2006). Under these assumptions, the existence of smooth solutions at each step to the Bellman equation (9.26) was further shown in Abu-Khalaf *et al.* (2006). Also shown was the asymptotic stability of $(f + gu_j + kd^i)$ at each step. In fact, the inner loop yields $V_j^{i+1}(x) \geq V_j^i(x), \forall x$ while the outer loop yields $V_{j+1}(x) \leq V_j(x), \forall x$ until convergence to V^*.

Note that this algorithm relies on successive solutions of Bellman equations (9.26). As such, the discussion surrounding (9.4) shows that the algorithm finds the value $V_j^i(x(t))$ of successive control policy/disturbance policy pairs.

9.3 Actor–critic approximator structure for online policy iteration algorithm

The ZS game PI Algorithm 9.1 is a *sequential* algorithm that solves the HJI equation (9.13) and finds the Nash solution (u^*, d^*) based on sequential solutions of the Bellman equation (9.26). That is, while the disturbance policy is being updated, the feedback policy is held constant. Sequential learning seems a bit strange from the point of view of adaptive control theory, where all parameter updates are performed continually in either discrete time or continuous time. In this section, we develop a novel adaptive control structure that operates like standard adaptive controllers, in that all the parameters are tuned continuously through time, yet converges online to the optimal ZS game solution. Standard adaptive controllers do not converge to optimal control solutions.

In this section, PI Algorithm 9.1 is used to lay a rigorous foundation for an *adaptive control structure that solves the two-player zero-sum differential game in real time.* This learning structure turns out to have a single critic network and two actor networks, one for the control policy and one for the disturbance. Value function approximation is use to implement these loops using three neural networks. In the next section, this critic–double actor structure will be used to develop an adaptive control algorithm of novel form that converges to the ZS game solution. It is important to define the neural network structures and the NN estimation errors properly or such an adaptive algorithm cannot be developed. This is assured by using the structure inherent in the ZS game policy iteration Algorithm 9.1.

The PI Algorithm 9.1 itself is not implemented in this chapter. Instead, here one implements both loops in the algorithm, the outer feedback control update loop and the inner disturbance update loop, *simultaneously* in time using neural network learning implemented as differential equations for tuning the weights, while simultaneously keeping track of and learning the value $V(x(t), u, d)$ (9.3) of the current control and disturbance by solution of the Bellman equation (9.4) or (9.26). We call this *synchronous learning* for online ZS games. The results in this chapter are from Vamvoudakis and Lewis (2010b).

9.3.1 Value function approximation and critic neural network

This chapter uses non-linear approximator structures for value function approximation (VFA) (Bertsekas and Tsitsiklis, 1996; Werbos, 1974, 1989, 1991, 1993) thereby sacrificing some representational accuracy in order to make the representation manageable in practice. Sacrificing accuracy in the representation of the value function is not so critical, since the ultimate goal is to find an optimal policy and not necessarily an accurate value function. Based on the structure of the ZS game PI Algorithm 9.1, VFA for online two-player games requires three approximators, which are taken here as neural networks (NNs), one for the value function, one for the feedback control policy and one for the disturbance policy. These are motivated respectively by the need to solve (9.26), (9.28) and (9.27).

To solve (9.26), we use VFA, which here requires approximation in Sobolev norm (Adams and Fournier, 2003), that is, approximation of the value $V(x)$ as well as its gradient $\nabla V(x)$. The following definition describes uniform convergence, which is needed later.

Definition 9.1. (Uniform convergence.) A sequence of functions $\{f_n\}$ converges uniformly to f if $\forall \varepsilon > 0, \exists N(\varepsilon) : \sup\|f_n(x) - f(x)\| < \varepsilon, n > N(\varepsilon)$.

According to the Weierstrass higher-order approximation Theorem (Abu-Khalaf *et al.*, 2006; Finlayson, 1990; Hornik *et al.*, 1990) there exists a complete independent basis set $\{\varphi_i(x)\}$ such that the solution $V(x)$ to Bellman equation (9.4) an its gradient are uniformly approximated. That is, for some constants c_i

$$V(x) = \sum_{i=1}^{\infty} c_i \varphi_i(x) \tag{9.30}$$

Therefore, assume there exist NN weights W_1 such that the value function $V(x)$ is approximated as

$$V(x) = W_1^T \phi_1(x) + \varepsilon(x) \tag{9.31}$$

with $\phi_1(x) : \mathbb{R}^n \to \mathbb{R}^N$ the NN activation function vector containing N basis functions $\{\varphi_i(x) : i = 1, N\}$, with N the number of neurons in the hidden layer, and $\varepsilon(x)$ the NN approximation error. For approximation in Sobolev space, the NN activation functions $\{\varphi_i(x) : i = 1, N\}$ should be selected so that $\{\varphi_i(x) : i = 1, \infty\}$ provides a complete independent basis set such that $V(x)$ and its derivative are uniformly approximated, for example additionally

$$\frac{\partial V}{\partial x} = \left(\frac{\partial \phi_1(x)}{\partial x}\right)^T W_1 + \frac{\partial \varepsilon}{\partial x} = \nabla \phi_1^T W_1 + \nabla \varepsilon \tag{9.32}$$

Then, as the number of hidden-layer neurons $N \to \infty$ the approximation errors $\varepsilon \to 0, \nabla \varepsilon \to 0$ uniformly (Abu-Khalaf *et al.*, 2006; Abu-Khalaf and Lewis, 2005). In addition, for fixed N, the NN approximation errors $\varepsilon(x)$ and $\nabla \varepsilon$ are bounded by constants locally (Hornik *et al.*, 1990). We refer to this NN with weights W_1 that performs VFA as the *critic* NN.

Standard usage of the Weierstrass high-order approximation theorem uses polynomial approximation. However, non-polynomial basis sets have been considered in the literature (e.g. Hornik *et al.*, 1990; Sandberg, 1998). The NN approximation literature has considered a variety of activation functions including sigmoids, tanh, radial basis functions, etc.

Using the NN VFA, considering fixed feedback and disturbance policies $u(x(t))$, $d(x(t))$, Bellman equation (9.4) becomes

$$\begin{aligned} H(x, W_1, u, d) = Q(x) + u^T Ru - \gamma^2 ||d||^2 + W_1^T \nabla\phi_1(f(x) \\ + g(x)u(x) + k(x)d(x)) = \varepsilon_H \end{aligned} \tag{9.33}$$

where the residual error is

$$\begin{aligned} \varepsilon_H &= -(\nabla\varepsilon)^T (f + gu + kd) \\ &= -(C_1 - W_1)^T \nabla\phi_1(f + gu + kd) - \sum_{i=N+1}^{\infty} c_i \nabla\varphi_i(x)(f + gu + kd) \end{aligned} \tag{9.34}$$

Under the Lipschitz assumption on the dynamics, this residual error is bounded locally.

Proposition 9.1 has been shown in Abu-Khalaf and Lewis (2005, 2008). Define $|v|$ as the magnitude of a scalar v, $||x||$ as the vector norm of a vector x and $|| \, ||_2$ as the induced matrix two-norm. Define matrix C_1 by $V(x) = \sum_{i=1}^{N} c_i\varphi_i(x) \equiv C_1\phi_1(x)$.

Proposition 9.1. For any policies $u(x(t))$, $d(x(t))$ the least-squares solution to (9.33) exists and is unique for each N. Denote this solution as W_1 and define

$$V_1(x) = W_1^T \phi_1(x) \tag{9.35}$$

Then, as $N \to \infty$:

a. $\sup|\varepsilon_H| \to 0$
b. $||W_1 - C_1||_2 \to 0$
c. $\sup|V_1 - V| \to 0$
d. $\sup||\nabla V_1 - \nabla V|| \to 0$ ■

This result shows that $V_1(x)$ converges uniformly in the Sobolev norm $W^{1,\infty}$ (Adams and Fournier, 2003) to the exact solution $V(x)$ to Bellman equation (9.4) as $N \to \infty$, and the weights W_1 converge to the first N of the weights, C_1, which exactly solve (9.4).

The effect of the approximation error on the HJI equation (9.13) is

$$\begin{aligned} Q(x) + W_1^T \nabla\varphi_1(x)f(x) - \frac{1}{4} W_1^T \nabla\varphi_1(x)g(x)R^{-1}g^T(x)\nabla\varphi_1^T W_1 \\ + \frac{1}{4\gamma^2} W_1^T \nabla\varphi_1(x)kk^T \nabla\varphi_1^T W_1 = \varepsilon_{HJI} \end{aligned} \tag{9.36}$$

where the residual error due to the function approximation error is

$$\varepsilon_{HJI} \equiv -\nabla\varepsilon^T f + \tfrac{1}{2} W_1^T \nabla\phi_1 g R^{-1} g^T \nabla\varepsilon + \tfrac{1}{4} \nabla\varepsilon^T g R^{-1} g^T \nabla\varepsilon$$
$$- \tfrac{1}{2\gamma^2} W_1^T \nabla\phi_1 k k^T \nabla\varepsilon - \tfrac{1}{4\gamma^2} \nabla\varepsilon^T k k^T \nabla\varepsilon \qquad (9.37)$$

It was also shown in Abu-Khalaf and Lewis (2008) that this error converges uniformly to zero as the number of hidden-layer units N increases. That is, $\forall \varepsilon > 0, \exists N(\varepsilon) : \sup\|\varepsilon_{HJI}\| < \varepsilon,\ N > N(\varepsilon)$.

9.3.2 Tuning and convergence of the critic neural network

This section addresses the tuning and convergence of the critic NN weights when fixed feedback control and disturbance policies are prescribed. Therefore, the focus is on solving the non-linear Bellman equation (9.4) (e.g. (9.26)) for a fixed feedback policy u and fixed disturbance policy d.

In fact, this amounts to the *design of an observer for the value function.* Therefore, this procedure is consistent with indirect adaptive control approaches which first design an identifier for the system state and unknown dynamics, and then use this observer in the design of a feedback control.

The ideal weights of the critic NN, W_1, which provide the best approximate solution for (9.33) are unknown. Therefore, the output of the critic NN is

$$\hat{V}(x) = \hat{W}_1^T \phi_1(x) \qquad (9.38)$$

where $\hat{W}_1$ are the current estimated values of W_1. The approximate Bellman equation is then

$$H(x, \hat{W}_1, u, d) = \hat{W}_1^T \nabla\phi_1 (f + gu + kd) + Q(x) + u^T R u - \gamma^2 \|d\|^2 = e_1 \qquad (9.39)$$

with e_1 a residual equation error. In view of Proposition 9.1, define the critic weight estimation error

$$\tilde{W}_1 = W_1 - \hat{W}_1$$

Then,

$$e_1 = -\tilde{W}_1^T \nabla\phi_1 (f + gu) + \varepsilon_H$$

Given any feedback control policy u, it is desired to select $\hat{W}_1$ to minimize the squared residual error

$$E_1 = \frac{1}{2} e_1^T e_1$$

Then $\hat{W}_1(t) \to W_1$ and $e_1 \to \varepsilon_H$. Select the tuning law for the critic weights as the normalized gradient descent algorithm:

$$\dot{\hat{W}}_1 = -a_1 \frac{\partial E_1}{\partial \hat{W}_1} = -a_1 \frac{\sigma_1}{(1+\sigma_1^T \sigma_1)^2} \left[\sigma_1^T \hat{W}_1 + Q(x) + u^T R u - \gamma^2 \|d\|^2 \right] \quad (9.40)$$

where $\sigma_1 = \nabla\phi_1(f + gu + kd)$. This is a nonstandard modified Levenberg–Marquardt algorithm where $(\sigma_1^T\sigma_1 + 1)^2$ is used for normalization instead of $(\sigma_1^T\sigma_1 + 1)$. This is required in the theorem proofs, where one needs both appearances of $\sigma_1/(1 + \sigma_1^T\sigma_1)$ in (9.40) to be bounded (Ioannou and Fidan, 2006; Tao, 2003).

Note that, from (9.33)

$$Q(x) + u^T Ru - \gamma^2 \|d\|^2 = -W_1^T \nabla\varphi_1(f + gu + kd) + \varepsilon_H \quad (9.41)$$

Substituting (9.41) in (9.40) and, with the notation

$$\bar{\sigma}_1 = \frac{\sigma_1}{(\sigma_1^T\sigma_1 + 1)}, \quad m_s = 1 + \sigma_1^T\sigma_1 \quad (9.42)$$

we obtain the dynamics of the critic weight estimation error as

$$\dot{\tilde{W}}_1 = -a_1\bar{\sigma}_1\bar{\sigma}_1^T\tilde{W}_1 + a_1\bar{\sigma}_1\frac{\varepsilon_H}{m_s} \quad (9.43)$$

To guarantee convergence of $\hat{W}_1$ to W_1, the next Persistence of Excitation (PE) assumption and associated technical lemmas are required.

Persistence of excitation (PE) assumption. Let the signal $\bar{\sigma}_1$ be persistently exciting over the interval $[t, t+T]$, that is there exist constants $\beta_1 > 0$, $\beta_2 > 0, T > 0$ such that, for all t

$$\beta_1 I \le S_0 \equiv \int_t^{t+T} \bar{\sigma}_1(\tau)\bar{\sigma}_1^T(\tau)d\tau \le \beta_2 I \quad (9.44)$$

with I the identity matrix of appropriate dimensions.

The PE assumption is needed in adaptive control if one desires to perform system identification using for example recursive least-squares (RLS) (Ioannou and Fidan, 2006; Tao, 2003). It is needed here because one effectively desires to identify the critic parameters that approximate $V(x)$.

The properties of tuning algorithm (9.40) are given in the subsequent results. They are proven in Chapter 7. See Technical Lemmas 7.1, 7.2 which are reproduced here.

Technical Lemma 9.1. Consider the error dynamics system with output defined as

$$\begin{aligned} \dot{\tilde{W}}_1 &= -a_1\bar{\sigma}_1\bar{\sigma}_1^T\tilde{W}_1 + a_1\bar{\sigma}_1\frac{\varepsilon_H}{m_s} \\ y &= \bar{\sigma}_1^T\tilde{W}_1 \end{aligned} \quad (9.45)$$

The PE condition (9.44) is equivalent to the uniform complete observability (UCO) (Lewis *et al.*, 1999) of this system, that is there exist constants $\beta_3 > 0,\ \beta_4 > 0,\ T > 0$ such that, for all t

$$\beta_3 I \le S_1 \equiv \int_t^{t+T} \Phi^T(\tau,t)\overline{\sigma}_1(\tau)\overline{\sigma}_1^T(\tau)\Phi(\tau,t)d\tau \le \beta_4 I \tag{9.46}$$

with $\Phi(t_1,t_0), t_0 \le t_1$ the state transition matrix of (9.45) and I the identity matrix of appropriate dimensions. ■

Technical Lemma 9.2. Consider the error dynamics system (9.45). Let the signal $\overline{\sigma}_1$ be persistently exciting. Then:

a. The system (9.45) is exponentially stable. In fact if $\varepsilon_H = 0$ then $||\tilde{W}(kT)|| \le e^{-akT}||\tilde{W}(0)||$ with decay factor

$$\alpha = -\frac{1}{T}\ln\left(\sqrt{1-2a_1\beta_3}\right) \tag{9.47}$$

b. Let $||\varepsilon_H|| \le \varepsilon_{\max}$ and $||y|| \le y_{\max}$. Then $||\tilde{W}_1||$ converges exponentially to the residual set

$$\tilde{W}_1(t) \le \frac{\sqrt{\beta_2 T}}{\beta_1}\{[y_{\max} + \delta\beta_2 a_1(\varepsilon_{\max} + y_{\max})]\} \tag{9.48}$$

where δ is a positive constant of the order of 1. ■

The next result shows that the tuning algorithm (9.40) is effective under the PE condition, in that the weights $\hat{W}_1$ converge to the actual unknown weights W_1 that solve the approximate Bellman equation (9.33) in a least-squares sense for the given feedback and disturbance policies $u(x(t)),\ d(x(t))$. That is, (9.38) converges close to the actual value function (9.31) of the current policies. The proof is in Chapter 7 (see Theorem 7.1).

Theorem 9.1. Consider system (9.1) with value function (9.3). Let $u(x(t)),\ d(x(t))$ be any bounded stabilizing policies. Let tuning for the critic NN be provided by (9.40) and assume that $\overline{\sigma}_1$ is persistently exciting. Let the residual error in (9.33) be bounded $||\varepsilon_H|| < \varepsilon_{\max}$. Then the critic parameter error converges exponentially with decay factor given by (9.47) to the residual set

$$\tilde{W}_1(t) \le \frac{\sqrt{\beta_2 T}}{\beta_1}\{[1 + 2\delta\beta_2 a_1]\varepsilon_{\max}\} \tag{9.49}$$

■

Remark 9.1. Note that, as $N \to \infty, \varepsilon_H \to 0$ uniformly (Abu-Khalaf and Lewis, 2005, 2008). This means that $\varepsilon_{\max}$ decreases as the number of hidden-layer neurons in (9.38) increases.

Remark 9.2. This theorem requires the assumption that the feedback policy $u(x(t))$ and the disturbance policy $d(x(t))$ are bounded, since the policies appear in (9.34). In the upcoming Theorems 9.2 and 9.3 this restriction is removed.

9.3.3 Action and disturbance neural networks

It is important to define the NN structures and the NN estimation errors properly for the control and disturbance, otherwise an adaptive algorithm that learns the optimal game solution online cannot be developed. To determine a rigorously justified form for the actor and the disturbance NN, consider one step of the policy iteration Algorithm 9.1. Suppose that the solution $V(x)$ to the Bellman equation (9.26) for given control and disturbance policies is smooth and is given by (9.31). Then, according to (9.30) and (9.27), (9.28) one has for the policy and the disturbance updates

$$u = -\frac{1}{2}R^{-1}g^T(x)\sum_{i=1}^{\infty} c_i \nabla\varphi_i(x) \tag{9.50}$$

$$d = \frac{1}{2\gamma^2}k^T(x)\sum_{i=1}^{\infty} c_i \nabla\varphi_i(x) \tag{9.51}$$

for some unknown coefficients c_i. Then one has the following result. The following proposition is proved in Abu-Khalaf and Lewis (2005) for constrained inputs. Non-constrained inputs are easier to prove.

Proposition 9.2. Let the least-squares solution to (9.33) be W_1 and define

$$u_1 = -\frac{1}{2}R^{-1}g^T(x)\nabla V_1(x) = -\frac{1}{2}R^{-1}g^T(x)\nabla\phi_1^T(x)W_1 \tag{9.52}$$

$$d_1 = \frac{1}{2\gamma^2}k^T(x)\nabla V_1(x) = \frac{1}{2\gamma^2}k^T(x)\nabla\phi_1^T(x)W_1 \tag{9.53}$$

with V_1 defined in (9.35). Then, as $N \to \infty$:

a. $\sup\|u_1 - u\| \to 0$
b. $\sup\|d_1 - d\| \to 0$
c. There exists a number of NN hidden-layer neurons N_0 such that u_1 and d_1 stabilize the system (9.1) for $N > N_0$.

In light of this result, the ideal feedback and disturbance policy updates are taken as (9.52), (9.53) with W_1 unknown. Therefore, define the feedback policy in the form of an action NN that computes the control input in the structured form

$$\hat{u}(x) = -\frac{1}{2}R^{-1}g^T(x)\nabla\phi_1^T\hat{W}_2 \tag{9.54}$$

where $\hat{W}_2$ denotes the current estimated values of the ideal NN weights W_1. Define the actor NN estimation error as

$$\tilde{W}_2 = W_1 - \hat{W}_2 \tag{9.55}$$

Likewise, define the disturbance in the form of a disturbance NN that computes the disturbance input in the structured form

$$\hat{d}(x) = \frac{1}{2\gamma^2} k^T(x) \nabla \phi_1^T \hat{W}_3 \tag{9.56}$$

where $\hat{W}_3$ denotes the current estimated values of the ideal NN weights W_1. Define the disturbance NN estimation error as

$$\tilde{W}_3 = W_1 - \hat{W}_3 \tag{9.57}$$

9.4 Online solution of two-player zero-sum games using neural networks

This section presents our main results. An online adaptive PI algorithm is given for online solution of the ZS game problem that involves simultaneous, or synchronous, tuning of critic, actor and disturbance NNs. That is, the weights of all three NNs are tuned at the same time. This approach is an extreme version of generalized policy iteration (GPI), as introduced for discrete-time systems in Sutton and Barto (1998) and Chapter 2 of this book. In the standard policy iteration Algorithm 9.1, the critic and actor NNs are tuned sequentially, for example one at a time, with the weights of the other NNs being held constant. By contrast, we *tune all NN simultaneously in real time*. This is more in keeping with standard operation of adaptive control systems.

Definition 9.2 and Assumption 9.1 complete the machinery required for the main results.

Definition 9.2. Lewis *et al.* (1999) (UUB) A time signal $\zeta(t)$ is said to be uniformly ultimately bounded (UUB) if there exists a compact set $S \subset \mathbb{R}^n$ so that for all $\zeta(0) \in S$ there exists a bound B and a time $T(B, \zeta(0))$ such that $||\zeta(t)|| \leq B$ for all $t \geq t_0 + T$.

Assumption 9.1.

a. For each feedback control and disturbance policy the Bellman equation (9.26) has a smooth local solution $V(x) \geq 0$.
b. $f(\cdot)$ is Lipschitz, and $g(\cdot)$, $k(\cdot)$ are bounded by constants

$$||f(x)|| < b_f||x||, ||g(x)|| < b_g, \quad ||k(x)|| < b_k$$

c. The NN approximation error and its gradient are bounded locally so that

$$||\varepsilon|| < b_\varepsilon$$
$$||\nabla\varepsilon|| < b_{\varepsilon_x}$$

d. The NN activation functions and their gradients are bounded locally so that

$$||\phi_1(x)|| < b_\phi$$
$$||\nabla\phi_1(x)|| < b_{\phi_x}$$

e. The critic NN weights are bounded by known constants

$$||W_1|| < W_{\max}$$

■

The discussion in Section 9.1 justifies Assumption 9.1a. Assumption 9.1d is satisfied, for example by sigmoids, tanh and other standard NN activation functions.

The main theorems of Chapter 9 are now given. They present a novel adaptive control structure based on policy iteration that solves the ZS game online by measuring data along the system trajectories in real time. Three adaptive structures are needed, one to learn the value function and one to learn the Nash control policy of each player. The next result provides the tuning laws for the critic and two actor NNs that guarantee convergence in real time to the ZS game Nash equilibrium solution, while also guaranteeing closed-loop stability. The practical notion of uniform ultimate boundedness (UUB) from Chapter 7 is used.

Theorem 9.2. System stability and convergence of NN weights

Let the dynamics be given by (9.1), the critic NN be given by (9.38), the control input be given by actor NN (9.54) and the disturbance input be given by disturbance NN (9.56). Let tuning for the critic NN be provided by

$$\dot{\hat{W}}_1 = -a_1 \frac{\sigma_2}{(\sigma_2^T\sigma_2 + 1)^2}\left[\sigma_2^T\hat{W}_1 + Q(x) - \gamma^2||\hat{d}||^2 + \hat{u}^T R\hat{u}\right] \tag{9.58}$$

where $\sigma_2 = \nabla\phi_1(f + g\hat{u} + k\hat{d})$. Let the actor NN be tuned as

$$\dot{\hat{W}}_2 = -a_2\left\{(F_2\hat{W}_2 - F_2\overline{\sigma}_2^T\hat{W}_1) - \frac{1}{4}\overline{D}_1(x)\hat{W}_2 m^T(x)\hat{W}_1\right\} \tag{9.59}$$

and the disturbance NN be tuned as

$$\dot{\hat{W}}_3 = -a_3\left\{(F_4\hat{W}_3 - F_3\overline{\sigma}_2^T\hat{W}_1) + \frac{1}{4\gamma^2}\overline{E}_1(x)\hat{W}_3 m^T(x)\hat{W}_1\right\} \tag{9.60}$$

where $\overline{D}_1(x) \equiv \nabla\phi_1(x)g(x)R^{-1}g^T(x)\nabla\phi_1^T(x)$, $\overline{E}_1(x) \equiv \nabla\phi_1(x)kk^T\nabla\phi_1^T(x)$, $m \equiv \sigma_2/(\sigma_2^T\sigma_2 + 1)^2$, and $F_1 > 0, F_2 > 0, F_3 > 0, F_4 > 0$ are tuning parameters. Let Assumptions 9.1 hold and let $Q(x) > 0$. Suppose that $\overline{\sigma}_2 = \sigma_2/(\sigma_2^T\sigma_2 + 1)$ is persistently exciting. Let the tuning parameters F_1, F_2, F_3, F_4 in (9.59), and (9.60) be selected to make the matrix D_{22} in the proof is positive definite. Then there exists an N_0 such that, for the number of hidden-layer units $N > N_0$ the closed-loop system state, the critic NN error $\tilde{W}_1$, the actor NN error $\tilde{W}_2$, and the disturbance NN error $\tilde{W}_3$ are UUB.

Proof: See appendix. ■

Remark 9.3. See the comments following equation (9.37). Let $\varepsilon > 0$ and let N_0 be the number of hidden-layer units above which $\sup\|\varepsilon_{HJI}\| < \varepsilon$. In the proof it is seen that the theorem holds for $N > N_0$.

Remark 9.4. The theorem shows that PE is needed for proper identification of the value function by the critic NN, and that nonstandard tuning algorithms are required for the actor and the disturbance NN to guarantee stability.

Remark 9.5. The assumption $Q(x) > 0$ is sufficient but not necessary for this result. If this condition is replaced by zero state observability, the proof still goes through, however, it is tedious and does not add insight. The method used would be the technique used in the proof of technical Lemma 7.2 Part a, or the standard methods of Ioannou and Fidan (2006) and Tao (2003).

Theorem 9.3. Exponential convergence and Nash equilibrium

Suppose the hypotheses of Theorem 9.2 hold. Then Theorem 9.1 holds so that exponential convergence of $\hat{W}_1$ to the approximate optimal critic value W_1 is obtained. Then:

a. $H(x, \hat{W}_1, \hat{u}_1, \hat{d}_1)$ is UUB. That is, $\hat{W}_1$ converges to the approximate HJI solution, the value of the ZS game. Here,

$$\hat{u}_1 = -\frac{1}{2}R^{-1}g^T(x)\nabla\phi_1^T(x)\hat{W}_1 \tag{9.61}$$

$$\hat{d}_1 = \frac{1}{2\gamma^2}k^T(x)\nabla\phi_1^T(x)\hat{W}_1 \tag{9.62}$$

b. $\hat{u}(x), \hat{d}(x)$ (see (9.54) and (9.56)) converge to the approximate Nash equilibrium solution of the ZS game

Proof. Consider the UUB weights $\tilde{W}_1$, $\tilde{W}_2$ and $\tilde{W}_3$ as proved in Theorem 9.2. Also $\varepsilon_{\max}$ in Theorem 9.1 is practically bounded by the bound in the proof of Theorem 9.2.

a. The approximate HJI equation is

$$H(x, \hat{W}_1) = Q(x) + \hat{W}_1^T\nabla\varphi_1(x)f(x) - \frac{1}{4}\hat{W}_1^T D_1\hat{W}_1 + \frac{1}{4\gamma^2}\hat{W}_1^T E_1\hat{W}_1 - \hat{\varepsilon}_{HJI} \tag{9.63}$$

After adding zero we have

$$\begin{aligned} H(x, \hat{W}_1) = {} & \tilde{W}_1^T\nabla\varphi_1(x)f(x) - \frac{1}{4}\tilde{W}_1^T D_1\hat{W}_1 - \frac{1}{2}\tilde{W}_1^T D_1\hat{W}_1 \\ & + \frac{1}{4\gamma^2}\tilde{W}_1^T E_1\tilde{W}_1 + \frac{1}{2\gamma^2}\tilde{W}_1^T E_1\hat{W}_1 - \varepsilon_{HJI} \end{aligned} \tag{9.64}$$

But

$$\hat{W}_1 = -\tilde{W}_1 + W_1 \tag{9.65}$$

After taking norms in (9.65) and letting $||W_1|| < W_{max}$ one has

$$||\hat{W}_1|| = || -\tilde{W}_1 + W_1|| \leq ||\tilde{W}_1|| + ||W_1|| \leq ||\tilde{W}_1|| + W_{max} \tag{9.66}$$

Now (9.64) becomes, by taking into account (9.66)

$$\begin{aligned} ||H(x,\hat{W}_1)|| \leq\ & ||\tilde{W}_1|| ||\nabla\varphi_1(x)|| \; ||f(x)|| + \frac{1}{4}||\tilde{W}_1||^2||\overline{D}_1|| \\ & + \frac{1}{2}||\tilde{W}_1|| ||\overline{D}_1|| \left(||\tilde{W}_1|| + W_{max}\right) \\ & + \frac{1}{4\gamma^2}||\tilde{W}_1||^2||\overline{E}_1|| \\ & + \frac{1}{2\gamma^2}||\tilde{W}_1|| ||\overline{E}_1|| \left(||\tilde{W}_1|| + W_{max}\right) + ||\varepsilon_{HJI}|| \end{aligned} \tag{9.67}$$

Let Assumption 9.1 hold and also $\sup||\varepsilon_{HJI}|| < \varepsilon$. Then (9.67) becomes

$$\begin{aligned} ||H(x,\hat{W}_1)|| \leq\ & b_{\phi x} b_f ||\tilde{W}_1|| ||x|| + \frac{1}{4}||\tilde{W}_1||^2||\overline{D}_1|| \\ & + \frac{1}{2}||\tilde{W}_1|| ||\overline{D}_1|| \left(||\tilde{W}_1|| + W_{max}\right) \\ & + \frac{1}{4\gamma^2}||\tilde{W}_1||^2||\overline{E}_1|| \\ & + \frac{1}{2\gamma^2}||\tilde{W}_1|| ||\overline{E}_1|| \left(||\tilde{W}_1|| + W_{max}\right) + \varepsilon \end{aligned} \tag{9.68}$$

All the signals on the right-hand side of (9.68) are UUB. So $||H(x,\hat{W}_1)||$ is UUB and convergence to the approximate HJI solution is obtained.

b. According to Theorem 9.1 and equations (9.52), (9.53) and (9.54), (9.56), $||\hat{u} - u_1||$ and $||\hat{d} - d_1||$ are UUB because $||\hat{W}_2 - W_1||$ and $||\hat{W}_3 - W_1||$ are UUB. Therefore, the pair $\hat{u}(x), \hat{d}(x)$ gives the approximate Nash equilibrium solution of the zero-sum game.

This completes the proof. ■

The theorems make no mention of finding the minimum nonnegative solution to the HJI. However, they do guarantee convergence to a solution $(u(x), d(x))$ such that $(f(x) + g(x)u(x) + k(x)d(x))$ is stable. This is only accomplished by the minimal nonnegative HJI solution. Practical implementation, in view of the policy iteration Algorithm 9.1, would start with initial weights of zero in the disturbance NN (9.56). NN usage suggests starting with the initial control NN weights in (9.54) randomly selected and nonzero.

The ZS game policy iteration Algorithm 9.1 requires initialization with a stabilizing initial control policy. The synchronous online adaptive controller does not require this.

Note that the dynamics is required to be known to implement this algorithm in that $\sigma_2 = \nabla\phi_1(f + g\hat{u} + k\hat{d}), \overline{D}_1(x), \overline{E}_1(x)$, and (9.54) and (9.56) depend on $f(x), g(x), k(x)$.

9.5 Simulations

9.5.1 Online solution of generalized ARE for linear quadratic ZS games

Consider the continuous-time F16 aircraft longitudinal plant with quadratic cost function used in Stevens and Lewis (2003). The system state vector is $x = [\alpha \;\; q \;\; \delta_e]$, where α denotes the angle of attack, q is the pitch rate and δ_e is the elevator deflection angle. The control input is the elevator actuator voltage and the disturbance is wind gusts on the angle of attack. One has the dynamics $\dot{x} = Ax + Bu + Dd$ given by

$$\dot{x} = \begin{bmatrix} -1.01887 & 0.90506 & -0.00215 \\ 0.82225 & -1.07741 & -0.17555 \\ 0 & 0 & -1 \end{bmatrix} x + \begin{bmatrix} 0 \\ 0 \\ 1 \end{bmatrix} u + \begin{bmatrix} 1 \\ 0 \\ 0 \end{bmatrix} d$$

where Q and R in the cost function are identity matrices of appropriate dimensions. Select $\gamma = 5$. In this linear quadratic game the solution of the HJI equation is given by the solution of the game algebraic Riccati equation (GARE)

$$0 = A^T S + SA + Q - SBR^{-1}B^T S + \frac{1}{\gamma^2} SDD^T S$$

Since the value is quadratic in the LQ game case, the critic NN basis set $\phi_1(x)$ was selected as the quadratic vector in the state components $x \otimes x$, with $\otimes$ the Kronecker product. Redundant terms were removed to leave $n(n+1)/2 = 6$ components. Solving the GARE gives the parameters of the optimal critic as

$$W_1^* = [1.6573 \quad 1.3954 \quad -0.1661 \quad 1.6573 \quad -0.1804 \quad 0.4371]^T$$

which are the components of the GARE solution matrix S.

The synchronous ZS game synchronous adaptive algorithm is implemented with parameter tuning in Theorem 9.2. We selected $a_1 = a_2 = a_3 = 1$, $F_1 = I$, $F_2 = 10I$, $F_3 = I$, $F_4 = 10I$, where I is the identity matrix of appropriate dimensions. PE was ensured by adding a small exponentially decaying probing noise to the control and the disturbance input.

Figure 9.2 shows the critic parameters, denoted by

$$\hat{W}_1 = [W_{c1} \quad W_{c2} \quad W_{c3} \quad W_{c4} \quad W_{c5} \quad W_{c6}]^T$$

converging to the optimal values given by W_1^*. After 300 s the critic parameters converged to

$$\hat{W}_1(t_f) = [1.7090 \quad 1.3303 \quad -0.1629 \quad 1.7354 \quad -0.1730 \quad 0.4468]^T$$

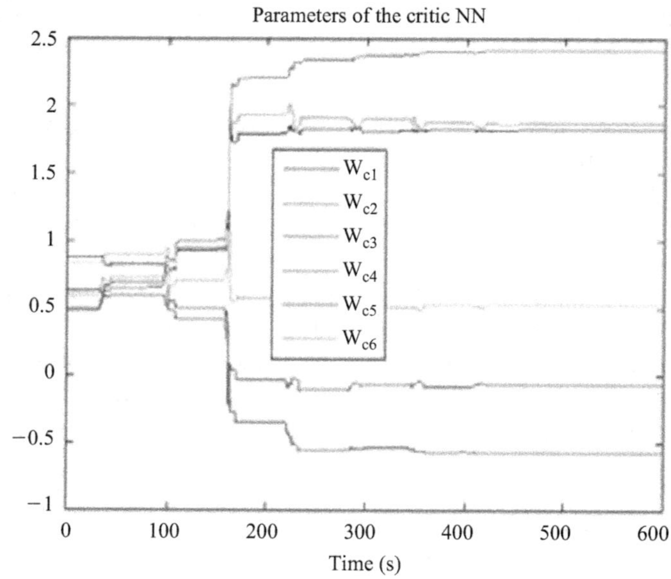

Figure 9.2 Convergence of the critic parameters to the parameters of the optimal critic

which are nearly the optimal values above. Similarly, the actor parameters after 300 s converge to the values of $\hat{W}_2(t_f) = [1.7090 \quad 1.3303 \quad -0.1629 \quad 1.7354 \quad -0.1730 \quad 0.4468]^T$. The disturbance parameters after 300s converge to the values of $\hat{W}_3(t_f) = [1.7090 \quad 1.3303 \quad -0.1629 \quad 1.7354 \quad -0.1730 \quad 0.4468]^T$. Then, the actor NN is given as

$$\hat{u}_2(x) = -\frac{1}{2}R^{-1}\begin{bmatrix}0\\0\\1\end{bmatrix}^T \begin{bmatrix}2x_1 & 0 & 0\\ x_2 & x_1 & 0\\ x_3 & 0 & x_1\\ 0 & 2x_2 & 0\\ 0 & x_3 & x_2\\ 0 & 0 & 2x_3\end{bmatrix} \hat{W}_2(t_f)$$

Then, the disturbance NN is given as

$$\hat{d}(x) = \frac{1}{2\gamma^2}\begin{bmatrix}1\\0\\0\end{bmatrix}^T \begin{bmatrix}2x_1 & 0 & 0\\ x_2 & x_1 & 0\\ x_3 & 0 & x_1\\ 0 & 2x_2 & 0\\ 0 & x_3 & x_2\\ 0 & 0 & 2x_3\end{bmatrix}^T \hat{W}_3(t_f)$$

The evolution of the system states is presented in Figure 9.3. One can see that the states are bounded, and they go to zero as the exponential probing noise decays.

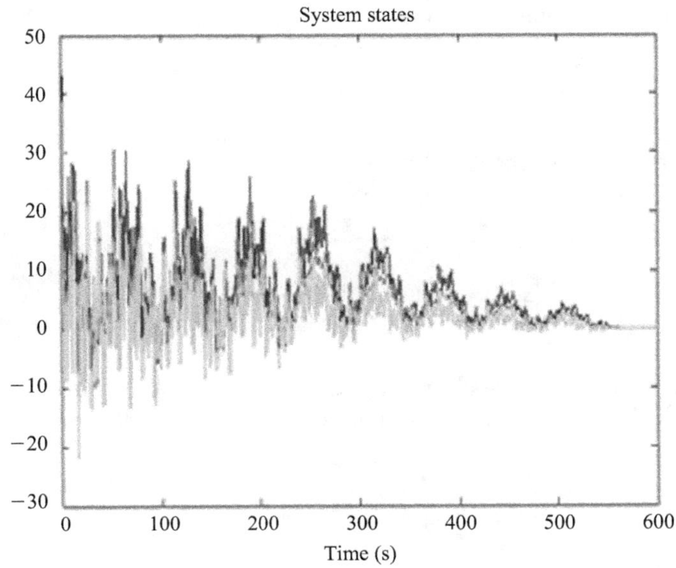

Figure 9.3 Evolution of the system states for the duration of the experiment

To select appropriate values of the tuning parameters a_1, a_2, a_3, F_1, F_2, F_3, F_4 in Theorem 9.2 one may have to run several simulations, changing their values until good convergence is obtained. In practice it is observed that tuning the parameters through several simulations is not difficult.

In conclusion, the synchronous ZS game adaptive control algorithm successfully solved the GARE online in real time by measuring data along the system trajectories. At the same time, system stability was ensured.

9.5.2 Online solution of HJI equation for non-linear ZS game

Consider the affine-in-control-input non-linear system $\dot{x} = f(x) + g(x)u + k(x)d$, $x \in \mathbb{R}^2$, where

$$f(x) = \begin{bmatrix} -x_1 + x_2 \\ -x_1^3 - x_2^3 + 0.25x_2(\cos(2x_1) + 2)^2 - 0.25x_2\dfrac{1}{\gamma^2}(\sin(4x_1) + 2)^2 \end{bmatrix}$$

$$g(x) = \begin{bmatrix} 0 \\ \cos(2x_1) + 2 \end{bmatrix}, k(x) = \begin{bmatrix} 0 \\ (\sin(4x_1) + 2) \end{bmatrix}$$

One selects $Q = \begin{bmatrix} 1 & 0 \\ 0 & 1 \end{bmatrix}, R = 1, \gamma = 8$. The optimal value function that solves the HJI equation (9.13) is

$$V^*(x) = \frac{1}{4}x_1^4 + \frac{1}{2}x_2^2$$

The optimal control signal (9.11) is

$$u^*(x) = -\frac{1}{2}(\cos(2x_1) + 2)x_2$$

and the worst-case disturbance (9.12) is

$$d^*(x) = \frac{1}{2\gamma^2}(\sin(4x_1) + 2)x_2$$

This example was constructed using the converse optimal procedure of Nevistic and Primbs (1996).

To implement the synchronous ZS game adaptive controller using the tuning algorithms in Theorem 9.2, one selects the critic NN vector activation function as

$$\varphi_1(x) = [x_1^2 \;\; x_2^2 \;\; x_1^4 \;\; x_2^4]$$

We selected $a_1 = a_2 = a_3 = 1$, $F_1 = I$, $F_2 = 10I$, $F_3 = I$, $F_4 = 10I$ where I is the identity matrix of appropriate dimensions.

Figure 9.4 shows the critic parameters, denoted by $\hat{W}_1 = [W_{c1} \;\; W_{c2} \;\; W_{c3} \;\; W_{c4}]^T$ when using the synchronous ZS game adaptive controller. After convergence at about 50 s have

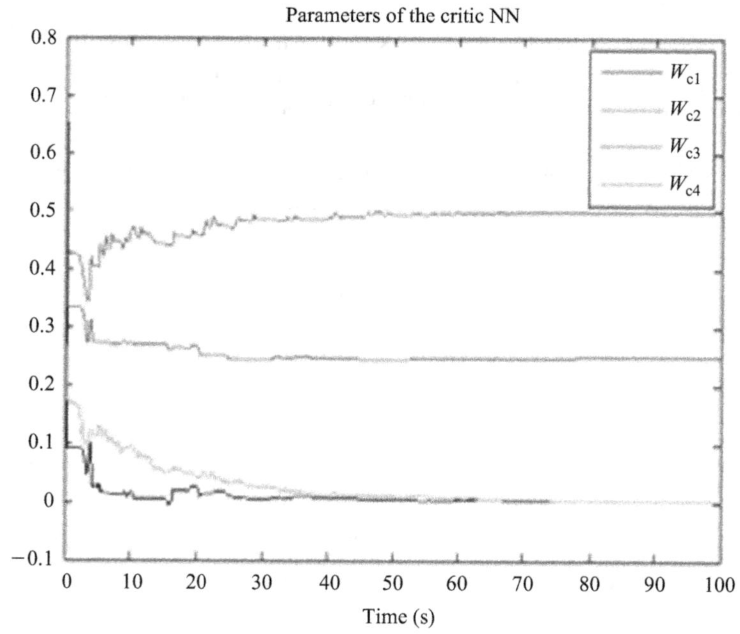

Figure 9.4 Convergence of the critic parameters

$$\hat{W}_1(t_f) = [-0.0006 \quad 0.4981 \quad 0.2532 \quad 0.0000]^T$$

which are very close to the correct values given above in $V^*(x)$.

Similarly, the actor parameters after 80 s converge to the values of $\hat{W}_2(t_f) = [-0.0006 \quad 0.4981 \quad 0.2532 \quad 0.0000]^T$, and the disturbance parameters after 50 s converge to the values of $\hat{W}_3(t_f) = [-0.0006 \quad 0.4981 \quad 0.2532 \quad 0.0000]^T$. Thus, the actor NN converged to the optimal control

$$\hat{u}_2(x) = -\frac{1}{2}R^{-1}\begin{bmatrix} 0 \\ \cos(2x_1) + 2 \end{bmatrix}^T \begin{bmatrix} 2x_1 & 0 \\ 0 & 2x_2 \\ 4x_1^3 & 0 \\ 0 & 4x_2^3 \end{bmatrix}^T \hat{W}_2(t_f)$$

and the disturbance NN converged to the worst-case disturbance

$$\hat{d}(x) = \frac{1}{2\gamma^2}\begin{bmatrix} 0 \\ \sin(4x_1) + 2 \end{bmatrix}^T \begin{bmatrix} 2x_1 & 0 \\ 0 & 2x_2 \\ 4x_1^3 & 0 \\ 0 & 4x_2^3 \end{bmatrix}^T \hat{W}_3(t_f)$$

The evolution of the system states is presented in Figure 9.5. As the probing noise decays to zero, the states converge to zero.

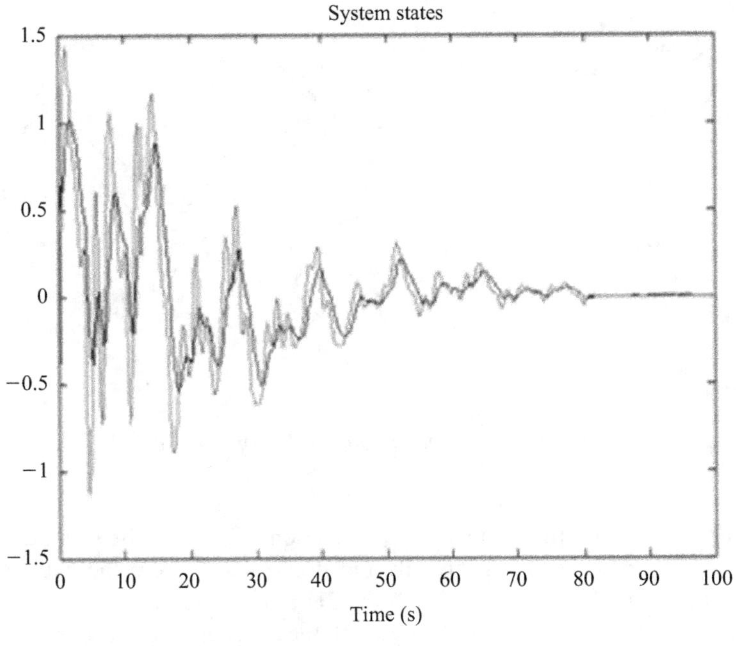

Figure 9.5 Evolution of the system states

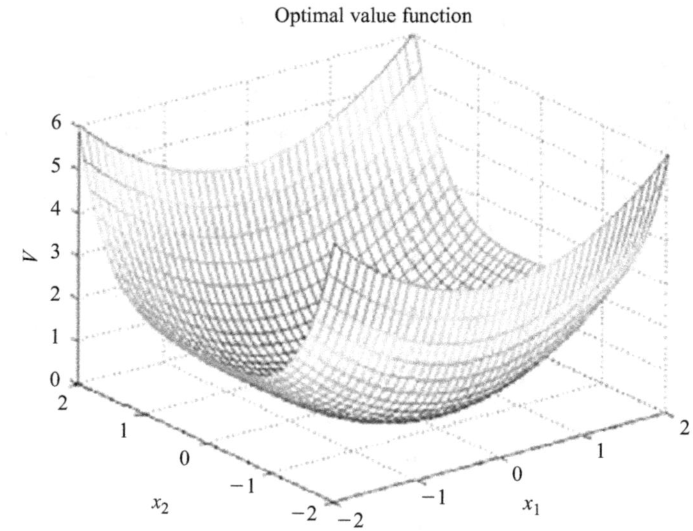

Figure 9.6 Optimal value function

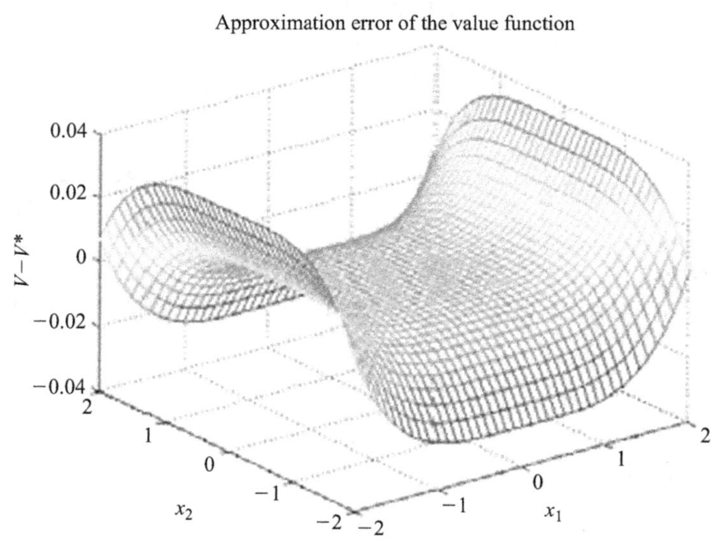

Figure 9.7 3D plot of the approximation error for the value function

Figure 9.6 shows the optimal value function. The identified value function given by $\hat{V}_1(x) = \hat{W}_1^T \phi_1(x)$ is virtually indistinguishable from the exact solution and so is not plotted. Figure 9.7 shows the 3D plot of the difference between the approximated value function and the optimal one. This error is close to zero. Good approximation of the actual value function is evolved. Figure 9.8 shows the 3D plot

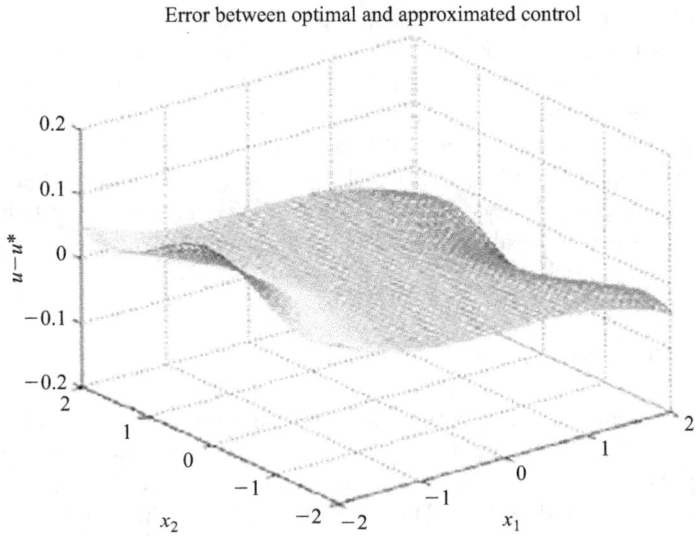

Figure 9.8 3D plot of the approximation error for the control

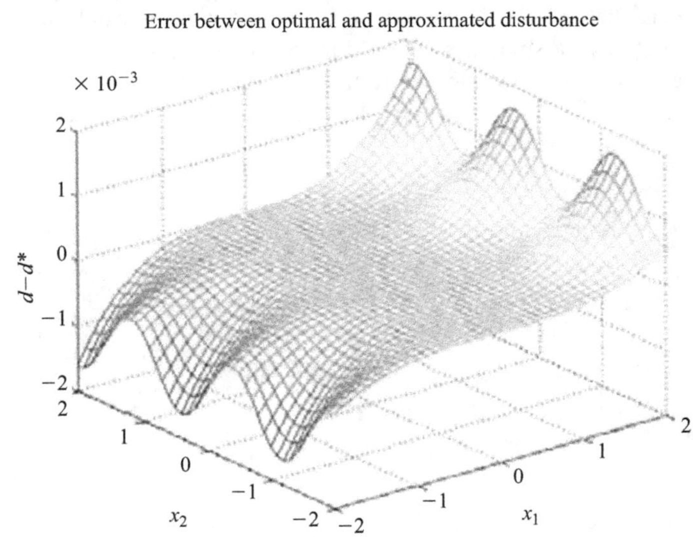

Figure 9.9 3D plot of the approximation error for the disturbance

of the difference between the approximated control, by using the online algorithm, and the optimal one. This error is close to zero. Finally Figure 9.9 shows the 3D plot of the difference between the approximated disturbance, by using the online algorithm, and the optimal one. This error is close to zero.

See the discussion above about selecting appropriate values of the tuning parameters a_1, a_2, a_3, F_1, F_2, F_3, F_4 in Theorem 9.2. The selection of the NN activation functions is made by choosing polynomials of even powers, since the value must serve as a Lyapunov function. If one had selected more polynomial basis functions and polynomials of higher powers in this example, their coefficients would go to zero as the ZS game adaptive controller learns the optimal solution.

In conclusion, the synchronous ZS game adaptive controller converged to an approximate smooth solution to the HJI equation (9.13) online in real time by measuring data along the system trajectories. At the same time, system stability was ensured. Note that solving the HJI for this non-linear system would be a formidable task.

9.6 Conclusion

This chapter proposed a new optimal adaptive algorithm which solves online the continuous-time zero-sum game problem for affine-in-the-input non-linear systems. The adaptive contoler has a three network structure that comes from a ZS game policy iteration algorithm. This algorithm relies on the simultaneous tuning of the actor, critic, and disturbance neural networks and it is termed synchronous ZS game adaptive controller. The ZS game adaptive controller converges to an approximate smooth solution of the HJI equation without explicitly solving this equation. Note that the HJI is intractable to solve for general non-linear systems and it may not have a smooth solution. Proofs using Lyapunov techniques guarantee convergence and closed-loop stability.

Chapter 10
Synchronous online learning for multiplayer non–zero-sum games

This chapter shows how to solve multiplayer non–zero-sum (NZS) games online using novel adaptive control structures based on reinforcement learning. For the most part, interest in the control systems community has been in the (non-cooperative) zero-sum games, which provide the solution of the H-infinity robust control problem (Başar and Olsder, 1999). However, dynamic team games may have some cooperative objectives and some selfish objectives among the players. This cooperative/non-cooperative balance is captured in the NZS games, as detailed herein.

In this chapter we are interested in feedback policies with full state information. We provide methods for *online gaming*, that is for solution of TV-player infinite-horizon NZS games *online*, through learning the Nash equilibrium in real time. The dynamics are non-linear in continuous time and are assumed known. A novel adaptive control structure is given that is based on reinforcement learning techniques, whereby each player's control policies are tuned online using data generated in real time along the system trajectories. Also tuned by each player are 'critic' approximator structures whose function is to identify the values of the current control policies for each player. Based on these value estimates, the players' policies are continuously updated. This is a sort of indirect adaptive control algorithm, yet, due to the simple form dependence of the control policies on the learned value, it is affected online as direct ('optimal') adaptive control.

This chapter proposes an algorithm for non-linear continuous-time systems with known dynamics to solve the N-player non–zero-sum (NZS) game problem where each player wants to optimize his *own performance index* (Başar and Olsder, 1999). The number of parametric approximator structures that are used is 2N. Each player maintains a critic approximator neural network (NN) to learn his optimal value and a control actor NN to learn his optimal control policy. Parameter update laws are given to tune the N-critic and N-actor NNs simultaneously online to converge to the solution to the coupled HJ equations, while also guaranteeing closed-loop stability. Rigorous proofs of performance and convergence are given. For the sake of clarity, we restrict ourselves to two-player differential NZS games in the actual proof. The proof technique can be directly extended using further careful bookkeeping to multiple players.

The chapter is organized as follows. Section 10.1 sets the background for multiplayer NZS games (Başar and Olsder, 1999), defining multiplayer Nash equilibrium and introducing the all-important coupled NZS Bellman equations. We derive the coupled NZS Hamilton–Jacobi (HJ) equations whose solution gives the Nash NZS game policies. It is necessary to develop policy iteration (PI) techniques for solving multiplayer games, for these PI algorithms give the controller structure needed for the online adaptive learning techniques presented in this work. Therefore, Section 10.2 presents a PI algorithm to solve the coupled non-linear HJ equations by successive solutions of the NZS Bellman equations. Based on the structure of this PI algorithm, Section 10.3 develops a novel adaptive control structure for online learning of the solution to the two-player NZS game problem in real time. Value function approximation is used as developed in Werbos (1989, 1991, 1992, 2009) and Bertsekas and Tsitsiklis (1996). Based on the PI algorithm in Section 10.2, suitable approximator structures (based on NNs) are developed for the value function and the control inputs of the players. Each player tunes two approximator structures – a critic network to evaluate its current performance, and an action network to update its control policy. A rigorous mathematical analysis is carried out for two players that generalizes to the multiplayer case. It is found that the actor and critic NNs require novel nonstandard tuning algorithms to guarantee stability and convergence to the Nash equilibrium. A persistence of excitation condition (Ioannou and Fidan, 2006; Tao, 2003; Lewis *et al.*, 1995) is needed to guarantee proper convergence to the optimal value functions. A Lyapunov analysis technique is used. Section 10.4 presents simulation examples that show the effectiveness of the synchronous online game algorithm in learning the optimal Nash equilibrium values in real time for both linear and non-linear systems.

10.1 *N*-player differential game for non-linear systems

10.1.1 Background on non–zero-sum games

Consider the *N*-player non-linear time-invariant differential game (Başar and Olsder, 1999) with dynamics

$$\dot{x} = f(x) + \sum_{j=1}^{N} g_j(x)u_j \tag{10.1}$$

where state $x(t) \in \mathbb{R}^n$, and the N players have control inputs $u_j(t) \in \mathbb{R}^{m_j}$. Assume that $f(0) = 0$ and $f(x)$, $g_j(x)$ are locally Lipschitz. The performance indices or cost functionals associated with each player are the infinite-horizon costs

$$\begin{aligned} J_i(x(0), u_1, u_2, \ldots u_N) &= \int_0^\infty \Big(Q_i(x) + \sum_{j=1}^{N} u_j^T R_{ij} u_j\Big)\, dt \\ &\equiv \int_0^\infty r_i(x(t), u_1, u_2, \ldots u_N)\, dt; \quad i \in N \end{aligned} \tag{10.2}$$

where functions $Q_i(x) \geq 0$ is generally non-linear, and $R_{ii} > 0, R_{ij} \geq 0$ are symmetric matrices.

We seek optimal controls among the set of feedback control policies with complete state information.

Definition 10.1. (Admissible policies.) Feedback control policies $u_i = \mu_i(x)$ are defined as admissible with respect to (10.2) on a set $\Omega \subset \mathbb{R}^n$, denoted by $\mu_i \in \Psi(\Omega)$ if $\mu_i(x)$ is continuous on Ω, $\mu_i(0) = 0$, $\mu_i(x)$ stabilizes (10.1) on Ω and (10.2) is finite $\forall x_0 \in \Omega$.

Given admissible feedback policies/strategies $u_i(t) = \mu_i(x)$, the value is

$$V_i(x(0), \mu_1, \mu_2, \ldots, \mu_N) = \int_0^\infty (Q_i(x) + \sum_{j=1}^N u_j^T R_{ij} \mu_j) d\tau$$

$$\equiv \int_0^\infty r_i(x(t), \mu_1, \mu_2, \ldots, \mu_N) d\tau; \quad i \in N \tag{10.3}$$

Define the N-player game

$$V_i^*(x(t), \mu_1, \mu_2, \ldots \mu_N) = \min_{\mu_i} \int_t^\infty (Q_i(x) + \sum_{j=1}^N \mu_j^T R_{ij} u_j) d\tau; \quad i \in N \tag{10.4}$$

By assuming that all the players have the same hierarchical level, we focus on the so-called Nash equilibrium that is given by the following definition.

Definition 10.2. Başar and Olsder (1999) (Nash equilibrium strategies) An *N-tuple* of strategies $\{\mu_1^*, \mu_2^*, \ldots, \mu_N^*\}$ with $\mu_1^* \in \Omega_i, i \in N$ is said to constitute a Nash equilibrium solution for an *N-player* finite game in extensive form, if the following N inequalities are satisfied for all $\mu_1^* \in \Omega_i, i \in N$

$$J_1^* \triangleq J_i(\mu_1^*, \mu_2^*, \mu_i^*, \ldots, \mu_N^*) \leq J_i(\mu_1^*, \mu_2^*, \mu_i, \ldots, \mu_N^*), \quad i \in N \tag{10.5}$$

The *N-tuple* of quantities $\{J_1^*, J_2^*, \ldots, J_N^*\}$ is known as a Nash equilibrium outcome of the N-player game.

This definition means that if any single player changes his policy from the Nash policy, he will pay an increased cost. Thus, no player has any incentive to unilaterally change his policy from the Nash policy.

Differential equivalents to each value function are given by the following coupled Bellman equations (different form from those given in Başar and Olsder (1999))

$$0 = r(x, u_1, \ldots, u_N) + (\nabla V_i)^T (f(x) + \sum_{j=1}^N g_j(x) u_j), \quad V_i(0) = 0, \quad i \in N \tag{10.6}$$

where $\nabla V_i = \partial V_i / \partial x \in \mathbb{R}^{n_i}$ is the gradient vector (e.g. transposed gradient).

Then, suitable nonnegative definite solutions to (10.6) are the values evaluated using the infinite integral (10.4) along the system trajectories. Define the Hamiltonian functions

$$H_i(x, \nabla V_i, u_1, \ldots, u_N) = r(x, u_1, \ldots u_N) + (\nabla V_i)^T (f(x) \sum_{j=1}^N g_j(x) u_j), \; i \in N \tag{10.7}$$

According to the stationarity conditions, associated feedback control policies are given by

$$\frac{\partial H_i}{\partial u_i} = 0 \Rightarrow \mu_i(x) = \frac{1}{2} R_{ii}^{-1} g_i^T(x) \nabla V_i, \quad i \in N \tag{10.8}$$

It is not difficult to show that Bellman equations (10.6) are non-linear Lyapunov equations, for their solutions V_i serve as Lyapunov functions for the closed-loop systems with control inputs given by (10.8) (see Lewis *et al.*, 2012).

Substituting (10.8) into (10.6) one obtains the N-coupled Hamilton–Jacobi (HJ) equations

$$\begin{aligned} 0 &= (\nabla V_i)^T \left(f(x) - \frac{1}{2} \sum_{j=1}^{N} g_j(x) R_{jj}^{-1} g_j^T(x) \nabla V_j \right) + Q_i(x) \\ &+ \frac{1}{4} \sum_{j=1}^{N} \nabla V_j^T g_j(x) R_{jj}^{-T} R_{ij} R_{jj}^{-1} g_j^T(x) \nabla V_j, \; V_i(0) = 0 \end{aligned} \tag{10.9}$$

These coupled HJ equations are in 'closed-loop' form. The equivalent 'open-loop' form is

$$\begin{aligned} 0 &= \nabla V_i^T f(x) + Q_i(x) - \frac{1}{2} \nabla V_i^T \sum_{j=1}^{N} g_j(x) R_{jj}^{-1} g_j^T(x) \nabla V_j \\ &+ \frac{1}{4} \sum_{j=1}^{N} \nabla V_j^T g_j(x) R_{jj}^{-T} R_{ij} R_{jj}^{-1} g_j^T(x) \nabla V_j, \; V_i(0) = 0 \end{aligned} \tag{10.10}$$

To solve the NZS game, one may solve the coupled HJ equations (10.9) for V_i; then the Nash equilibrium control inputs are given by (10.8).

In linear systems of the form $\dot{x} = Ax + \sum_{j=1}^{N} B_j u_j$, (10.9) becomes the N-coupled generalized algebraic Riccati equations

$$0 = P_i A_c + A_c^T P_i + Q_i + \frac{1}{4} \sum_{j=1}^{N} P_j B_j R_{jj}^{-T} R_{ij} R_{jj}^{-1} B_j^T P_j, \quad i \in N \tag{10.11}$$

where $A_c = A - \frac{1}{2}\sum_{i=1}^{N} B_i R_{ii}^{-1} B_i^T P_i$. It is shown in Başar and Olsder (1999) that if there exist solutions to (10.11) further satisfying the conditions that for each $i \in N$ the pair $(A - \frac{1}{2} \sum_{j \in N, j \neq i} B_j R_{jj}^{-1} B_j^T P_j, \;\; B_i)$ is stabilizable and the pair $(A - \frac{1}{2}\sum_{j \in N, j \neq i} B_j R_{jj}^{-1} B_j^T P_j, \sqrt{Q_i + \frac{1}{4}\sum_{j \in N, j \neq i} P_j B_j R_{jj}^{-T} R_{ij} R_{jj}^{-1} B_j^T P_j})$ is detectable, then the N-tuple of the stationary feedback policies $\mu_i^*(x) = -K_i x = -\frac{1}{2} R_{ii}^{-1} B_i^T P_i x$, $i \in N$ provides a Nash equilibrium solution for the linear quadratic N-player differential game among feedback policies with full state information. Furthermore the resulting system dynamics, described by $\dot{x} = A_c x$, $x(0) = x_0$ are asymptotically stable. The corresponding result for the non-linear case (10.9) is shown in Lewis *et al.* (2012).

10.1.2 *Cooperation vs. non-cooperation in multiplayer dynamic games*

Differential games where each player minimizes his own performance index generally result in Nash equilibria. These are commonly called non-cooperative games, since each player minimizes his own cost without explicit regard for the actions of other players. However, Nash equilibrium may imply some sort of cooperation between players.

In fact, one may write the costs as

$$J_i = \frac{1}{N}\sum_{j=1}^{N} J_j + \frac{1}{N}\sum_{j=1}^{N}(J_i - J_j) \equiv \overline{J} + \tilde{J}_i, \quad i \in N \tag{10.12}$$

where $\overline{J}$ is an overall cooperative 'team' cost (which can be viewed as a 'center of gravity' cost) and $\tilde{J}_i$ a 'conflict' cost for player i. If $\overline{J} = 0$ one has a zero-sum game. The most studied case is the two-player zero-sum game when $J_1 = -J_2$. Such games are known as convex–concave games, result in saddle-point Nash equilibria, and have been extensively studied in control systems due to their relation to the H_∞ control problem (Başar and Olsder, 1999). However, general dynamic team games may have some cooperative objectives and some selfish objectives among the players. This interplay between cooperative and non-cooperative objectives is captured in NZS games, as detailed in (10.12). Therefore, this work is interested in general N-player games that may or may not be zero-sum.

Though NZS games may contain non-cooperative components, note that the solution to each player's coupled HJI equations (10.9) requires knowledge of all the other players' strategies (10.8). This is in the spirit of rational opponents (Başar and Olsder, 1999) whereby the players share information, yet independently minimize their own cost. Then, from the definition of Nash equilibrium each player benefits by remaining at the equilibrium policy.

10.2 Policy iteration solution for non–zero-sum games

To solve the NZS game, one may solve the coupled HJ equations (10.9) for V_i, then the Nash equilibrium control inputs are given by (10.8). The coupled non-linear HJ equations (10.9) are difficult or impossible to solve. An iterative offline solution technique is given by the following policy iteration algorithm. It solves the coupled HJ equations by iterative solution of the coupled Bellman equations (10.6), which are linear in the value gradient.

Algorithm 10.1. Policy iteration (PI) for N-player games

1. Start with stabilizing initial policies $\mu_1^0(x), \ldots, \mu_N^0(x)$.

2. (*Policy evaluation*) Given the N-tuple of policies $\mu_1^k(x), \ldots, \mu_N^k(x)$, solve for the N-tuple of costs $V_1^k(x(t)), V_2^k(x(t)) \ldots V_N^k(x(t))$ using the Bellman equations

$$0=r(x,\mu_1^k,\ldots,\mu_N^k)+(\nabla V_i^k)^T\left(f(x)+\sum_{j=1}^{N}g_j(x)\mu_j^i\right),\ V_i^k(0)=0,\ i\in N \quad (10.13)$$

3. (*Policy improvement*) Update the N-tuple of control policies using

$$\mu_i^{k+1} = \arg\min_{u_i\in\Psi(\Omega)}[H_i(x, \nabla V_i, u_i, \ldots, u_N)] \quad i \in N \quad (10.14)$$

which explicitly is

$$\mu_i^{k+1}(x) = -\frac{1}{2}R_{ii}^{-1}g_i^T(x)\nabla V_i^k \quad i \in N \quad (10.15)$$

■

A linear two-player version of Algorithm 10.1 is given in Gajic and Li (1988) and can be considered as an extension of Kleiman's Algorithm (Kleinman, 1968) to two-player games.

Section 10.3 uses PI Algorithm 10.1 to motivate a novel adaptive control structure for online solution of the N-player game in real time. Then it is proven that the 'optimal adaptive' control algorithm converges online to the solution of coupled HJs (10.10), while guaranteeing closed-loop stability.

10.3 Online solution for two-player non–zero-sum games

10.3.1 Value function approximation and critic neural networks for solution of Bellman equations

This work uses non-linear approximator structures for value function approximation (VFA) (Werbos, 1989, 1991, 1992, 2009; Bertsekas and Tsitsiklis, 1996) to solve (10.13). We show how to solve the two-player NZS game presented in Section 10.2, the approach can easily be extended to more than two players.

Consider the non-linear time-invariant affine-in-the-input dynamical system given by

$$\dot{x} = f(x) + g(x)u(x) + k(x)d(x) \quad (10.16)$$

where state $x(t) \in \mathbb{R}^n$, first control input $u(x) \in \mathbb{R}^m$ and second control input $d(x) \in \mathbb{R}^q$. Assume that $f(0) = 0$ and $f(x)$, $g(x)$, $k(x)$ are locally Lipschitz, $\|f(x)\| < b_f\|x\|$.

Assumption 10.1. For admissible feedback control policies the Bellman equations (10.13) have locally smooth solutions $V_1'(x) \geq 0,\ \ V_2'(x) \geq 0,\ \forall x \in \Omega$.

If so, according to the Weierstrass higher-order approximation theorem (Abu-Khalaf and Lewis, 2005; Finlayson, 1990; Hornik *et al.*, 1990), there exist complete independent basis sets such that the solutions to $V_1'(x(t)),\ V_2'(x(t))$ (10.6), (10.13) and their gradients are uniformly approximated. Specifically, the basis sets are dense in the Sobolev norm $W^{1,\infty}$ (Adams and Fournier, 2003).

Therefore, assume there exist constant neural-network (NN) weights W_1 and W_2 such that the value functions $V_1'(x)$ and $V_2'(x)$ are approximated on a compact set Ω as

$$V_1'(x) = W_1^T \phi_1(x) + \varepsilon_1(x) \tag{10.17}$$

$$V_2'(x) = W_2^T \phi_2(x) + \varepsilon_2(x) \tag{10.18}$$

with $\phi_1(x) : \mathbb{R}^n \to \mathbb{R}^K$ and $\phi_2(x) : \mathbb{R}^n \to \mathbb{R}^K$ the NN activation function basis set vectors, K the number of neurons in the hidden layer and $\varepsilon_1(x)$ and $\varepsilon_2(x)$ the NN approximation errors. From the approximation literature, the basis functions can be selected as sigmoids, tanh, polynomials, etc.

The value function derivatives are also uniformly approximated, for example, additionally, $\forall x \in \Omega$

$$\frac{\partial V_i'}{\partial x} = \left(\frac{\partial \phi_i(x)}{\partial x}\right)^T W_i + \frac{\partial \varepsilon_i}{\partial x} = \nabla \phi_i^T W_i + \nabla \varepsilon_i, \quad i = 1, 2 \tag{10.19}$$

Then, as the number of hidden-layer neurons $K \to \infty$, the approximation errors $\varepsilon_i \to 0$, $\nabla \varepsilon_i \to 0$, $i = 1, 2$ uniformly (Abu-Khalaf and Lewis, 2005; Finlayson, 1990). In addition, for fixed K, the NN approximation errors $\varepsilon_i(x), \varepsilon_2(x)$ and $\nabla \varepsilon_i, \nabla \varepsilon_2$ are bounded on a set Ω by constants if Ω is compact. We refer to the NN with weights W_1 and W_2 that perform VFA as the *critic* NNs for each player.

Using the NN VFA (10.17), (10.18) considering fixed feedback policies u and d the Hamiltonians (10.7) become

$$\begin{aligned} H_1(x, W_1, u, d) = {} & Q_1(x) + u^T R_{11} u + d^T R_{12} d \\ & + W_1^T \nabla \phi_1 (f(x) + g(x)u + k(x)d) = \varepsilon_{H_1} \end{aligned} \tag{10.20}$$

$$\begin{aligned} H_2(x, W_2, u, d) = {} & Q_2(x) + u^T R_{21} u + d^T R_{22} d \\ & + W_2^T \nabla \phi_2 (f(x) + g(x)u + k(x)d) = \varepsilon_{H_2} \end{aligned} \tag{10.21}$$

The residual errors for the two players are

$$\varepsilon_{H_i} = -(\nabla \varepsilon_i)^T (f + gu + kd), \quad i = 1, 2 \tag{10.22}$$

Substituting W_1, W_2 into the HJ equations (10.10), one obtains

$$Q_1(x) - \frac{1}{4}W_1^T\nabla\phi_1(x)g(x)R_{11}^{-1}g^T(x)\nabla\phi_1^T(x)W_1 + W_1^T\nabla\phi_1(x)f(x)$$
$$+\frac{1}{4}W_2^T\nabla\phi_2(x)k(x)R_{22}^{-T}R_{12}R_{22}^{-1}k^T(x)\nabla\phi_2^T(x)W_2$$
$$-\frac{1}{2}W_1^T\nabla\phi_1(x)k(x)R_{22}^{-1}k^T(x)\nabla\phi_2^T(x)W_2 = \varepsilon_{HJ_1},\ V_1(0)=0 \tag{10.23}$$

$$Q_2(x) - \frac{1}{4}W_2^T\nabla\phi_2(x)k(x)R_{22}^{-1}k^T(x)\nabla\phi_2^T(x)W_2 + W_2^T\nabla\phi_2(x)f(x)$$
$$+\frac{1}{4}W_1^T\nabla\phi_1(x)g(x)R_{11}^{-T}R_{21}R_{11}^{-1}g^T(x)\nabla\phi_1^T(x)W_1$$
$$-\frac{1}{2}W_2^T\nabla\phi_2(x)g(x)R_{11}^{-1}g^T(x)\nabla\phi_1^T(x)W_1 = \varepsilon_{HJ_2},\ V_2(0)=0 \tag{10.24}$$

where the residual errors due to function approximation error for the first player is

$$\varepsilon_{HJ_1} \equiv \frac{1}{2}W_1\nabla\phi_1^T g(x)R_{11}^{-1}g^T(x)\nabla\varepsilon_1 + \frac{1}{4}\nabla\varepsilon_1^T g(x)R_{11}^{-1}g^T(x)\nabla\varepsilon_1$$
$$-\frac{1}{4}\nabla\varepsilon_2^T k(x)R_{22}^{-T}R_{22}^{-1}k^T(x)\nabla\varepsilon_2 - \frac{1}{4}W_2\nabla\phi_2^T k(x)R_{22}^{-T}R_{12}R_{22}^{-1}k^T(x)\nabla\varepsilon_2$$
$$-\nabla\varepsilon_1^T f(x) + \frac{1}{2}W_1\nabla\phi_1^T k(x)R_{22}^{-T}k^T(x)\nabla\varepsilon_2 + \frac{1}{2}\nabla\varepsilon_1^T k(x)R_{22}^{-1}k^T(x)\nabla\varepsilon_2$$
$$+\frac{1}{2}W_2\nabla\phi_2^T k(x)R_{22}^{-1}k^T(x)\nabla\varepsilon_1$$

and ε_{HJ_2} for the second player is similarly defined.

The following proposition follows as in Abu-Khalaf *et al.* (2006) and Abu-Khalaf and Lewis (2005).

Proposition 10.1. For any admissible feedback policies $u(x(t))$, $d(x(t))$ the least-squares pair solution to (10.20) and (10.21) exists and is unique for each K. Denote these solutions as W_1 and W_2 and define

$$V_1(x) = W_1^T\phi_1(x) \tag{10.25}$$

$$V_2(x) = W_2^T\phi_2(x) \tag{10.26}$$

Then, as $K \to \infty$:

a. $\sup_{x\in\Omega}|\varepsilon_{H_1}| \to 0, \quad i = 1, 2$
b. $\sup_{x\in\Omega}|\varepsilon_{HJ_i}| \to 0, \quad i = 1, 2$

c. $\sup_{x\in\Omega} |V_i - V_i'| \to 0, \quad i = 1,2$

d. $\sup_{x\in\Omega} |\nabla V_i - \nabla V_i'| \to 0, \quad i = 1,2$

This result shows that $V_1(x), V_2(x)$ converge uniformly in Sobolev norm $W^{1,\infty}$ (Abu-Khalaf and Lewis, 2005) to the exact solution $V_1'(x), V_2'(x)$ to (10.6). Therefore, $\forall \bar{\varepsilon}_1 > 0 \exists K(\bar{\varepsilon}_1) : \sup_{x\in\Omega} \|\varepsilon_{HJ_1}\| < \bar{\varepsilon}_1, K > K(\bar{\varepsilon}_1)$ and

$$\forall \bar{\varepsilon}_2 > 0 \exists K(\bar{\varepsilon}_2) : \sup_{x\in\Omega} \|\varepsilon_{HJ_2}\| < \bar{\varepsilon}_2,\ K > K(\bar{\varepsilon}_2)$$

The ideal NN weights W_1, W_2 are unknown. Assuming current NN weight estimates $\hat{W}_1$ and $\hat{W}_2$, the outputs of the two critic NN are given by

$$\hat{V}_1(x) = \hat{W}_1^T \phi_1(x) \tag{10.27}$$

$$\hat{V}_2(x) = \hat{W}_2^T \phi_2(x) \tag{10.28}$$

The approximate Bellman equations are then

$$\begin{aligned} H_1(x, \hat{W}_1, u_1, d_2) &= Q_1(x) + u_1^T R_{11} u_1 + d_2^T R_{12} d_2 \\ &\quad + \hat{W}_1^T \nabla\phi_1(f(x) + g(x)u_1 + k(x)d_2) = e_1 \end{aligned} \tag{10.29}$$

$$\begin{aligned} H_2(x, \hat{W}_2, u_1, d_2) &= Q_2(x) + u_1^T R_{21} u_1 + d_2^T R_{22} d_2 \\ &\quad + \hat{W}_2^T \nabla\phi_2(f(x) + g(x)u_1 + k(x)d_2) = e_2 \end{aligned} \tag{10.30}$$

It is desired to select $\hat{W}_1$ and $\hat{W}_2$ to minimize the square residual error

$$E_1 = \frac{1}{2} e_1^T e_1 + \frac{1}{2} e_2^T e_2$$

Then $\hat{W}_1(t) \to W_1, \hat{W}_2(t) \to W_2$ and $e_1 \to \varepsilon_{H_1}, e_2 \to \varepsilon_{H_2}$. Therefore, select the tuning laws for the critic weights as the normalized gradient descent algorithms

$$\dot{\hat{W}}_1 = -a_1 \frac{\partial E_1}{\partial \hat{W}_1} = -a_1 \frac{\sigma_1}{(1 + \sigma_1^T \sigma_1)^2} \left[\sigma_1^T \hat{W}_1 + Q_1(x) + u_1^T R_{11} u_1 + d_2^T R_{12} d_2\right] \tag{10.31}$$

$$\dot{\hat{W}}_2 = -a_2 \frac{\partial E_1}{\partial \hat{W}_2} = -a_2 \frac{\sigma_1}{(1 + \sigma_2^T \sigma_2)^2} \left[\sigma_2^T \hat{W}_2 + Q_2(x) + u_1^T R_{21} u_1 + d_2^T R_{22} d_2\right] \tag{10.32}$$

where $\sigma_i = \nabla\phi_i(f(x) + g(x)u_1 + k(x)d_2)$, $i = 1,2$. Note that these are modified gradient descent algorithms with the normalizing terms in the denominators raised to the power 2.

Persistence of excitation (PE) assumption. Let the signals $\overline{\sigma}_1, \overline{\sigma}_2$ be persistently exciting over the interval $[t, t+T]$, *that is* there exist constants $\beta_1 > 0, \beta_2 > 0, \beta_3 > 0, \beta_4 > 0, T > 0$ such that, for all t

$$\beta_1 I \leq S_0 \equiv \int_t^{t+T} \overline{\sigma}_1(\tau)\overline{\sigma}_1^T(\tau)d\tau \leq \beta_2 I \tag{10.33}$$

$$\beta_3 I \leq S_1 \equiv \int_t^{t+T} \overline{\sigma}_2(\tau)\overline{\sigma}_2^T(\tau)d\tau \leq \beta_4 I \tag{10.34}$$

where $\overline{\sigma}_i = \sigma_i/(\sigma_i^T\sigma_i + 1)$, $i = 1, 2$ and I the identity matrix of appropriate dimensions.

The PE assumption is needed in adaptive control if one desires to perform system identification using for example RLS (Ioannou and Fidan, 2006; Tao, 2003). It is needed in the upcoming Theorem 10.1 because one effectively desires to identify the critic parameters that approximate the solutions $V_1(x)$ and $V_2(x)$ to the Bellman equations (10.6).

The properties of tuning algorithms (10.31) and (10.32) and their exponential convergence to residual sets are given in the following theorem that was proven for a single player in Chapter 7.

Theorem 10.1. Let $u(x(t))$, $d(x(t))$ be any admissible bounded feedback policies. Let tuning for the critic NNs be provided by (10.31), (10.32) and assume that $\overline{\sigma}_1$ and $\overline{\sigma}_2$ are persistently exciting. Let the residual errors (10.22) be bounded by $||\varepsilon_{H_1}|| < \varepsilon_{\max_1}$ and $||\varepsilon_{H_2}|| < \varepsilon_{\max_2}$. Then the critic parameter errors converge exponentially to the residual sets

$$\tilde{W}_1(t) \leq \frac{\sqrt{\beta_2 \mathrm{T}}}{\beta_1}\left\{[1 + 2\delta_1\beta_2 a_1]\varepsilon_{\max_1}\right\} \tag{10.35}$$

$$\tilde{W}_2(t) \leq \frac{\sqrt{\beta_4 \mathrm{T}}}{\beta_3}\left\{[1 + 2\delta_2\beta_4 a_2]\varepsilon_{\max_2}\right\} \tag{10.36}$$

where δ_1, δ_2 are positive constants of the order of 1. ∎

Proof: Follows as in Chapter 7 (Theorem 7.1). ∎

10.3.2 Action neural networks and online learning algorithm

This section develop an optimal adaptive control algorithm that solves the two-player game problem *online* using data measured along the system trajectories. The technique given here generalizes directly to the *N*-player game. A Lyapunov technique is used to derive novel parameter tuning algorithms for the values and control policies that guarantee closed-loop stability as well as convergence to the approximate game solution of (10.10).

A suitable 'actor–critic' control structure is developed based on PI Algorithm 10.1. The adaptive control structure relies on each player maintaining two approximator structures, one for its current value estimate (cf. (10.13)) and one for its current policy (cf. (10.14), (10.15)). This structure is based on reinforcement learning precepts.

Define

$$u_1(x) = -\frac{1}{2}R_{11}^{-1}g^T(x)\nabla V_1(x) = -\frac{1}{2}R_{11}^{-1}g^T(x)\nabla\phi_1^T(x)W_1 \tag{10.37}$$

$$d_2(x) = -\frac{1}{2}R_{22}^{-1}g^T(x)\nabla V_2(x) = -\frac{1}{2}R_{22}^{-1}g^T(x)\nabla\phi_2^T(x)W_2 \tag{10.38}$$

with V_1 and V_2 defined in terms of the least-squares solutions to (10.20), (10.21) as given in Proposition 10.1. According to Proposition 10.1, as $K \to \infty$ (10.37) and (10.38) converge uniformly to (10.8). The next result follows as in Abu-Khalaf *et al.* (2006) and Abu-Khalaf and Lewis (2005).

Proposition 10.2. There exists a number of hidden-layer units K_0 such that $u_1(x)$ and $d_2(x)$ are admissible for $K > K_0$.

In light of this result, the ideal control policy updates are taken as (10.37) and (10.38) with W_1 and W_2 unknown. Based on this structure, define the control policies in the form of two action neural networks which compute the control inputs in the structured form

$$u_3(x) = -\frac{1}{2}R_{11}^{-1}g^T(x)\ \nabla\phi_1^T\hat{W}_3 \tag{10.39}$$

$$d_4(x) = -\frac{1}{2}R_{22}^{-1}k^T(x)\ \nabla\phi_2^T\hat{W}_4 \tag{10.40}$$

where $\tilde{W}_3$ and $\tilde{W}_4$ denote the current estimated values of the ideal NN weights W_1 and W_2, respectively. Define the critic and the actor NN estimation errors respectively as

$$\tilde{W}_1 = W_1 - \hat{W}_1, \tilde{W}_2 = W_2 - \hat{W}_2, \tilde{W}_3 = W_1 - \hat{W}_3, \tilde{W}_4 = W_2 - \hat{W}_4 \tag{10.41}$$

Assumptions 10.2. For a given compact set $\Omega \subset \mathbb{R}^n$:

a. $g(\cdot)$, $k(\cdot)$ are bounded by constants

$$\|g(x)\| < b_g, \quad \|k(x)\| < b_k$$

b. The NN approximation errors and their gradients are bounded so that

$$||\varepsilon_1|| < b_{\varepsilon 1}, ||\nabla\varepsilon_1|| < b_{\varepsilon_{1x}}, ||\varepsilon_2|| < b_{\varepsilon 2}, ||\nabla\varepsilon_2|| < b_{\varepsilon_{2x}}$$

c. The NN activation functions and their gradients are bounded so that

$$||\phi_1(x)|| < b_{\phi_1}, ||\nabla\phi_1(x)|| < b_{\phi_{1x}}, ||\phi_2(x)|| < b_{\phi_2}, ||\nabla\varphi_2(x)|| < b_{\phi_{2x}}$$

d. The critic NN weights are bounded by known constants

$$||W_1|| < W_{1\max}, ||W_2|| < W_{2\max}$$

The main theorems of Chapter 10 are now given. They present a novel adaptive control structure based on policy iteration that solves the NZS game online by measuring data along the system trajectories in real time. Two adaptive structures are needed for each player, one to learn the value function and one to learn the Nash control policy. The next result provides the tuning laws for these actor and critic neural networks that guarantee convergence in real time to the two-player non–zero-sum game Nash equilibrium solution, while also guaranteeing closed-loop stability. The practical notion of uniform ultimate boundedness (UUB) from Chapter 7 is used.

Theorem 10.2. Stability and bounded neural-network weight errors. Let the dynamics be given by (10.16), and consider the two-player game formulation in Section 10.2. Let the critic networks be given by (10.27) and (10.28), the control inputs be given by action networks (10.39) and (10.40). Let tuning for the critic NNs be provided by

$$\dot{\hat{W}}_1 = -a_1 \frac{\sigma_3}{(\sigma_3^T\sigma_3 + 1)^2}\left[\sigma_3^T \hat{W}_1 + Q_1(x) + u_3^T R_{11} u_3 + d_4^T R_{12} d_4\right] \tag{10.42}$$

$$\dot{\hat{W}}_2 = -a_2 \frac{\sigma_4}{(\sigma_4^T\sigma_4 + 1)^2}\left[\sigma_4^T \hat{W}_2 + Q_2(x) + u_3^T R_{21} u_3 + d_4^T R_{22} d_4\right] \tag{10.43}$$

where $\sigma_3 = \nabla\phi_1(f + gu_3 + kd_4)$ and $\sigma_4 = \nabla\phi_2(f + gu_3 + kd_4)$. Let the first actor NN (first player) be tuned as

$$\begin{aligned}\dot{\hat{W}}_3 &= -a_3\Big\{(F_2\hat{W}_3 - F_1\bar{\sigma}_3^T\hat{W}_1)\\ &\quad -\frac{1}{4}(\nabla\phi_1 g(x) R_{11}^{-T} R_{21} R_{11}^{-1} g^T(x) \nabla\phi_1^T \hat{W}_3 m_2^T \hat{W}_2 + \overline{D}_1(x)\hat{W}_3 m_1^T \hat{W}_1)\Big\}\end{aligned} \tag{10.44}$$

and the second actor NN (second player) be tuned as

$$\begin{aligned}\hat{W}_4 &= -a_4\Big\{(F_4\hat{W}_4 - F_3\bar{\sigma}_4^T\hat{W}_2) - \frac{1}{4}(\nabla\phi_2 k(x) R_{22}^{-1} R_{12} R_{22}^{-1} g^T(x)\\ &\quad \times \nabla\phi_2^T \hat{W}_4 m_1^T \hat{W}_1 + \overline{D}_2(x)\hat{W}_4 m_2^T \hat{W}_2)\Big\}\end{aligned} \tag{10.45}$$

where

$$\overline{D}_1(x) \equiv \nabla\phi_1(x)g(x)R_{11}^{-1}g^T(x)\ \nabla\phi_1^T(x), \quad \overline{D}_2(x) \equiv \nabla\phi_2(x)kR_{22}^{-1}k^T\nabla\phi_2^T(x),$$

$m_1 \equiv \sigma_3/(\sigma_3^T\sigma_3 + 1)^2, m_2 \equiv \sigma_4/(\sigma_4^T\sigma_4 + 1)^2$ and $F_1 > 0, F_2 > 0, F_3 > 0, F_4 > 0$ are tuning parameters. Let Assumption 10.2 hold, and also assume $Q_1(x) > 0$ and $Q_2(x) > 0$. Suppose that $\overline{\sigma}_3 = \sigma_3/(\sigma_3^T\sigma_3 + 1)$ and $\overline{\sigma}_4 = \sigma_4/(\sigma_4^T\sigma_4 + 1)$ are persistently exciting. Let the tuning parameters be selected as detailed in the proof. Then there exists a K_0 such that, for the number of hidden-layer units $K > K_0$ the closed-loop system state, the critic NN errors $\tilde{W}_1$ and $\tilde{W}_2$, the first actor NN error $\tilde{W}_3$ and the second actor NN error $\tilde{W}_4$ are UUB. ■

Proof: See Appendix A. ■

Remark 10.1. The tuning parameters F_1, F_2, F_3, F_4 in (10.44), and (10.45) must be selected to make the matrix M in (A.68) positive definite (more discussion is presented in the appendix).

Remark 10.2. NN usage suggests starting the tuning algorithm online with the initial control NN weights in (10.39) and (10.40) randomly selected and non-zero.

Remark 10.3. The assumption $Q_1(x) > 0$ and $Q_2(x) > 0$ is sufficient but not necessary.

Remark 10.4. Standard neuroadaptive controllers for single players require only one NN, namely the control action NN (Lewis *et al.*, 1999). However, *optimal* neuroadaptive controllers based on reinforcement learning (e.g. through PI Algorithm) require also a critic NN that identifies the value function for each player. In the optimal control (single player) case this requires two NN, as described in Chapter 7. That is, the price paid for an adaptive controller that converges to an *optimal control* solution is a doubling in the number of NN, which approximately doubles the computational complexity.

Theorem 10.3. Nash solution of the game. Suppose the hypotheses of Theorem 10.2 hold. Then:

a. $H_1(x, \hat{W}_1, \hat{u}_1, \hat{d}_2)$ and $H_2(x, \hat{W}_2, \hat{u}_1, \hat{d}_2)$ are UUB, where

$$\hat{u}_1 = -\frac{1}{2}R_{11}^{-1}g^T(x)\nabla\phi_1^T(x)\hat{W}_1 \tag{10.46}$$

$$\hat{d}_2 = -\frac{1}{2}R_{22}^{-1}k^T(x)\nabla\phi_2^T(x)\hat{W}_2 \tag{10.47}$$

   That is, $\hat{W}_1$ and $\hat{W}_2$ converge to the approximate coupled HJ solution.

b. $u_3(x)$, $d_4(x)$ (see (10.39) and (10.40)) converge to the approximate Nash solution of the game. ■

Proof: Consider the weights $\hat{W}_1, \hat{W}_2, \hat{W}_3$ and $\hat{W}_4$ to be UUB as proved in Theorem 10.2.

a. The approximate coupled HJ equations are

$$
\begin{aligned}
H_1(x, \hat{W}_1, \hat{u}_1, \hat{d}_2) &\equiv H_1(x, \hat{W}_1, \hat{W}_2) \\
&= Q_1(x) - \frac{1}{4}\hat{W}_1^T \nabla\phi_1(x) g(x) R_{11}^{-1} g^T(x) \nabla\phi_1^T(x) \hat{W}_1 \\
&\quad + \hat{W}_1^T \nabla\phi_1(x) f(x) \\
&\quad - \frac{1}{2}\hat{W}_1{}^T \nabla\phi_1(x) k(x) R_{22}^{-1} k^T(x) \nabla\phi_2^T(x) \hat{W}_2 \\
&\quad + \frac{1}{4}\hat{W}_2^T \nabla\phi_2(x) k(x) R_{22}^{-T} R_{12} R_{22}^{-1} k^T(x) \nabla\phi_2^T(x) \hat{W}_2 - \varepsilon_{HJ_1}
\end{aligned}
$$

and

$$
\begin{aligned}
H_2(x, \hat{W}_2, \hat{u}_1, \hat{d}_2) &\equiv H_2(x, \hat{W}_1, \hat{W}_2) \\
&= Q_2(x) - \frac{1}{2}\hat{W}_2^T \nabla\phi_2(x) g(x) R_{11}^{-1} g^T(x) \nabla\phi_1^T(x) \hat{W}_1 \\
&\quad + \frac{1}{4}\hat{W}_1^T \nabla\phi_1(x) g(x) R_{11}^{-T} R_{21} R_{11}^{-1} g^T(x) \nabla\phi_1^T(x) \hat{W}_1 - \varepsilon_{HJ_2} \\
&\quad + \frac{1}{4}\hat{W}_2^T \nabla\phi_2(x) k(x) R_{22}^{-1} k^T(x) \nabla\phi_2^T(x) \hat{W}_2 + \hat{W}_2^T \nabla\phi_2(x) f(x)
\end{aligned}
$$

After adding zero we have

$$
\begin{aligned}
H_1(x, \hat{W}_1, \hat{W}_2) &= \tilde{W}_1^T \nabla\phi_1(x) f(x) + \frac{1}{4}\tilde{W}_1^T \overline{D}_1 \tilde{W}_1 - \frac{1}{2} W_1^T \overline{D}_1 \tilde{W}_1 \\
&\quad + \frac{1}{4}\hat{W}_2^T \nabla\phi_2(x) k(x) R_{22}^{-T} R_{12} R_{22}^{-1} k^T(x) \nabla\phi_2^T(x) \hat{W}_2 \\
&\quad + \frac{1}{4} W_2^T \nabla\phi_2(x) k(x) R_{22}^{-1} R_{12} R_{22}^{-1} k^T(x) \nabla\phi_2^T(x) W_2 \\
&\quad + \frac{1}{2}\hat{W}_1^T \nabla\phi_1(x) k(x) R_{22}^{-1} k^T(x) \nabla\phi_2^T(x) \hat{W}_2 \\
&\quad - \frac{1}{2} W_1^T \nabla\phi_1(x) k(x) R_{22}^{-1} k^T(x) \nabla\phi_2^T(x) W_2 - \varepsilon_{HJ_1} \qquad (10.48)
\end{aligned}
$$

and

$$
\begin{aligned}
H_2(x,\hat{W}_1,\hat{W}_2) = {} & \tilde{W}_2^T \nabla\phi_2(x)f(x) + \frac{1}{2}\hat{W}_2^T \nabla\phi_2(x)g(x)R_{11}^{-1}g^T(x)\nabla\phi_1^T(x)\hat{W}_1 \\
& -\frac{1}{2}W_2^T\nabla\phi_2(x)g(x)R_{11}^{-T}g^T(x)\nabla\phi_1^T(x)W_1 \\
& -\frac{1}{4}\hat{W}_1^T\nabla\phi_1(x)g(x)R_{11}^{-1}R_{21}R_{11}^{-1}g^T(x)\nabla\phi_1^T(x)\hat{W}_1 \\
& +\frac{1}{2}W_1^T\nabla\phi_1(x)g(x)R_{11}^{-1}R_{21}R_{11}^{-1}g^T(x)\nabla\phi_1^T(x)W_1 \\
& +\frac{1}{4}\tilde{W}_2^T\nabla\phi_2(x)k(x)R_{22}^{-1}k^T(x)\nabla\phi_2^T(x)\tilde{W}_2 \\
& -\frac{1}{2}W_2^T\nabla\phi_2(x)k(x)R_{22}^{-1}k^T(x)\nabla\phi_2^T(x)\tilde{W}_2 - \varepsilon_{HJ_2}
\end{aligned} \tag{10.49}
$$

But

$$
\hat{W}_1 = -\tilde{W}_1 + W_1 \text{ and } \hat{W}_2 = -\tilde{W}_2 + W_2 \tag{10.50}
$$

After taking norms in (10.50) and letting $||W_1|| < W_{1\max}$ and $||W_2|| < W_{2\max}$ one has

$$
\|\hat{W}_1\| = \|-\tilde{W}_1 + W_1\| \le \|\tilde{W}_1\| + \|W_1\| \le \|\tilde{W}_1\| + W_{1\max} \tag{10.51}
$$

and

$$
\|\hat{W}_2\| = \|-W_2 + W_2\| \le \|\tilde{W}_2\| + \|W_2\| \le \|W_2\| + W_{2\max} \tag{10.52}
$$

Now (10.48) becomes, by taking into account (10.51), (10.52), Asumption 10.2, and

$$
\begin{aligned}
\sup_{x\in\Omega}||\varepsilon_{HJ_1}|| < \overline{\varepsilon}_1 ||H_1(x,\hat{W}_1,\hat{W}_2)|| \le {} & b_{\phi_{1x}}b_f||x||\|\tilde{W}_1\| + \frac{1}{4}\|\tilde{W}_1\|^2\|\overline{D}_1\| \\
& +\frac{1}{4}\left(||\tilde{W}_2|| + W_{2\max}\right)^2\|\nabla\phi_2(x)k(x)R_{22}^{-T}R_{12}R_{22}^{-1}k^T(x)\nabla\phi_2^T(x)\| \\
& +\frac{1}{4}W_{2\max}^2\|\nabla\phi_2(x)k(x)R_{22}^{-T}R_{12}R_{22}^{-1}k^T(x)\nabla\phi_2^T(x)\| \\
& +\frac{1}{2}\left(\|\tilde{W}_1\| + W_{1\max}\right)\left(\|\tilde{W}_2\| + W_{2\max}\right)\|\nabla\phi_1(x)k(x)R_{22}^{-1}k^T(x) \\
& \times \nabla\phi_2^T(x)\| + \frac{1}{2}W_{1\max}\|\tilde{W}_1\|\|\overline{D}_1\| \\
& +\frac{1}{2}W_{1\max}W_{2\max}\|\nabla\phi_1(x)k(x)R_{22}^{-1}k^T(x)\nabla\phi_2^T(x)\| + \overline{\varepsilon}_1
\end{aligned} \tag{10.53}
$$

and same for (10.49) with $\sup_{x\in\Omega}\|\varepsilon_{HJ_2}\| < \bar{\varepsilon}_2$

$$\begin{aligned}
||H_2(x,\hat{W}_1,\hat{W}_2)|| \le\ & b_{\phi_{1x}} b_f ||x|| ||\tilde{W}_2|| \\
& + \frac{1}{2}\left(||\tilde{W}_2|| + W_{2\max}\right)\left(||\tilde{W}_1|| + W_{1\max}\right) \\
& \times ||\nabla\phi_2(x)g(x)R_{11}^{-1}g^T(x)\nabla\phi_1^T(x)|| \\
& + \frac{1}{2}W_{2\max}W_{1\max}||\nabla\phi_2(x)g(x)R_{11}^{-1}g^T(x)\nabla\phi_1^T(x)|| \\
& + \frac{1}{4}\left(||\tilde{W}_1|| + W_{1\max}\right)^2 ||\nabla\phi_1(x)g(x)R_{11}^{-T}R_{21}R_{11}^{-1}g^T(x) \\
& \times \nabla\phi_1^T(x)|| \\
& + \frac{1}{4}W_{1\max}{}^2||\nabla\phi_1(x)g(x)R_{11}^{-T}R_{21}R_{11}^{-1}g^T(x)\nabla\phi_1^T(x)|| \\
& + \frac{1}{4}||\tilde{W}_2||^2||\nabla\phi_2(x)k(x)R_{22}^{-1}k^T(x)\nabla\phi_2^T(x)|| \\
& + \frac{1}{2}\tilde{W}_{2\max}||\tilde{W}_2||||\nabla\phi_2(x)k(x)R_{22}^{-1}k^T(x)\nabla\phi_2^T(x)|| + \bar{\varepsilon}_2
\end{aligned} \tag{10.54}$$

All the signals on the right-hand side of (10.53) and (10.54) are UUB. So $H_1(x,\hat{W}_1,\hat{u}_1,\hat{d}_2)$ and $H_2(x,\hat{W}_2,\hat{u}_1,\hat{d}_2)$ are UUB and convergence to the approximate coupled HJ solution is obtained.

b. According to Theorem 10.1, 10.2 and equations (10.39), (10.40) $||u_3 - u_1||$ and $||d_4 - d_2||$ are UUB because $||\hat{W}_3 - W_1||$ and $||\hat{W}_4 - W_2||$ are UUB. Therefore, the pair $u_3(x)$, $d_4(x)$ gives the approximate Nash equilibrium solution of the game.

This completes the proof. ■

Remark 10.5. The theorems show that persistence of excitation is needed for proper identification of the value functions by the critic NNs, and that nonstandard tuning algorithms are required for the actor NNs to guarantee stability.

Remark 10.6. Theorems 10.2 and 10.3 show UUB of key quantities. According to the definition of vector D in (A.69) and to (10.53), (10.54), the error bounds depend on the NN approximation errors, Bellman equation and HJ equation residuals, and the bounds $W_{1\max}$, $W_{2\max}$ on the unknown NN weights. By the Weierstrass approximation theorem and Proposition 10.1, all of these go to zero uniformly as the number of NN hidden layers increases except for $W_{1\max}$, $W_{2\max}$. The question of obtaining improved errors bounds in neuroadaptive control has a long and studied literature. Methods have been developed for removing $W_{1\max}$, $W_{2\max}$ from the UUB error bounds. See for instance the work of Ge *et al.* (e.g. Ge and Wang, 2004). Those results could be used to extend the tuning algorithms provided here, but do not constitute a part of the novel results presented herein.

The bounds in the theorems are, in fact, conservative. The simulation results show that the values and Nash solutions are very closely identified.

10.4 Simulations

Here, we present simulations of non-linear and linear systems to show that the game can be solved online by learning in real time, using the adaptive control algorithm of Theorem 10.2, which has two adaptive structures for each player, one to learn the Valkue function and one to learn the Nash control policy. PE is needed to guarantee convergence to the Nash solution. In these simulations, probing noise is added to the two control inputs to ensure PE until convergence is obtained. Then, the probing noise is turned off.

10.4.1 Non-linear system

Consider the following affine in control input non-linear system, with a quadratic cost:

$$\dot{x} = f(x) + g(x)u + k(x)d, x \in \mathbb{R}^2$$

where

$$f(x) = \begin{bmatrix} x_2 \\ -x_2 - \frac{1}{2}x_1 + \frac{1}{4}x_2(\cos(2x_1) + 2)^2 + \frac{1}{4}x_2(\sin(4x_1^2) + 2)^2 \end{bmatrix}$$

$$g(x) = \begin{bmatrix} 0 \\ \cos(2x_1) + 2 \end{bmatrix}, \; k(x) = \begin{bmatrix} 0 \\ \sin(4x_1^2) + 2 \end{bmatrix}$$

Select $Q_1 = 2Q_2$, $R_{11} = 2R_{22}$ and $R_{12} = 2R_{21}$, where Q_2, R_{22} and R_{21} are identity matrices.

This example was constructed as in Nevistic and Primbs (1996). The optimal value function for the first player is

$$V_1^*(x) = \frac{1}{2}x_1^2 + x_2^2$$

and for the second player is

$$V_2^*(x) = \frac{1}{4}x_1^2 + \frac{1}{2}x_2^2$$

The optimal control signal for the first player is $u^*(x) = -2(\cos(2x_1) + 2)x_2$ and the optimal control signal for the second player is

$$d^*(x) = -(\sin(4x_1^2) + 2)x_2$$

One selects the NN vector activation function for the critics as $\phi_1(x) = \phi_2(x) \equiv [x_1^2 \; x_1x_2 \; x_2^2]$ and uses the adaptive control algorithm presented in Theorem 10.2. Each player maintains two NN, a critic NN to estimate its current value and an action NN to estimate its current control policy. Select $a_1 = a_2 = a_3 = a_4 = 1$ and $F_4 = F_3 = F_2 = F_1 = 100I$ for the constants in the tuning laws in Theorem 10.2, where I is an identity matrix of appropriate dimensions. Several simulations were run to decide on these suitable values of these parameters.

Figure 10.1 shows the critic parameters for the first player, denoted by $\hat{W}_1 = [W_{c1} \quad W_{c2} \quad W_{c3}]^T$ by using the proposed game algorithm. After convergence at about 150 s one has $\hat{W}_1(t_f) = [0.5015 \quad 0.0007 \quad 1.0001]^T$ Figure 10.2 shows the

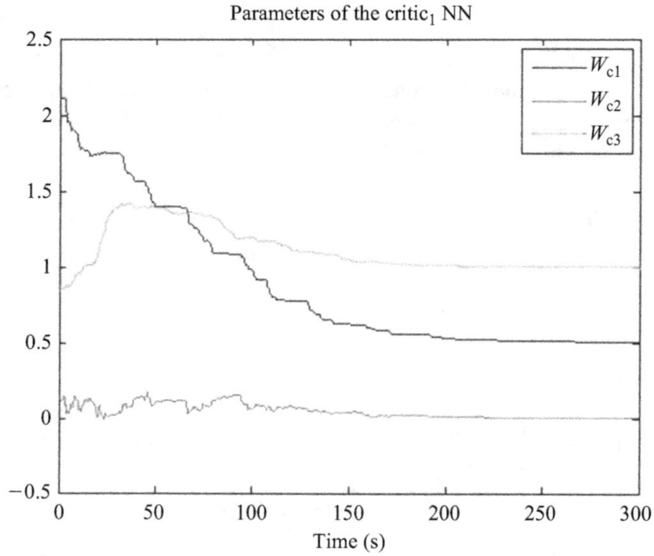

Figure 10.1 Convergence of the critic parameters for the first player

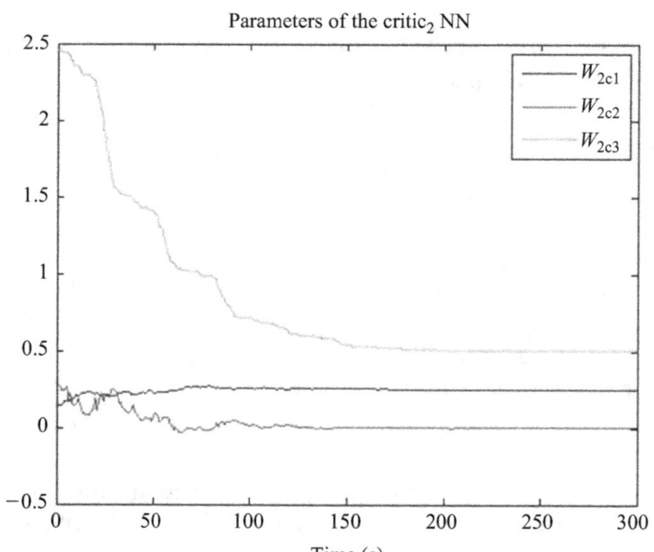

Figure 10.2 Convergence of the critic parameters for the second player

critic parameters for the second player $\hat{W}_2 = [W_{2c1} \quad W_{2c2} \quad W_{2c3}]^T$. After convergence at about 150 s one has $\hat{W}_2(t_f) = [0.2514 \quad 0.0006 \quad 0.5001]^T$. These are all close to the optimal value function parameters.

The actor parameters for the first player after 150 s converge to the values of $\hat{W}_1(t_f) = [0.5015 \quad 0.0007 \quad 1.0001]^T$, and the actor parameters for the second player converge to the values of $\hat{W}_4(t_f) = \left[0.2514 \quad 0.0006 \quad 0.5001\right]^T$. Therefore, the actor NN for the first player

$$\hat{u}(x) = -\frac{1}{2}R11 - 1\begin{bmatrix} 0 \\ \cos(2x_1) + 2 \end{bmatrix}^T \begin{bmatrix} 2x_1 & 0 \\ x_2 & x_1 \\ 0 & 2x_2 \end{bmatrix}^T \begin{bmatrix} 0.5015 \\ 0.0007 \\ 1.0001 \end{bmatrix}$$

also converged to the optimal control, and similarly for the second player

$$\hat{d}(x) = -\frac{1}{2}R_{22}^{-1}\begin{bmatrix} 0 \\ \sin(4x_1^2) + 2 \end{bmatrix}^T \begin{bmatrix} 2x_1 & 0 \\ x_2 & x_1 \\ 0 & 2x_2 \end{bmatrix}^T \begin{bmatrix} 0.2514 \\ 0.0006 \\ 0.5001 \end{bmatrix}$$

The evolution of the system states is presented in Figure 10.3. After the probing noise is turned off, the states converge to zero. Figure 10.4 shows the 3D plot of the difference between the approximated value function for player 1 and the optimal one. The results are similar for the second player. These errors are close to zero. Good approximations of the actual value functions are evolved. Figure 10.5 shows the 3D plot of the difference between the approximated control for the first player, by using the online algorithm, and the optimal one. This error is close to zero. Similarly for the second player.

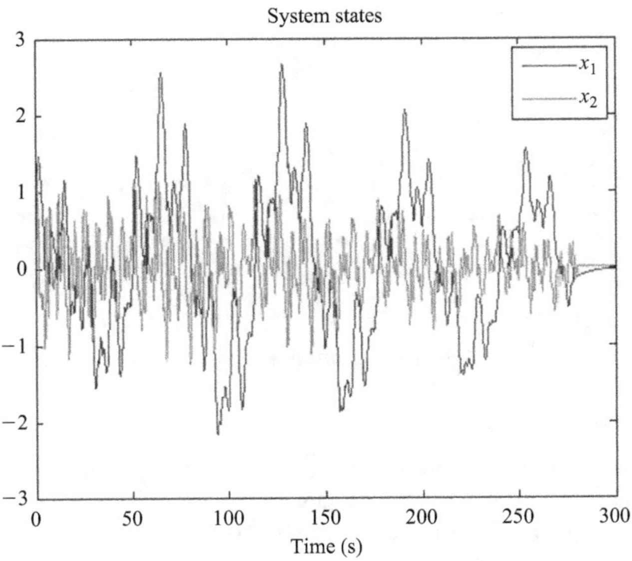

Figure 10.3 Evolution of the system states

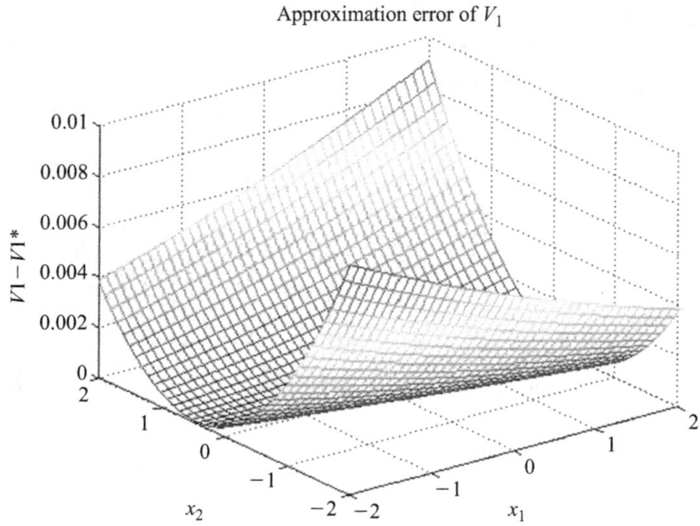

Figure 10.4 3D plot of the approximation error for the value function of player 1

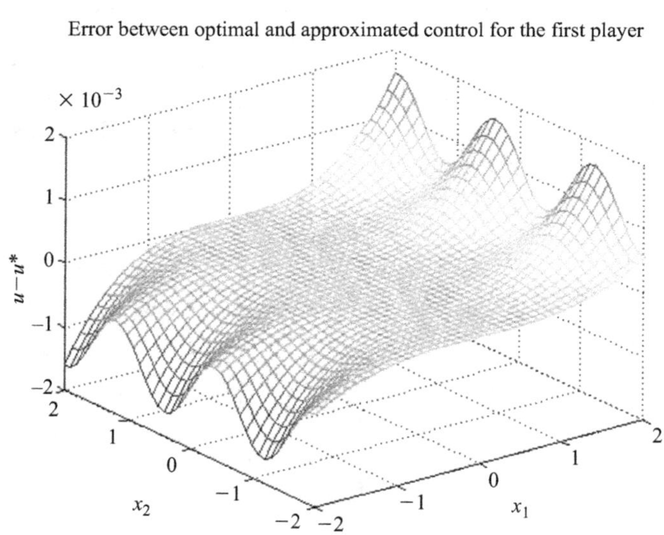

Figure 10.5 3D plot of the approximation error for the control of player 1

10.4.2 Linear system

The aim of this example is to illustrate the online algorithm with a simple example that was simulated in previous work in Jungers *et al.* (2007) and Limebeer *et al.* (1994) to solve the coupled Riccati equations (10.11).

Suppose

$$\dot{x}(t) = 2x(t) + u_1(t) + 3u_2(t)$$

with

$$J_1 = \int_0^\infty (-9x^2 + 0.64u_1^2 - u_2^2)\, dt$$

and

$$J_2 = \int_0^\infty (9x^2 + u_1^2)\, dt$$

According to Limebeer *et al.* (1994) the solution of the coupled Riccati is given by $P_1 = -1.4145$ and $P_2 = 1.5718$.

Select $a_1 = a_2 = a_3 = a_4 = 1$, $F_4 = F_3 = F_2 = F_1 = 5$ for the constants in the tuning laws. In this example, exponentially decaying probing noise was added to the control input of each player to guarantee PE. By using the optimal adaptive control algorithm proposed in Theorem 10.2 the solution of the coupled Riccati equations is found online to be $P_1 = -1.4250$ and $P_2 = 1.5754$. Figure 10.6 shows the evolution of the system state, which reaches zero as the probing noise decays. Figures 10.7 and 10.8 show the convergence of the critic neural networks for each player to $P_1 = -1.4250$ and $P_2 = 1.5754$.

10.4.3 Zero-sum game with unstable linear system

Consider the unstable linear system

$$\dot{x} = \begin{bmatrix} 0 & 0.25 \\ 1 & 0 \end{bmatrix} x + \begin{bmatrix} 1 \\ 0 \end{bmatrix} u + \begin{bmatrix} 1 \\ 0 \end{bmatrix} d = Ax + Bu + Dd$$

with cost

$$J_1 = \int_0^\infty \left(x^T \begin{bmatrix} 1 & 0 \\ 0 & 1 \end{bmatrix} x + u^T u - 25 d^T d \right) dt$$

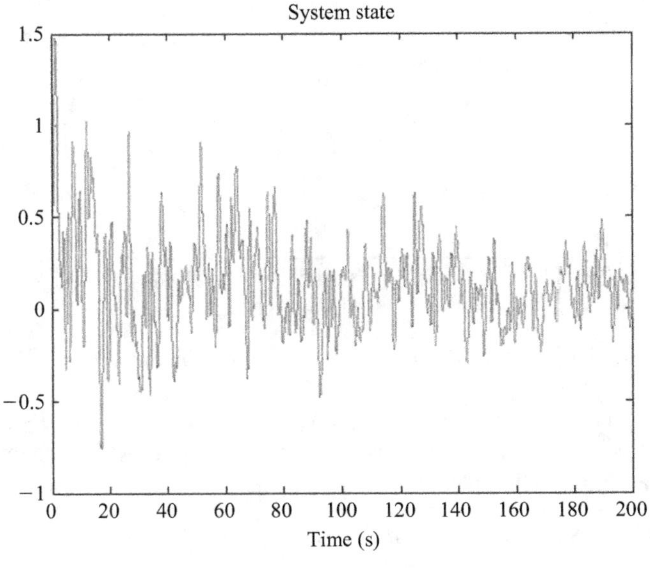

Figure 10.6 Evolution of the system state

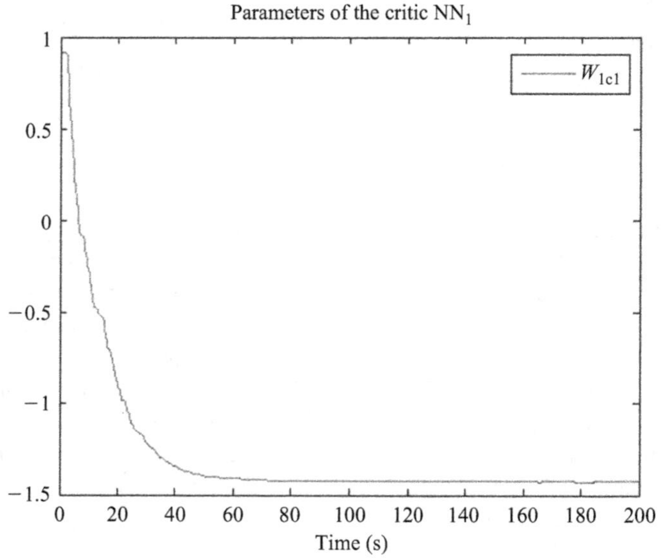

Figure 10.7 Convergence of the critic NN for the first player

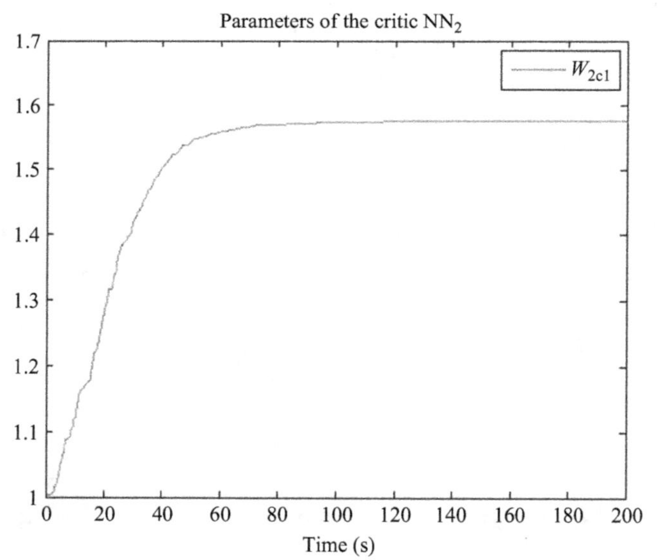

Figure 10.8 Convergence of the critic NN for the second player

and $J_2 = -J_1$. That is, this is a zero-sum game with $Q_1 = -Q_2$, $R_{11} = -R_{21}$, $R_{22} = -R_{12}$.

In this linear case the solution is given by the solution of the game algebraic Riccati equation (GARE) (Chapter 9)

$$0 = A^T S + SA + H^T H - SBR^{-1}B^T S + \frac{1}{\gamma^2} SDD^T S$$

with

$$\gamma \equiv \sqrt{R_{22}} = 5, \quad R \equiv R_{11} = 1 \quad \text{and} \quad Q \equiv Q_1 = \begin{bmatrix} 1 & 0 \\ 0 & 1 \end{bmatrix}.$$

Solving the GARE gives the parameters of the optimal critic as $W^* = [1.9439 \quad 1.3137 \quad 1.9656]^T$ that are the components of the Riccati solution matrix S.

Since the value is quadratic in the LQR game case, the critic NN basis sets $\phi_1(x) = \phi_2(x)$ were selected as the quadratic vector in the state components $x \otimes x$ with $\otimes$ the Kronecker product. Redundant terms were removed to leave $n(n+1)/2 = 3$ components.

The two-player game algorithm is implemented as in Theorem 10.2. Select $a_1 = a_2 = a_3 = a_4 = 1, F_4 = F_3 = F_2 = F_1 = 5I$ for the constants in the tuning laws, where I is an identity matrix of appropriate dimensions. PE was ensured by adding a small exponentially decreasing probing noise to the two control inputs. Figure 10.9 shows the critic parameters for the first player, denoted by $\hat{W}_1 = [W_{c1} \quad W_{c2} \quad W_{c3}]^T$ and Figure 10.10 the critic parameters for the second player, $\hat{W}_2 = [W_{2c1} \quad W_{2c2} \quad W_{2c3}]^T$. It is clear that $\hat{W}_1 = -\hat{W}_2$. In fact, after 400 s the critic parameters for the first player converged to $\hat{W}_1(t_f) = [1.9417 \quad 1.3138 \quad 1.9591]^T$ and the second to $\hat{W}_2(t_f) = [-1.9417 \quad -1.3138 \quad -1.9591]^T$. These are the optimal value function parameters.

Finally, Figure 10.11 shows the evolution of the states. They converge to zero as the probing noise decays to zero.

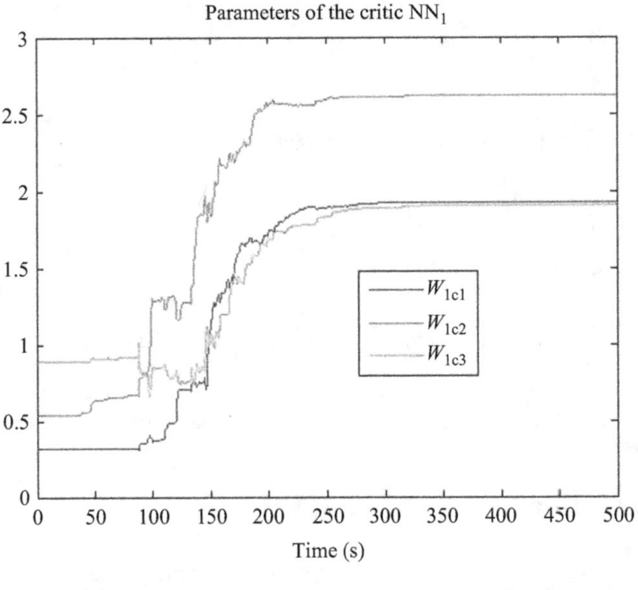

Figure 10.9 Convergence of the critic NN for the first player

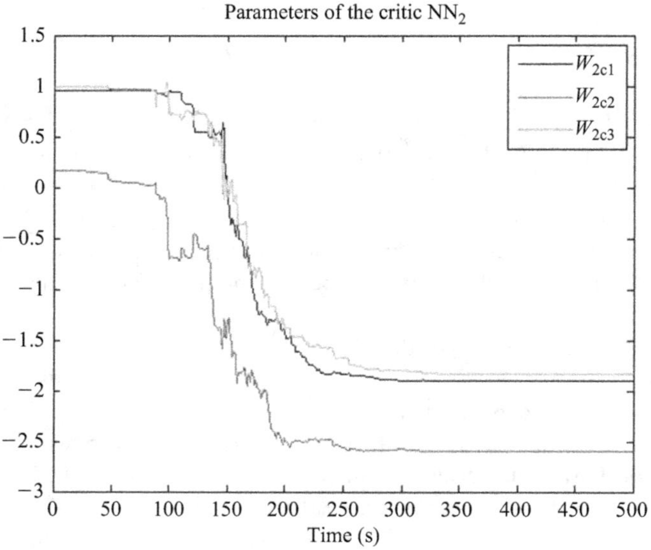

Figure 10.10 Convergence of the critic NN for the second player

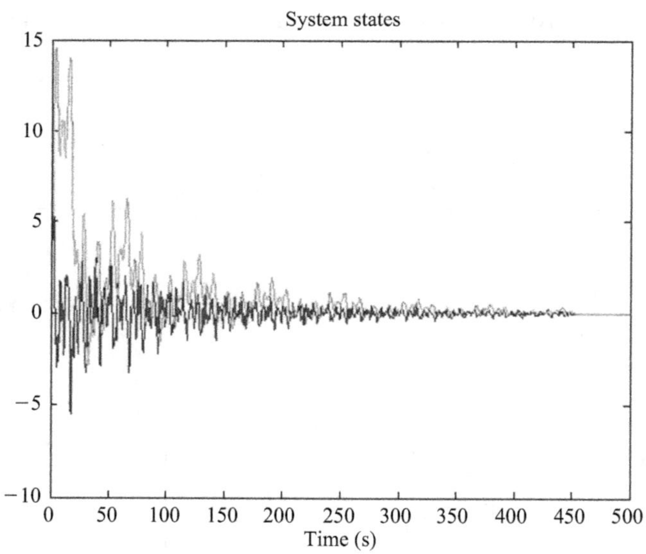

Figure 10.11 Evolution of the system states

10.5 Conclusion

In this chapter we presented multiplayer non–zero-sum games. The Nash equilibrium solution is found by solving a set of coupled non-linear Hamilton–Jacobi equations. A reinforcement learning policy iteration algorithm was given to solve these nonlinear HJ equations based on repeated solution of linear coupled Bellman equations. This PI algorithm was used to develop a novel adaptive control architecture that solves NZS games online in real time using data measured along the system trajectories. This adaptive controller has two approximation structures for each player, one to estimate its current value function and one to estimate its Nash control policy. Parameter tuning algorithms were given that learn the NZS game solution online while at the same time guaranteeing system stability. Simulation examples showed that the NZS adaptive controller converges online to the Nash solutions of the NZS game in real time by measuring data along the system trajectories.

Chapter 11
Integral reinforcement learning for zero-sum two-player games

In this chapter we present a continuous-time adaptive dynamic programming (ADP) procedure that uses the idea of integral reinforcement learning (IRL) to find online the Nash-equilibrium solution for the two-player zero-sum (ZS) differential game. We consider continuous-time (CT) linear dynamics of the form $\dot{x} = Ax + B_1 w + B_2 u$, where $u(t)$, $w(t)$ are the control actions of the two players, and an infinite-horizon quadratic cost. This work is from Vrabie and Lewis (2010).

Most continuous-time reinforcement learning algorithms provide an online heuristic approach to the solution of optimal control problems under the assumption that the controlled system is not affected by disturbances. However, there are numerous control applications in which the presence of disturbance signals is certain, and these have a negative effect on the performance of the control system. In these cases, the optimal control problem is formulated with the purpose of finding all admissible controllers that minimize the H-infinity norm. These methods are known as H-infinity controllers (Van Der Schaft, 1992; Başar and Bernard, 1995; Zames, 1981). Such control policies counteract, in an optimal sense, the effects of the worst case disturbance that might affect the system.

The background on ZS games and their applications to H-infinity control are given in Section 9.1. It is known that the solution of the H-infinity problem is the saddle point solution of a two-player zero-sum differential game. For the case when the system has linear dynamics and the cost index is quadratic and has infinite horizon, it is also known that finding the Nash-equilibrium solution to the game problem depends on calculating a solution of a generalized Riccati equation with sign indefinite quadratic term (Başar and Olsder, 1999; Başar and Bernard, 1995; Doyle *et al.*, 1989; Stoorvogel, 1992; Zhou and Khargonekar, 1988); that is the game algebraic Riccati equation (GARE)

$$0 = A^T P + PA + C^T C - P\left(B_2 B_2^T - \frac{1}{\gamma^{*2}} DD^T\right)P \tag{11.1}$$

Suboptimal H-infinity controllers can be determined such that the H-infinity norm is less than a given prescribed bound that is larger than the minimum H-infinity norm (Başar and Olsder, 1999; Van Der Schaft, 1992; Başar and Bernard, 1995). This means that for any $\gamma > \gamma^*$ one can find a suboptimal H-infinity state-feedback controller, which admits a performance level of at least γ, by solving (11.1), where γ^* is replaced by γ. To simplify the mathematical notation, for a given γ, in the following we will denote $B_1 = \gamma^{-1}D$. Thus, the GARE of our concern will be written as

$$0 = A^T P + PA + C^T C - P(B_2 B_2^T - B_1 B_1^T)P \tag{11.2}$$

The solution of the GARE has been approached in Başar and Olsder (1999), Başar and Bernard (1995), Damm (2004), while for its non-linear generalization, that is the Hamilton–Jacobi–Isaacs (HJI) equation (Section 9.1), iterative solutions have been presented in Van Der Schaft (1992), Abu-Khalaf *et al.* (2006), Abu-Khalaf and Lewis (2008). In all cases the solution is determined in an iterative manner by means of a Newton-type of algorithm. These algorithms construct sequences of cost functions that are monotonically convergent to the solution of interest. In all cases exact knowledge of the system dynamics (A, B_1, B_2) is required and the solution is obtained by means of offline computation.

Adaptive dynamic programming (ADP) is a class of methods that provide online solutions to optimal control problems by making use of measured information from the system and using computation in a forward-in-time fashion, as opposed to the backward in time procedure that is characterizing the classical dynamic programming approach (Lewis *et al.*, 2012). These methods were initially developed for systems with finite state and action spaces and are based on Sutton's temporal difference learning (Sutton, 1988), Werbos' Heuristic Dynamic Programming (HDP) (Werbos, 1991, 1992, 2009) and Watkins's Q-learning (Watkins, 1989).

To our knowledge, there exists only one ADP procedure that provides solutions to the HJI equation (Wei and Zhang, 2008). This algorithm involves calculation of two sequences of cost functions, the upper and lower performance indices, which converge to the saddle point solution of the game. The adaptive critic structure required for learning the saddle point solution comprised of four action networks and two critic networks. The requirement for full knowledge of the system dynamics (A, B_1, B_2) is still present.

The result presented in this chapter is a reinforcement learning approach to the saddle point solution of a two player zero-sum differential game and comes from Vrabie and Lewis (2010). We use the idea of integral reinforcement learning (IRL) introduced in Chapter 3, which allows calculation of the value function associated with the pair of behavior policies of the two players. By virtue of this online ADP method, exact knowledge of part of the system dynamics is not required. Specifically, the system drift matrix A need not be known.

The online ADP algorithm presented in this chapter to solve the two-player differential zero-sum game is built on the mathematical result given in Lanzon *et al.* (2008). This algorithm involves solving a sequence of Riccati equations in order to construct a monotonically increasing sequence of matrices that converges to the equilibrium solution of the game. Every matrix in this sequence is determined by solving a Riccati equation with sign definite quadratic term of the sort associated with the optimal control problem (Lewis *et al.*, 2012). In this chapter, we use ADP techniques and the idea of IRL to find the solution of these optimal control problems in an online fashion and using reduced information on the system dynamics.

While working in the framework of control applications we will refer to the two players $u(t)$, $w(t)$ in the dynamics $\dot{x} = Ax + B_1 w + B_2 u$, as the controller and disturbance, respectively, in contrast to the regular nomenclature used in game theory of 'pursuer' and 'evader'. We are assuming that both players have perfect instantaneous state information. A player's policy is a function that allows computation of the player's control signal based on the state information. Both players compete and learn in real time using data measured along the system trajectories. The two policies that characterize the Nash-equilibrium solution of the game are thus determined based on online measured data from the system.

In the ADP approach described in this chapter only one of the two players is actively learning and improving its policy. The algorithm is built on interplay between a *learning* phase, performed by the *controller* that is learning in order to optimize its behavior, and a policy *update* step, performed by the *disturbance* that is gradually increasing its detrimental effect. The controller learns online in order to maximize its performance while the policy of the disturbance remains constant. The disturbance player then updates its action policy only after the controller has learned its optimal behavior in response to the present policy of his opponent. The update of the disturbance policy uses the information about the policy of his opponent, and makes way for further improvement for the controller policy. For learning the control policy in this chapter we use the online continuous-time HDP or value iteration procedure developed in Chapter 6.

We begin our investigation by providing in Section 11.1 the formulation of the two-player zero-sum game problem and reviewing the iterative mathematical algorithm from Lanzon *et al.* (2008) that solves its underlying GARE. Next, we briefly review the online HDP (value iteration) algorithm based on IRL from Chapter 6 that provides the solution to the single-player optimal control problem. Section 11.2 presents the online ADP algorithm for solution of the two-player zero-sum game in real time. It is shown that this algorithm has a novel adaptive critic structure. Section 11.3 gives simulation results obtained for the load-frequency control of a power system. It is shown that the proposed ADP algorithm solves the GARE online without knowing the system A matrix, thus finding in real time the best control policy to counteract the worst case load disturbance.

11.1 Zero-sum games for linear systems

11.1.1 Background

Consider the linear time-invariant system described by the equation

$$\dot{x} = Ax + B_1 w + B_2 u \tag{11.3}$$

where $x \in \mathbb{R}^n$, $u \in \mathbb{R}^m$, $w \in \mathbb{R}^q$. The control actions of the control player and disturbance player are $u(t)$, $w(t)$, respectively.

The controller player, which computes the signal u, desires to minimize the quadratic performance index

$$V(x_0, u, w) = \int_0^\infty (x^T C^T C x + u^T u - w^T w)\, dt \tag{11.4}$$

while the disturbance player, which computes the signal w, desires to maximize it. The goal is to determine the saddle point stabilizing solution. Details are given in Section 9.1.

The following assumptions, related to the solvability of the two-player game, are made:

- (C, A) is detectable, that is all unobservable modes of (C, A) are asymptotically stable,
- (A, B_2) is stabilizable, that is all uncontrollable modes of (A, B_2) are asymptotically stable,
- there exists a unique positive definite stabilizing solution of the GARE (11.2).

Denoting by Π the unique positive definite stabilizing solution of (11.2), the saddle point control solution and value of the Nash game are

$$\begin{aligned} u &= -B_2^T \Pi x \\ w &= B_1^T \Pi x \\ V(x_0, u, w) &= x_0^T \Pi x_0 \end{aligned} \tag{11.5}$$

We say the solution Π of the GARE is stabilizing if the closed-loop system $A + B_1 B_1^T \Pi - B_2 B_2^T \Pi$ is asymptotically stable.

We shall use the notations $u = Kx$ and $w = Lx$ for the state-feedback control and disturbance policies, respectively. We say that K is the gain of the control policy and L is the gain of the disturbance policy. The meaning of the saddle point solution of the Nash differential game is that, for any state-feedback control policy $\tilde{u} = \tilde{K}x$ and any state-feedback disturbance policy $\tilde{w} = \tilde{L}x$, different than the Nash policies in (11.5), the value of the game satisfies

$$V(x_0, \tilde{u}, w) \geq V(x_0, u, w) \geq V(x_0, u, \tilde{w}) \tag{11.6}$$

11.1.2 Offline algorithm to solve the game algebraic Riccati equation

Given real matrices A, B_1, B_2, C with compatible dimensions, define the map $F : \mathbb{R}^{n \times n} \to \mathbb{R}^{n \times n}$

$$F(P) = A^T P + PA + C^T C - P(B_2 B_2^T - B_1 B_1^T)P \tag{11.7}$$

The following iterative method for solving the game algebraic Riccati equation (GARE) (11.2) was introduced in Lanzon *et al.* (2008). We use the notation

$$A_u^{i-1} = A + B_1 B_1^T p_u^{i-1} - B_2 B_2^T p_u^{i-1}$$

Algorithm 11.1. Offline iterations on the control policy to solve GARE (Lanzon *et al.*, 2008)

1. *Initialize.*

$$P_u^0 = 0 \tag{11.8}$$

2. *Iterate for i = 1, 2, ...* Solve for the unique positive definite Z_u^i

$$0 = (A_u^{i-1})^T Z_u^i + Z_u^i A_u^{i-1} - Z_u^i B_2 B_2^T Z_u^i + F(P^{-1}) \tag{11.9}$$

3. *Update.*

$$P_u^i = P_u^{i-1} + Z_u^i \tag{11.10}$$

■

This is an offline algorithm for solving the zero-sum (ZS) GARE that requires full knowledge of the system dynamics (A, B_1, B_2). The sub-index 'u' is used in this algorithm to underline the fact that the controller player is the one that will be learning online to find solution of the sequence of Riccati equations (11.9).

The convergence of this algorithm to the unique positive definite solution of the GARE (11.2) was given by the following result.

Theorem 11.1. (Lanzon *et al.*, 2008) Given real matrices A, B_1, B_2, C with compatible dimensions, such that (C, A) is detectable and (A, B_2) is stabilizable, define the map F as in (11.7). Suppose that there exists a stabilizing solution $\Pi > 0$ of (11.2). Then:

i. There exist two square matrix series $P_u^i \in \mathbb{R}^{n\times n}$ and $Z_u^i \in \mathbb{R}^{n\times n}$ for all $i \in \mathbb{N}$ satisfying Algorithm 11.1.
ii. The elements of these two series, defined recursively, have the following properties:
   a. $(A + B_1B_1^TP_u^i, B_2)$ is stabilizable for all $i \in \mathbb{N}$.
   b. $Z_u^i \geq 0, \quad \forall i \in \mathbb{N}$.
   c. $F(P_u^{i+1}) = Z_u^i B_1 B_1^T Z_u^i, \quad \forall i \in \mathbb{N}$.
   d. $A + B_1B_1^TP_u^i - B_2B_2^TP_u^{i+1}$ is Hurwitz $\forall i \in \mathbb{N}$.
   e. $\Pi \geq P_u^{i+1} \geq P_u^i \geq 0, \quad \forall i \in \mathbb{N}$.
iii. Let $P_u^\infty = \lim_{x\to\infty} P_u^i \geq 0$. Then $P_u^\infty = \Pi$. ■

We now introduce two propositions that provide equivalent formulations for Algorithm 11.1. We introduce them here in order to bring meaning to every step of this iterative algorithm. These propositions will provide a compact mathematical formulation for Algorithm 11.1. Using these results we will then show that the iterative algorithm that solves the two-player ZS game involves solving a sequence of single-player games, namely solving a sequence of optimal control problems (Lewis *et al.*, 2012).

Proposition 11.2. The iteration between (11.9) and (11.10) in Algorithm 11.1 can be written as

$$\begin{aligned} &P_u^i(A + B_1B_1^TP_u^{i-1}) + (A + B_1B_1^TP_u^{i-1})^TP_u^i - P_u^iB_2B_2^TP_u^i \\ &\quad - P_u^{i-1}B_1B_1^Tp_u^{i-1} + C^TC = 0 \end{aligned} \tag{11.11}$$

Proof: The result is obtained by writing compactly the two equations and making use of the definition of the map F. Explicitly, (11.11) becomes

$$0 = (A_u^{i-1})^T(P_u^i - P_u^{i-1}) + (P_u^i - P_u^{i-1})A_u^{i-1} - (P_u^i - P_u^{i-1})B_2B_2^T(P_u^i - P_u^{i-1})$$
$$+ A^TP_u^{i-1} + P_u^{i-1}A + C^TC - P_u^{i-1}(B_2B_2^T - B_1B_1^T)P_u^{i-1}$$

Unfolding the parentheses, using the notation $A_u^{i-1} = A + B_1B_1^TP_u^{i-1} - B_2B_2^TP_u^{i-1}$, and cancelling the equivalent terms we obtain

$$0 = P_u^{i-1}B_1B_1^T(P_u^i - P_u^{i-1}) - P_u^{i-1}B_2B_2^T(P_u^i - P_u^{i-1}) + (P_u^i - P_u^{i-1})B_1B_1^TP_u^{i-1}$$
$$- (P_u^i - P_u^{i-1})B_2B_2^TP_u^{i-1} - P_u^iB_2B_2^TP_u^i + P_u^iB_2B_2^TP_u^{i-1} + P_u^{i-1}B_2B_2^TP_u^i$$
$$- P_u^{i-1}B_2B_2^TP_u^{i-1} + A^TP_u^i + P_u^iA + C^TC - P_u^{i-1}(B_2B_2^T - B_1B_1^T)P_u^{i-1}$$

After more cancelations and rearranging we obtain

$$0 = P_u^{i-1}B_1B_1^TP_u^i + P_u^iB_1B_1^TP_u^{i-1} + A^TP_u^i + P_u^iA - P_u^iB_2B_2^TP_u^i$$
$$- P_u^{i-1}B_1B_1^TP_u^{i-1} + C^TC$$

which is the result (11.11). ■

Proposition 11.3. The iteration between (11.9) and (11.10) in Algorithm 11.1 can be written as

$$0 = (P_u^i - P_u^{i-1})A_u^{i-1} + (A_u^{i-1})^T(P_u^i - P_u^{i-1}) - (P_u^i - P_u^{i-1})B_2B_2^T(P_u^i - P_u^{i-1})$$
$$+ (P_u^{i-1} - P_u^{i-2})B_1B_1^T(P_u^{i-1} - P_u^{i-2}) \tag{11.12}$$

■

This results directly from (11.9), (11.10) and Theorem 11.1.

It is important to notice at this point that the result given in Proposition 11.3 includes three instances of the index of the sequence $\{P_u^i\}_{i\geq 0}$, namely $P_u^{i-2}, P_u^{i-1}, P_u^i$. For this reason it can only be used for calculating the values $\{P_u^i\}_{i\geq 2}$ provided the first two elements in the sequence are available.

The next two propositions formulate optimal control problems associated with the Riccati equations (11.11) and (11.12). This is important since they attach meaning to Algorithm 11.1, enhancing both its reinforcement learning perspective and its game theoretical reasoning.

Proposition 11.4. Solving the Riccati equation (11.11) is equivalent to finding the solution of the following optimal control problem:

'For the system $\dot{x} = A + B_1 w_{i-1} + B_2 u_i$ let the state-feedback disturbance policy gain be $L_u^{i-1} = B_i^T P_u^{i-1}$ such that $w_{i-1} = B_1^T P_u^{i-1} x$. Determine the state-feedback control policy u_i such that the infinite-horizon quadratic cost index $\int_0^\infty [x^T C^T C x - w_{i-1}^T w_{i-1} + u_i^T u_i]\, dt$ is minimized.'

Let $x_0^T P_u^i x_0 = \min_{u_i} \int_0^\infty [x^T (C^T C - P_u^{i-1} B_1 B_1^T P_u^{i-1}) x + u_i^T u_i]\, dt$. Then the optimal state-feedback control policy is $K_u^i = -B_2^T P_u^i$, so that the optimal state-feedback control is $u_i = -B_2^T P_u^i x$.

Proposition 11.5. Solving the Riccati equation (11.12) is equivalent to finding the solution of the following optimal control problem:

'For the system $\dot{x} = A + B_1 w_{i-1} + B_2 (u_{i-1} + \hat{u}_i)$ let the state-feedback disturbance policy be $w_{i-1} = B_1^T P_u^{i-1} x$ and the base state-feedback control policy be $u_{i-1} = -B_2^T P_x^{i-1} x$. Determine the *correction* for the state-feedback control policy, $\hat{u}_i$, which minimizes the infinite-horizon quadratic cost index

$$\hat{J} = \int_0^\infty [x^T (P_u^{i-1} - P_u^{i-2}) B_1 B_1^T (P_u^{i-1} P_u^{i-2}) x + \hat{u}_i^T \hat{u}_i]\, dt$$

Let $x_0^T Z_u^i x_0 = \min_{\hat{u}_i} \hat{J}$. Then the optimal control policy $u_i = u_{i-1} + \hat{u}_i$ is $u_i = -B_2^T (P_u^{i-1} + Z_u^i) x$.

Algorithm 11.1 can be used as the backbone for an online approach that finds the saddle point solution of the ZS differential game in real time using data measure along the system trajectories. In the next section we describe this online algorithm.

Because Algorithm 11.1 reaches solution of the ZS game my means of building a sequence of solutions for standard Riccati equations, in the next subsection we provide a brief review of the continuous-time heuristic dynamic programming (CT-HDP) method from Chapter 6 that finds, in an online fashion, the solution of a Riccati equation with a sign definite quadratic term.

11.1.3 Continuous-time HDP algorithm to solve Riccati equation

The goal of this section is to briefly review the value iteration or HDP algorithm from Chapter 6 that uses reinforcement learning ideas to solve online in real time the standard ARE with sign definite quadratic term

$$A^T P + PA + Q - PBR^{-1} B^T P = 0 \tag{11.13}$$

We take note that the notation used in this subsection is general and not specifically related to the notations used in the previous subsections. The continuous-time value iteration or HDP algorithm introduced in Vrabie *et al.* (2008) solves the optimal control ARE (11.13) in real time using data measured along the system trajectories. This CT-HDP algorithm was presented as Algorithm 6.2 and is reproduced as follows.

Algorithm 11.2. Continuous-time heuristic dynamic programming (CT-HDP) (Vrabie *et al.*, 2008)

Initialize. Let $P_0 = 0$ and let K_0 be a state-feedback control policy (not necessarily stabilizing). *Iterate for* $i = 0, 1, 2, \ldots$ between

$$x_t^T P_{i+1} x_t = \int_t^{t+T_0} x_\tau^T (Q + K_i^T R K_i) x_\tau \, d\tau + x_{t+T_0}^T P_i x_{t+T_0} \tag{11.14}$$

$$K_i + 1 = R^{-1} B^T P_{i+1} \tag{11.15}$$

until convergence, where x_τ denotes the state of the system described by $\dot{x} = (A + BK_i)x$ with initial condition x_t. The value of the state at time $t + T_0$ is denoted by x_{t+T_0}. ■

The online implementation of this algorithm was discussed in Chapter 6 in some detail. It is now briefly reviewed. The solution of (11.14) consists of the value of the matrix P_{i+1} that parameterizes the cost function at iteration step $i + 1$.Write the quadratic cost function at step i as

$$x_t^T P_i x_t = \bar{p}_i^T \bar{x}_t \tag{11.16}$$

where $\bar{x}_t$ denotes the Kronecker product (Brewer, 1978) quadratic polynomial basis vector with elements $\{x_i(t) x_j(t)\}_{i=1,n; j=i,n}$. Vector $\bar{p} = v(P)$ consists of the columns of P stacked on top of one another with redundant lower diagonal terms deleted. Here, the off-diagonal terms are taken as $2P_{jk}$, with P_{jk} the element of matrix P.

The *integral reinforcement* over the time interval $[t, t + T_0]$ is defined as

$$d(\bar{x}_t, K_i) = \int_t^{t+T_0} x^T(\tau)(Q + K_i{}^T R K_i) x(\tau) \, d\tau \tag{11.17}$$

Based on these notations and structures (11.14) is rewritten as

$$\bar{p}_{i+1}{}^T \bar{x}_t = d(\bar{x}_t, K_i) + \bar{p}_i{}^T \bar{x}_{t+T_0}. \tag{11.18}$$

This is a linear equation in which the vector of unknown parameters is $\bar{p}_{i+1}$. This is in the standard form of equations appearing in system identification (Ljung, 1999), and $\bar{x}_t$ acts as a regression vector. The right-hand side of this equation can be viewed as a target reinforcement function. It is given in terms of data measured along the state trajectories over the time interval $[t, t + T_0]$, namely the integral reinforcement over $[t, t + T_0]$ and the state value at $t + T_0, \bar{x}_{t+T_0}$.

The parameter vector $\overline{p}_{i+1}$ is found by finding the least-squares solution to (11.18) for numerous time steps $[t, t+T_0]$. This is standard in system identification, and can be accomplished by using batch least-squares or recursive least-squares algorithms (Ljung, 1999).

The online value iteration Algorithm 11.2 is an online data-based approach that uses reinforcement learning ideas to find the solution of the standard algebraic Riccati equation (11.13). This algorithm does not require explicit knowledge of the system drift dynamics matrix A, which does not appear in Algorithm 11.2.

11.2 Online algorithm to solve the zero-sum differential game

Algorithm 11.1 solves the ZS game, but it is an offline method that relies on iterations on nonstandard Riccati equations with sign indefinite terms. We now plan to formulate Algorithm 11.1 in terms of iterations on standard Riccati equations of the form (11.13) with sign definite quadratic term. This is presented as Algorithm 11.3. At every iteration step, these standard Riccati equations can be solved by means of the online value iteration (CT-HDP) Algorithm 11.2. The end result is the online Algorithm 11.4, which finds the saddle point solution of the differential game online in real time using data measured along the system trajectories. In this algorithm, neither of the two players uses any knowledge about the system drift dynamics A.

In the online algorithm given in this section, the solution of the game is found online while the game is played. We shall see that only one of the players is learning and optimizing his behavior strategy while the other is playing based on fixed policies. We say that *the learning player is leading the learning procedure*. His opponent is a passive player who decides on a fixed policy and maintains it during every learning phase. The passive player changes his behavior policy based only on information regarding his opponent's optimal strategy. It is noted that the passive player does not choose the best strategy that he could play to get a leading advantage in the sense of a Stackelberg game solution. Instead, he plays the strategy that his opponent, that is the learning player, chooses for him, because this will allow the learning process to terminate with the calculation of the desired Nash-equilibrium solution of the game.

That is, in the algorithm presented in this section, the reinforcement learning technique is employed only by the controller, while the passive player is the disturbance.

First, we give the formulation of Algorithm 11.1 as iterations on standard Riccati equations of the form (11.13). Throughout, we make use of the notation $A_u^{i-1} = A + B_1 B_1^T P_u^{i-1} - B_2 B_2^T P_u^{i-1}$.

Algorithm 11.3. Formulation of Algorithm 11.1 in terms of standard AREs

1. *Initialize*. Let $P_u^0 = 0$.
2. *Solve* online the Riccati equation

$$P_u^i A + A^T P_u^1 - P_u^1 B_2 B_2^T P_u^1 + C^T C = 0 \tag{11.19}$$

3. Let $Z_u^1 = P_u^1$.
4. For $i \geq 2$ solve online the Riccati equation

$$Z_u^i A_u^{i-1} + A_u^{i-1T} Z_u^i - Z_u^i B_2 B_2^T Z_u^i + Z_u^{i-1} B_1 B_u^{i-1} Z_u^{i-1} = 0 \tag{11.20}$$

5. $P_u^i = P_u^{i-1} + Z_u^i$. ■

At every step the Riccati equations can be solved online using the data-based approach of Algorithm 11.2. This does not require knowledge of the system drift matrix A.

The online algorithm is given next. It can be viewed as an extension of approximate dynamic programming (Werbos, 1991, 1992, 2009) to solve ZS games online for continuous-time systems.

Algorithm 11.4. Online solution of ZS games using integral reinforcement learning

1. *Initialize*. Let $P_u^0 = 0$.
2. Let $K_u^{0(0)}$ be zero, let $k=0$.
   a. Solve online

$$x_t^T P_u^{0(k+1)} x_t = \int_t^{t+T_0} x_\tau^T (C^T C + K_u^{0(k)T} R K_u^{0(k)}) x_\tau \, d\tau + x_{t+T_0}^T P_u^{0(k)} x_{t+T_0}$$

   b. Update $K_u^{0(k+1)} = -B_2^T P_u^{0(k+1)}$, $k = k + 1$.
   c. Until $\|P_u^{0(k)} - P_u^{0(k-1)}\| < \varepsilon$.
3. $P_u^1 = P_u^{0(k)}, Z_u^1 = P_u^1$.
4. For $i \geq 2$:
   a. Let $K_u^{i-1(0)}$ be zero, let $k=0$. Denote $A_u^{i-1(k)} = A + B_1 B_1^T P_u^{i-1} + B_2 K_u^{i-1(k)}$ and let x_τ be the solution over the interval $[t, t+T_0]$ of $\dot{x} = A_u^{i-1(k)} x$ with initial condition x_t.
   b. Solve online for the value of $Z_u^{i(k+1)}$ using value iteration (CT-HDP)

$$x_t^T Z_u^{i(k+1)} x_t = \int_t^{t+T_0} x_\tau^T (Z_u^{i-1} B_1 B_1^T Z_u^{i-1} + K_u^{i-1(k)T} K_u^{i-1(k)}) x_\tau \, d\tau + x_{t+T_0}^T Z_u^{i(k)} x_{t+T_0}$$

   c. Update $K_u^{i-1(k+1)} = K_u^{i-1(k)} - B_2^T Z_u^{i(k+1)}, k = k + 1$.
   d. Until $\|Z_u^{i(k)} - Z_u^{i(k+1)}\| < \varepsilon$.
5. $Z_u^i = Z_u^{i(k)}, P_u^i = P_u^{i-1} + Z_u^i$.
6. Until $\|P_u^i - P_u^{i-1}\| < \varepsilon_p$. ■

This is an online algorithm to solve the ZS game problem in real time using data measured along the system trajectories. It does not require knowledge of the system drift matrix A. From the perspective of two-player ZS games, this algorithm translates as follows:

1. Let the initial disturbance policy be zero, $w=0$.
2. Let $K_u^{0(0)}$ be an initial control policy for the system (11.3) with zero disturbance $w=0$, and let $k=0$.
   a. Update the value associated with the controller $K_u^{0(k)}$;
   b. update the control policy, $k=k+1$;
   c. until the controller with the highest value (i.e. minimum cost) has been obtained.
3. Update the disturbance policy using the gain of the control policy.
4. For $i \geq 2$:
   a. Let $K_u^{i-1(0)}$ be a control policy for the system (11.3) with disturbance policy $w = B_1^T P_u^{i-1} x$ and let $k=0$:
      i. find the added value associated with the change in the control policy;
      ii. update the control policy, $k=k+1$;
      iii. until the controller with the highest value has been obtained.
5. Go to step 3 until the control policy and disturbance policy have the same gain.

Concisely, the game is played as follows.

1. The game starts while the disturbance player does not play.
2. The controller player plays the game without opponent and uses reinforcement learning to find the optimal behavior that minimizes his costs; then informs his opponent on his new behavior policy.
3. The disturbance player starts playing using the behavior policy of his opponent.
4. The controller player corrects iteratively his own behavior using reinforcement knowledge such that his costs are again minimized; then informs his opponent on his new behavior policy.
5. Go to step 3 until the two policies are characterized by the same parameter values.

The two players execute successively steps 3 and 4 until the controller player can no longer lower his costs by changing his behavior policy. Then, the saddle point equilibrium has been obtained. This algorithm corresponds to a novel adaptive critic structure as given in Figure 11.1. This structure has one critic and two actor networks, one for each player. An important aspect revealed by the adaptive critic structure is the fact that this ADP algorithm uses three time scales:

- the continuous-time scale, represented by the full lines, which is connected with the continuous-time dynamics of the system and the continuous-time computations performed by the two players.
- a discrete time scale given by Δ_1, which is a multiple of the sample time T. This time scale is connected with the online learning procedure that is based on discrete time measured data.
- a slower discrete time scale characterized by the period Δ_2, that is a multiple of Δ_1. This time scale is connected with the update procedure of the disturbance policy, procedure that is performed once the controller policy has converged.

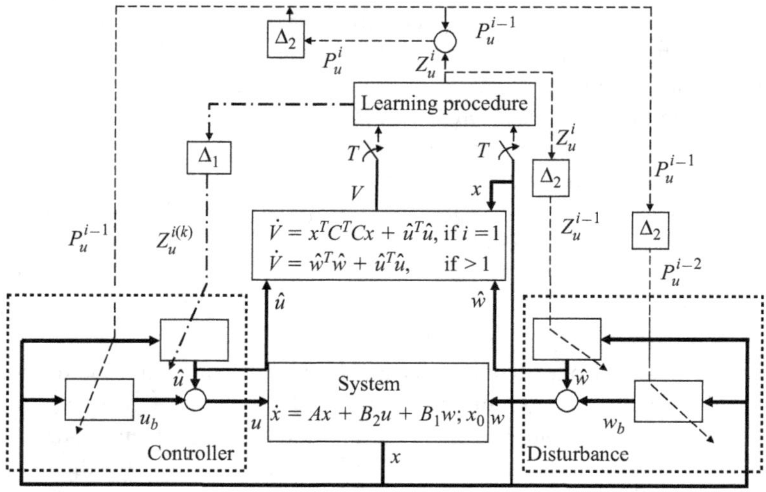

Figure 11.1 Adaptive critic structure for the online ADP game controller

The values of the time periods Δ_1 and Δ_2 can be variable and are controlled by the critic, which outputs the matrices $Z_u^{i(k)}$after every Δ_1 interval, and Z_u^i after every Δ_2 time interval. These periods are based on the sample time $T = T_0$. In fact, the integration period T_0 does not need to be constant, and can vary at each iteration step. In is not a sample time in the sense of standard sampled data systems.

One notes that both the control and disturbance signals are obtained as sums of two signals: a base signal and a correction signal.

$$\begin{aligned} u &= u_b + \hat{u} = -B_2^T P_u^{i-1} x - B_2^T Z_u^{i(k)} x \\ w &= w_b + \hat{w} = -B_1^T P_u^{i-2} x + B_1^T Z_u^{i-1} x \end{aligned} \tag{11.21}$$

The disturbance policy is constant over the time intervals Δ_2, and is equal with $B_1^T P_u^{i-1}$, being described by the parameters of the base policy of the controller given by the matrix P_u^{i-1}. The control policy is constant over the shorter time intervals Δ_1, and is equal with $-B_2^T(P_u^{i-1} + Z_u^{i(k)})$. The correction policy of the controller is the one that changes at every time interval Δ_1 while the base policy remains constant over the larger interval Δ_2.

11.3 Online load–frequency controller design for a power system

This section simulates the application of Algorithm 11.4 while finding the optimal controller for a power system.

Even though power systems are characterized by non-linearities, linear state-feedback control is regularly employed for load–frequency control about certain nominal operating points that are characterized by variations of the system load in a given range around a constant value. Although this assumption seems to simplify the design problem of load–frequency control, a new problem appears from the fact that the parameters of the actual plant are not precisely known and only the range of these parameters can be determined. For this reason it is particularly advantageous to apply model-free methods to obtain the optimal H-infinity controller for a given operating point of the power system. Algorithm 11.4 does not require knowledge of the system A matrix, so is particularly useful for such applications.

The plant considered is the linear model of the power system presented in Wang *et al.* (1993). The purpose of the design method is to determine the control strategy that results in minimal control cost in response to a maximal load change.

The state vector is

$$x = \begin{bmatrix} \Delta f & \Delta P_g & \Delta X_g & \Delta E \end{bmatrix}^T \tag{11.22}$$

where the state components are the incremental frequency deviation Δf (Hz), incremental change in generator output ΔP_g (p.u. MW), incremental change in governor value position ΔX_g (p.u. MW) and the incremental change in integral control ΔE. The matrices of the model of the plant used in this simulation are

$$A_{nom} = \begin{bmatrix} -0.0665 & 8 & 0 & 0 \\ 0 & -3.663 & 3.663 & 0 \\ -6.86 & 0 & -13.736 & -13.736 \\ 0.6 & 0 & 0 & 0 \end{bmatrix} \tag{11.23}$$

$$B_2 = [0 \quad 0 \quad 13.736 \quad 0]^T B_1 = [-8 \quad 0 \quad 0 \quad 0]^T$$

The cost function parameter, that is the C matrix, was chosen such that $Q = C^T C$ is the identity matrix of appropriate dimensions.

Knowing the system matrices one determines the saddle point of the zero-sum game by solving GARE (11.2) to obtain

$$P_u^{\infty} = \Pi = \begin{bmatrix} 0.6036 & 0.7398 & 0.0609 & 0.5877 \\ 0.7398 & 1.5438 & 0.1702 & 0.5978 \\ 0.0609 & 0.1702 & 0.0502 & 0.0357 \\ 0.5877 & 0.5978 & 0.0357 & 2.3307 \end{bmatrix} \tag{11.24}$$

For the purpose of demonstrating Algorithm 11.4, the closed-loop system was excited with an initial condition of one incremental change in integral control ΔE, the initial state of the system being $x_0 = [0 \quad 0 \quad 0 \quad 1]$. The simulation was conducted using data obtained from the system every 0.1 s. The value of the stop criterion s was 10^{-7}.

Algorithm 11.4 was run. To solve online for the values of the P_u^i matrix that parameterizes the cost function at step i, a least-squares problem of the sort described in Section 11.1.3 was set up before each iteration step 2.a or 4.a.i in the Algorithm 11.4. Since there are 10 independent elements in the symmetric matrix

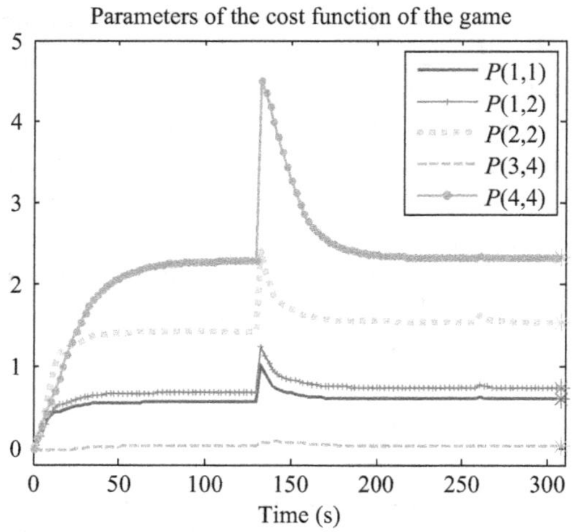

Figure 11.2 Convergence of the cost function of the game using the ADP method

P^i_u, the setup of the least-squares problem requires at least 10 measurements of the cost function associated with the given control policy, and measurements of the system's states at the beginning and the end of each time interval. Persistence of excitation is needed to solve this least-squares problem. A least-squares problem was solved after 25 data samples were acquired and thus the policy of the controller was updated every 2.5 s.

Figure 11.2 presents the evolution of the parameters of the value of the game, that is the elements of the kernel matrix P^i_u. It is clear that these cost function (i.e. critic) parameters converged to the optimal ones in (11.24).

The value of the game after five updates of the control policy is given by the matrix

$$P^5_u = \begin{bmatrix} 0.6036 & 0.7399 & 0.0609 & 0.5877 \\ 0.7399 & 1.5440 & 0.1702 & 0.5979 \\ 0.0609 & 0.1702 & 0.0502 & 0.0357 \\ 0.5877 & 0.5979 & 0.0357 & 2.3307 \end{bmatrix} \tag{11.25}$$

Comparing (11.25) with (11.24) one sees that the solution obtained using the ADP gaming method Algorithm 11.4 is very close to the exact solution (11.24), which was obtained offline by solving the GARE. Note that the solution of the GARE requires full knowledge of the system dynamics (A, B_1, B_2), whereas the solution obtained online using Algorithm 11.4 does not require the matrix A. This is important, for in LPC the A matrix is not exactly known. In fact, if the system parameters, and hence the matrix A, change during operation, Algorithm 11.4 will converge in real time to the new game solution corresponding to the new A matrix.

The number of iterations until convergence of the Z^i_u for $i = \overline{1,5}$are given in the vector $iter = [52 \quad 51 \quad 15 \quad 3 \quad 1]$. The large change in the parameters of the cost function, indicated in the figure at time 132.5 s, is determined by the first update of the policy of the disturbance player, that is after the sequence $Z_u^{1(k)}$ has converged to Z_u^1, and P_u^2 was calculated.

It is important to note that the convergence to the equilibrium of the game is strongly connected with the convergence of steps 2 and 4 of Algorithm 11.4. If convergence at these steps is not obtained, and the disturbance policy is yet updated, then there are no guarantees that the saddle point solution of the game will still be obtained.

As discussed in Lanzon *et al.* (2008), there are no guarantees that the closed-loop dynamics of the game, characterized by $A_u^{i-1} = A + B_1B_1^TP_u^{i-1} - B_2B_2^TP_u^{i-1}$, obtained after the disturbance has updated its policy, will be stable. For this reason, the iteration that was employed at the steps 2 and 4 of the Algorithm 11.4 was chosen to be the CT-HDP algorithm described in Algorithm 11.2, because this CT-HDP procedure does not require initialization with a stabilizing controller.

11.4 Conclusion

This chapter introduced an online data-based approach that makes use of reinforcement learning techniques, namely the integral reinforcement learning method, to determine in an online fashion the solution of the two-player zero-sum differential game with linear dynamics. The result, Algorithm 11.4, is based on a mathematical algorithm in Lanzon *et al.* (2008) that solves the GARE offline. Algorithm 11.4 involves iterations on Riccati equations to build a sequence of control and disturbance policies. These sequences converge monotonically to the state-feedback saddle point solution of the two-player zero-sum differential game.

The Riccati equations that appear at each step of Algorithm 11.4 are solved online using measured data by means of a continuous-time value iteration (i.e. CT-HDP) algorithm. This algorithm is based on the integral reinforcement learning techniques of Chapter 3. In this way, the H-infinity state-feedback optimal controller, or the solution of the differential game, can be obtained online without using exact knowledge on the drift dynamics of the system, that is matrix A in the system dynamics (11.3).

Appendix A
Proofs

Proofs for selected results from Chapter 4

Lemma 4.1. Solving for $V^{\mu^{(i)}}$ in (4.9) is equivalent with finding the solution of

$$0 = r(x, \mu^{(i)}(x)) + (\nabla V_x^{\mu^{(i)}})^T (f(x) + g(x)\mu^{(i)}(x)), \; V^{\mu^{(i)}}(0) = 0 \tag{4.12}$$

Proof: Since $\mu^{(i)} \in \Psi(\Omega)$, then $V^{\mu^{(i)}} \in C^1(\Omega)$, defined as $V^{\mu^{(i)}}(x(t)) = \int_t^{\infty} r(r(s), \mu^{(i)}(x(s)))\, ds$, is a Lyapunov function for the system $\dot{x}(t) = f(x(t)) + g(x(t))\mu^{(i)}(x(t)) \cdot V^{\mu^{(i)}} \in C^1(\Omega)$ satisfies

$$(\nabla\, V_x^{\mu^{(i)}})^T (f(x) + g(x)\mu^{(i)}(x)) = -r(x(t), \mu^{(i)}(x(t))) \tag{A.1}$$

with $r(x(t), \mu^{(i)}(x(t))) > 0; x(t) \neq 0$. Integrating (A.1) over the time interval $[t, t+T]$ one obtains

$$V^{\mu^{(i)}}(x(t)) = \int_t^{t+T} r(x(s), \mu^{(i)}(x(s)))\, ds + V^{\mu^{(i)}}(x(t+T)) \tag{A.2}$$

This means that the unique solution of (4.12), $V^{\mu^{(i)}}$, satisfies also (A.2).

To complete the proof, uniqueness of solution of (A.2) must be established. The proof is by contradiction. Thus, assume that there exists another cost function $V \in C^1(\Omega)$ that satisfies (4.12) with the end condition $V(0) = 0$. This cost function also satisfies $\dot{V}(x(t)) = -r(x(t), \mu^{(i)}(x(t)))$. Subtracting this from (A.2) we obtain

$$\left(\frac{d[V(x(t)) - V^{\mu^{(i)}}(x(t))]^T}{dx}\right)\dot{x} = \left(\frac{d[V(x(t)) - V^{\mu^{(i)}}(x(t))]^T}{dx}\right)(f(x(t)) \\ + g(x(t))\mu^{(i)}(x(t))) = 0 \tag{A.3}$$

which must hold for any x on the system trajectories generated by the stabilizing policy $\mu^{(i)}$.

Thus, $V(x(t)) = V^{\mu^{(i)}}(x(t)) + c$. As this relation must hold also for $x(t) = 0$ then $V(0) = V^{\mu^{(i)}}(0) + c \Rightarrow 0 = c$ and thus $V(x(t)) = V^{\mu^{(i)}}(x(t))$, that is (4.9) has a unique solution that is equal with the unique solution of (4.12). ■

Lemma 4.3. Let $\mu(x) \in \Psi(\Omega)$ such that $f(x) + g(x)\mu(x)$ is asymptotically stable. Given that the set $\{\phi_j\}_1^N$ is linearly independent then $\exists T > 0$ such that $\forall x(t) \in \Omega - \{0\}$, the set $\{\overline{\phi}_j(x(t), T) = \phi_j(x(t+T)) - \phi_j(x(t))\}_1^N$ is also linearly independent.

Proof: The proof is by contradiction.

The vector field $\dot{x} = f(x) + g(x)\mu(x)$ is asymptotically stable. Denote with $\eta(\tau; x(t), \mu), x(t) \in \Omega$ the system trajectories obtained using the policy $\mu(x)$ for any $x(t) \in \Omega$. Then, along the system trajectories, we have

$$\phi(x(t+T)) - \phi(x(t)) = \int_t^{t+T} \nabla\phi_x^T(f + g\mu)(\eta(\tau; x(t), \mu))\, d\tau \tag{A.4}$$

Suppose that the result is not true, then $\forall T > 0$ there exists a nonzero constant vector $c \in \mathbb{R}^N$ such that $\forall x(t) \in \Omega \quad c^T[\phi(x(t+T)) - \phi(x(t))] \equiv 0$. This implies that $\forall T > 0, c^T \int_t^{t+T} \nabla\phi_x^T(f + g\mu)(\phi(\tau; x(t), \mu))\, d\tau \equiv 0$ and thus, $\forall x(t) \in \Omega, c^T \nabla\phi_x^T (f + g\mu)(\varphi(\tau; x(t), \mu)) \equiv 0$. This means that $\{\nabla\phi_j^T(f + gu)\}_1^N$ is not linearly independent contradicting Lemma 4.2. Thus, $\exists T > 0$ such that $\forall x(t_0) \in \Omega$ the set $\{\overline{\phi}_j(x(t_0), T\}_1^N$ is also linearly independent. ■

Corollary 4.2. (Admissibility of $\mu_L^{(i)}(x)$) $\exists L_0$ such that $\forall L > L_0, \mu_L^{(i)} \in \Psi(\Omega)$.

Proof: Consider the function $V^{\mu^{(i)}}$, which is a Lyapunov function for the system (4.1) with control policy $\mu^{(i)}$. Taking derivative of $V^{\mu^{(i)}}$ along the trajectories generated by the controller $\mu_L^{(i+1)}(x)$ one obtains

$$\dot{V}^{\mu^{(i)}} = (\nabla V_x^{\mu^{(i)}})^T (f(x) + g(x)\mu_L^{(i+1)}(x)) \tag{A.5}$$

We also have that $\dot{V}^{\mu^{(i)}} = (\nabla V_x^{\mu^{(i)}})^T (f(x) + g(x)\mu^{(i)}(x)) = -Q(x) - (\mu^{(i)}(x))^T R\mu^{(i)}(x)$ and thus $(\nabla V_x^{\mu^{(i)}})^T f(x) = -(\nabla V_x^{\mu^{(i)}})^T g(x)\mu^{(i)}(x) - Q(x) - (\mu^{(i)}(x))^T R\mu^{(i)}(x)$. With this, (A.5) becomes

$$\dot{V}^{\mu^{(i)}} = -Q(x) - (\mu^{(i)}(x))^T R\mu^{(i)}(x) - (\nabla V_x^{\mu^{(i)}})^T g(x)(\mu^{(i)}(x) - \mu_L{}^{(i+1)}(x)) \tag{A.6}$$

Using the controller update $\mu^{(i+1)}(x) = -\frac{1}{2}R^{-1}g(x)^T \nabla V_x^{\mu^{(i)}}$ we can write $(\nabla V_x^{\mu^{(i)}})^T g(x) = 2R\mu^{(i+1)}(x)$. Thus (A.6) becomes

$$\begin{aligned}\dot{V}^{\mu^{(i)}} &= -Q - (\mu^{(i)})^T R\mu^{(i)} - 2\mu^{(i+1)} R(\mu^{(i)} - \mu_L^{(i+1)}) \\ &= -Q - (\mu^{(i)} - \mu^{(i+1)})^T R(\mu^{(i)} - \mu^{(i+1)}) - (\mu_L^{(i+1)})^T R\mu_L^{(i+1)} \\ &\quad + (\mu^{(i+1)} - \mu_L^{(i+1)})^T R(\mu^{(i+1)} - \mu_L^{(i+1)})\end{aligned} \tag{A.7}$$

Since $\sup_{x\in\Omega} |\mu_L^{(i)}(x) - \mu^{(i)}(x)| \to 0$ as $L \to \infty$ then there exists a L_0 such that $\forall L > L_0, \dot{V}^{\mu^{(i)}} < 0$, which means that $V^{\mu^{(i)}}$ is a Lyapunov function for the system with control policy $\mu_L^{(i+1)}(x)$, which proves the corollary. ■

Proofs for selected results from Chapter 5

Corollary 5.1. T'_μ: $X^P \to X^P$ is a contraction map on X^P.

Proof: The fixed point of $T'_\mu P = P_1^\mu \triangleq M^\mu + (A_d^{T_0})^T P A_d^{T_0}$ is the unique positive definite solution of the discrete-time Lyapunov equation

$$P^\mu = M^\mu + (A_d^{T_0})^T P^\mu A_d^{T_0} \tag{A.8}$$

Using the recursion k times, with $P_0^\mu = P$ we have

$$P_k^\mu = M^\mu + (A_d^{T_0})^T P_{k-1}^\mu A_d^{T_0} \tag{A.9}$$

Subtracting (A.8) from (A.9)

$$P^\mu - P_k^\mu = (A_d^{T_0})^T (P^\mu - P_{k-1}^\mu) A_d^{T_0} \tag{A.10}$$

Using the norm operator, (A.10) becomes

$$\|P^\mu - P_k^\mu\|_\rho \le \|A_d^{T_0}\|_\rho^2 \|P^\mu - P_{k-1}^\mu\|_\rho \tag{A.11}$$

Since $A_d^{T_0}$ is a discrete version of the closed loop continuous-time system stabilized by the state feedback control policy $\mu(x) = -K^\mu x$, then $0 < \rho(A_d^{T_0}) < 1$. Thus, T'_μ is a contraction map on

$$(X^P, \|\ \|_\rho)$$

■

Proofs for Chapter 7

Proof for Technical Lemma 7.2 Part a

This is a more complete version of results in Ioannou and Fidan (2006) and Tao (2003).

Set $\varepsilon_H = 0$ in (7.17). Take the Lyapunov function

$$L = \frac{1}{2}\tilde{W}_1^T a_1^{-1} \tilde{W}_1 \tag{A.12}$$

The derivative is

$$\dot{L} = -\tilde{W}_1^T \bar{\sigma}_1 \bar{\sigma}_1^T \tilde{W}_1$$

Integrating both sides

$$L(t+T) - L(t) = -\int_t^{t+T} \tilde{W}_1^T \sigma_1(\tau)\sigma_1^T(\tau)\tilde{W}_1 \, d\tau L(t+T)$$

$$= L(t) - \tilde{W}_1^T(t)\int_t^{t+T} \Phi^T(\tau,t)\sigma_1(\tau)\sigma_1^T(\tau)\Phi(\tau,t)\, d\tau \;\; \tilde{W}_1(t)$$

$$= L(t) - \tilde{W}_1^T(t)S_1\tilde{W}_1(t) \le (1-2a_1\beta_3)L(t)$$

So

$$L(t+T) \le (1-2a_1\beta_3)L(t) \tag{A.13}$$

Define $\gamma = (1-2a_1\beta_3)$. By using norms, we write (A.13) in terms of $\tilde{W}_1$ as

$$\frac{1}{2a_1}\|\tilde{W}(t+T)\|^2 \le \sqrt{(1-2a_1\beta_3)}\frac{1}{2a_1}\|\tilde{W}(t)\|^2$$
$$\|\tilde{W}(t+T)\| \le \sqrt{(1-2a_1\beta_3)}\|\tilde{W}(t)\|$$
$$\|\tilde{W}(t+T)\| \le \gamma\|\tilde{W}(t)\|$$

Therefore

$$\|\tilde{W}(k\mathrm{T})\| \le \gamma^k\|\tilde{W}(0)\| \tag{A.14}$$

that is $\tilde{W}(t)$ decays exponentially. To determine the decay time constant in continuous time, note

$$\|\tilde{W}(k\mathrm{T})\| \le e^{-akT}\|\tilde{W}(0)\| \tag{A.15}$$

where $e^{-akT} = \gamma^k$. Therefore the decay constant is

$$\alpha = -\frac{1}{T}\ln(\gamma) \Leftrightarrow \alpha = -\frac{1}{T}\ln(\sqrt{1-2a_1\beta_3}) \tag{A.16}$$

This completes the proof. ∎

Proof for Technical Lemma 7.2 Part b

Consider the system

$$\begin{cases} \dot{x}(t) = B(t)u(t) \\ y(t) = C^T(t)x(t) \end{cases} \tag{A.17}$$

The state and the output are

$$\begin{cases} x(t+T) = x(t) + \int_t^{t+T} B(\tau)u(\tau)\, d\tau \\ y(t+T) = C^T(t+T)x(t+T) \end{cases} \tag{A.18}$$

Let $C(t)$ be PE, so that

$$\beta_1 I \leq S_C \equiv \int_t^{t+T} C(\lambda)C^T(\lambda)\, d\lambda \leq \beta_2 I \tag{A.19}$$

Then,

$$\begin{aligned}
y(t+T) &= C^T(t+T)x(t) + \int_t^{t+T} C^T(t+T)B(\tau)u(\tau)\, d\tau \\
&\int_t^{t+T} C(\lambda)\left(y(\lambda) - \int_t^{\lambda} C^T(\lambda)B(\tau)u(\tau)\, d\tau\right) d\lambda \\
&= \int_t^{t+T} C(\lambda)C^T(\lambda)x(t)\, d\lambda \\
&\int_t^{t+T} C(\lambda)\left(y(\lambda) - \int_t^{\lambda} C^T(\lambda)B(\tau)u(\tau)\, d\tau\right) d\lambda = S_C x(t) \\
x(t) &= S_C^{-1}\left\{\int_t^{t+T} C(\lambda)\left(y(\lambda) - \int_t^{\lambda} C^T(\lambda)B(\tau)u(\tau)\, d\tau\right) d\lambda\right\}
\end{aligned}$$

Taking the norms in both sides yields

$$\begin{aligned}
\|x(t)\| &\leq \|S_C^{-1}\int_t^{t+T} C(\lambda)y(\lambda)\, d\lambda\| + \|S_C^{-1}\left\{\int_t^{t+T} C(\lambda)\left(\int_t^{\lambda} C^T(\lambda)B(\tau)u(\tau)\, d\tau\right) d\lambda\right\}\| \\
\|x(t)\| &\leq (\beta_1 \mathrm{I})^{-1}\left(\int_t^{t+T} C(\lambda)C^T(\lambda)\, d\lambda\right)^{1/2}\left(\int_t^{t+T} y(\lambda)^T y(\lambda)\, d\lambda\right)^{1/2} \\
&\quad + \|S_C^{-1}\|\left\{\int_t^{t+T} \|C(\lambda)C^T(\lambda)\| d\lambda \int_t^{t+T} \|B(\tau)u(\tau)\| d\tau\right\} \\
\|x(t)\| &\leq \frac{\sqrt{\beta_2 \mathrm{T}}}{\beta_1} y_{\max} + \frac{\delta\beta_2}{\beta_1}\int_t^{t+T} \|B(\tau)\| \cdot \|u(\tau)\| d\tau
\end{aligned} \tag{A.20}$$

where δ is a positive constant of the order of 1. Now consider

$$\dot{\tilde{W}}_1(t) = a_1 \overline{\sigma}_1 u \tag{A.21}$$

Note that setting $u = -y + (\varepsilon_H/m_s)$ with output given $y = \overline{\sigma}_1^T \tilde{W}_1$ turns (A.21) into (7.17). Set $B = a_1\overline{\sigma}_1, C = \overline{\sigma}_1, x(t) = \tilde{W}_1$ so that (A.17) yields (A.21). Then,

$$\|u\| \leq \|y\| + \left\|\frac{\varepsilon_H}{m_s}\right\| \leq y_{\max} + \varepsilon_{\max} \tag{A.22}$$

since $\|m_s\| \geq 1$. Then,

$$\begin{aligned}
N &\equiv \int_t^{t+T} \|B(\tau)\| \cdot \|u(\tau)\| d\tau = \int_t^{t+T} \|a_1\overline{\sigma}_1(\tau)\| \cdot \|u(\tau)\| d\tau \\
&\leq a_1(y_{\max} + \varepsilon_{\max})\int_t^{t+T} \|\overline{\sigma}_1(\tau)\| d\tau
\end{aligned}$$

$$\leq a_1(y_{\max} + \varepsilon_{\max}) \left[\int_t^{t+T} \|\overline{\sigma}_1(\tau)\|^2 \, d\tau \right]^{1/2} \left[\int_t^{t+T} 1 \, d\tau \right]^{1/2}$$

By using (A.19),

$$N \leq a_1(y_{\max} + \varepsilon_{\max})\sqrt{\beta_2 T} \tag{A.23}$$

Finally, (A.20) and (A.23) yield

$$\tilde{W}_1(t) \leq \frac{\sqrt{\beta_2 T}}{\beta_1} \{[y_{\max} + \delta\beta_2 a_1(\varepsilon_{\max} + y_{\max})]\} \tag{A.24}$$

This completes the proof. ■

Proof of Theorem 7.2. The convergence proof is based on Lyapunov analysis. We consider the Lyapunov function

$$L(t) = V(x) + \frac{1}{2} tr(\tilde{W}_1^T a_1^{-1} \tilde{W}_1) + \frac{1}{2} tr(\tilde{W}_2^T a_2^{-1} \tilde{W}_2) \tag{A.25}$$

With the chosen tuning laws one can then show that the errors $\tilde{W}_1$ and $\tilde{W}_2$ are UUB and convergence is obtained. This proceeds as follows:

Take the derivative of the Lyapunov function to obtain

$$\dot{L}(x) = \dot{V}(x) + \tilde{W}_1^T \alpha_1^{-1} \dot{\tilde{W}}_1 + \tilde{W}_2^T \alpha_2^{-1} \dot{\tilde{W}}_2 = \dot{L}_V(x) + \dot{L}_1(x) + \dot{L}_2(x) \tag{A.26}$$

First term is,

$$\dot{V}(x) = W_1^T \left(\nabla\phi_1 f(x) - \frac{1}{2}\overline{D}_1(x)\hat{W}_2 \right) + \nabla\varepsilon^T(x) \left(f(x) - \frac{1}{2} g(x) R^{-1} g^T(x) \nabla\phi_1^T \hat{W}_2 \right)$$

Then

$$\begin{aligned}
\dot{V}(x) &= W_1^T \left(\nabla\phi_1 f(x) - \frac{1}{2}\overline{D}_1(x)\hat{W}_2 \right) + \varepsilon_1(x) \\
&= W_1^T \nabla\phi_1 f(x) + \frac{1}{2} W_1^T \overline{D}_1(x)(W_1 - \hat{W}_2) - \frac{1}{2} W_1^T \overline{D}_1(x) W_1 + \varepsilon_1(x) \\
&= W_1^T \nabla\phi_1 f(x) = \frac{1}{2} W_1^T \overline{D}_1(x)\tilde{W}_2 - \frac{1}{2} W_1^T \overline{D}_1(x) W_1 + \varepsilon_1(x) \\
&= W_1^T \sigma_1 + \frac{1}{2} W_1^T \overline{D}_1(x)\tilde{W}_2 + \varepsilon_1(x)
\end{aligned}$$

where

$$\varepsilon_1(x) \equiv \dot{\varepsilon}(x) = \nabla\varepsilon^T(x) \left(f(x) - \frac{1}{2} g(x) R^{-1} g^T(x) \nabla\phi_1^T(x) \hat{W}_2 \right).$$

From the HJB equation

$$W_1^T \sigma_1 = -Q(x) - \frac{1}{4} W_1^T \overline{D}_1(x) W_1 + \varepsilon_{HJB}(x)$$

Then

$$\begin{aligned}\dot{L}_V(x) &= -Q(x) - \frac{1}{4} W_1^T \overline{D}_1(x) W_1 + \frac{1}{2} W_1^T \overline{D}_1(x) \tilde{W}_2 + \varepsilon_{HJB}(x) - \varepsilon_1(x) \\ &\equiv \dot{\bar{L}}_V(x) + \frac{1}{2} W_1^T \overline{D}_1(x) \tilde{W}_2 + \varepsilon_1(x)\end{aligned} \tag{A.27}$$

Second term is,

$$\begin{aligned}\dot{L}_1 &= \tilde{W}_1^T \alpha_1^{-1} \dot{\tilde{W}}_1 \\ &= \tilde{W}_1^T \alpha_1^{-1} \alpha_1 \frac{\sigma_2}{(\sigma_2^T \sigma_2 + 1)^2} \left(\sigma_2^T \hat{W}_1 + Q(x) + \frac{1}{4} \hat{W}_2^T \bar{D}_1 \hat{W}_2 \right) \\ &= \tilde{W}_1^T \frac{\sigma_2}{(\sigma_2^T \sigma_2 + 1)^2} (\sigma_2^T \hat{W}_1 + Q(x) + \frac{1}{4} \hat{W}_2^T \bar{D}_1(x) \hat{W}_2 - Q(x) \\ &\quad - \sigma_1^T W_1 \frac{1}{4} W_1^T \bar{D}_1(x) W_1 + \varepsilon_{HJB}(x)) \\ &= \tilde{W}_1^T \frac{\sigma_2}{(\sigma_2^T \sigma_2 + 1)^2} (\sigma_2^T(x) \hat{W}_1 - \sigma_1^T(x) W_1 + \frac{1}{4} \hat{W}_2^T \bar{D}_1(x) \hat{W}_2 \\ &\quad - \frac{1}{4} W_1^T \bar{D}_1(x) W_1 + \varepsilon_{HJB}(x)) \\ &= \tilde{W}_1^T \frac{\sigma_2}{(\sigma_2^T \sigma_2 + 1)^2} (-\tilde{W}_1^T (\nabla \phi_1^T(x))^T f(x) - \frac{1}{2} \hat{W}_2^T \bar{D}_1(x) \hat{W}_1 \\ &\quad + \frac{1}{2} W_1^T \bar{D}_1(x) W_1 + \frac{1}{4} \hat{W}_2^T \bar{D}_1(x) \hat{W}_2 - \frac{1}{4} W_1^T \bar{D}_1(x) W_1 + \varepsilon_{HJB}(x)) \\ &= \tilde{W}_1^T \frac{\sigma_2}{(\sigma_2^T \sigma_2 + 1)^2} (-f(x)^T \nabla \phi_1^T(x) \tilde{W}_1 + \frac{1}{2} \hat{W}_2^T \bar{D}_1(x) \tilde{W}_1 \\ &\quad + \frac{1}{4} \tilde{W}_2^T \bar{D}_1(x) \tilde{W}_2 + \varepsilon_{HJB}(x)) \\ \dot{L}_1 &= \tilde{W}_1^T \frac{\sigma_2}{(\sigma_2^T \sigma_2 + 1)^2} (-\sigma_2^T \tilde{W}_1 + \frac{1}{4} \tilde{W}_2^T \bar{D}_1(x) \tilde{W}_2 + \varepsilon_{HJB}(x)) \\ &= \dot{\bar{L}}_1 + \frac{1}{4} \tilde{W}_1^T \frac{\sigma_2}{(\sigma_2^T \sigma_2 + 1)^2} \tilde{W}_2^T \bar{D}_1(x) \tilde{W}_2\end{aligned} \tag{A.28}$$

where

$$\dot{\bar{L}}_1 = \tilde{W}_1^T \frac{\sigma_2}{(\sigma_2^T \sigma_2 + 1)^2} (-\sigma_2^T \tilde{W}_1 + \varepsilon_{HJB}(x)) = \tilde{W}_1^T \overline{\sigma}_2 \left(-\sigma_2^T \tilde{W}_1 + \frac{\varepsilon_{HJB}(x)}{m_s} \right)$$

Finally, by adding the terms (A.27) and (A.28)

$$\begin{aligned}
\dot{L}(x) &= -Q(x) - \frac{1}{4}W_1^T\overline{D}_1(x)W_1 + \frac{1}{2}W_1^T\overline{D}_1(x)\tilde{W}_2 + \varepsilon_{HJB}(x) + \varepsilon_1(x) \\
&\quad + \tilde{W}_1^T \frac{\sigma_2}{(\sigma_2^T\sigma_2+1)^2}\left(-\sigma_2^T\tilde{W}_1 + \frac{1}{4}\tilde{W}_2^T\overline{D}_1(x)\tilde{W}_2 + \varepsilon_{HJB}(x)\right) \\
&\quad + \tilde{W}_2^T\alpha_2^{-1}\dot{\tilde{W}}_2 \\
\dot{L}(x) &= \dot{\overline{L}}_V + \dot{\overline{L}}_1 + \varepsilon_1(x) - \tilde{W}_2^T\alpha_2^{-1}\dot{\hat{W}}_2 + \frac{1}{2}\tilde{W}_2^T\overline{D}_1(x)W_1 \\
&\quad + \frac{1}{4}\tilde{W}_2^T\overline{D}_1(x)W_1\frac{\overline{\sigma}_2^T}{m_s}\tilde{W}_1 - \frac{1}{4}\tilde{W}_2^T\overline{D}_1(x)W_1\frac{\overline{\sigma}_2^T}{m_s}W_1 \\
&\quad + \frac{1}{4}\tilde{W}_2^T\overline{D}_1(x)\tilde{W}_2\frac{\overline{\sigma}_2^T}{m_s}W_1 + \frac{1}{4}\tilde{W}_2^T\overline{D}_1(x)\hat{W}_2\frac{\overline{\sigma}_2^T}{m_s}\hat{W}_1
\end{aligned} \tag{A.29}$$

where $\overline{\sigma}_2 = (\sigma_2/\sigma_2^T\sigma_2 + 1)$ and $m_s = \sigma_2^T\sigma_2 + 1$.

To select the update law for the action neural network, write (A.29) as

$$\begin{aligned}
\dot{L}(x) &= \dot{\overline{L}}_V + \dot{\overline{L}}_1 + \varepsilon_1(x) - \tilde{W}_2^T\left[\alpha_2^{-1}\dot{\hat{W}}_2 - \frac{1}{4}\overline{D}_1(x)\hat{W}_2\frac{\overline{\sigma}_2^T}{m_s}\hat{W}_1\right] \\
&\quad + \frac{1}{2}\tilde{W}_2^T\overline{D}_1(x)W_1 + \frac{1}{4}\tilde{W}_2^T\overline{D}_1(x)W_1\frac{\overline{\sigma}_2^T}{m_s}\tilde{W}_1 - \frac{1}{4}\tilde{W}_2^T\overline{D}_1(x)W_1\frac{\overline{\sigma}_2^T}{m_s}W_1 \\
&\quad + \frac{1}{4}\tilde{W}_2^T\overline{D}_1(x)W_1\frac{\overline{\sigma}_2}{m_s}\tilde{W}_2
\end{aligned}$$

Now define the actor tuning law as

$$\dot{\hat{W}}_2 = -\alpha_2\left\{(F_2\hat{W}_2 - F_1\overline{\sigma}_2^T\hat{W}_1) - \frac{1}{4}\overline{D}_1(x)\hat{W}_2 m^T\hat{W}_1\right\} \tag{A.30}$$

This adds to $\dot{L}$ the terms

$$\begin{aligned}
\tilde{W}_2^T F_2\hat{W}_2 - \tilde{W}_2^T F_1\overline{\sigma}_2^T\hat{W}_1 &= \tilde{W}_2^T F_2(W_1 - \tilde{W}_2) - \tilde{W}_2^T F_1\overline{\sigma}_2^T(W_1 - \tilde{W}_1) \\
&= \tilde{W}_2^T F_2 W_1 - \tilde{W}_2^T F_2\tilde{W}_2 - \tilde{W}_2^T F_1\overline{\sigma}_2^T W_1 \\
&\quad + \tilde{W}_2^T F_1\overline{\sigma}_2^T\tilde{W}_1
\end{aligned}$$

Overall

$$\dot{L}(x) = -Q(x) - \frac{1}{4}W_1^T\overline{D}_1(x)W_1 + \varepsilon_{HJB}(x) + \tilde{W}_1^T\overline{\sigma}_2\left(-\overline{\sigma}_2^T\tilde{W}_1 + \frac{\varepsilon_{HJB}(x)}{m_s}\right)$$
$$+ \varepsilon_1(x) + \frac{1}{2}\tilde{W}_2^T\overline{D}_1(x)W_1 + \frac{1}{4}\tilde{W}_2^T\overline{D}_1(x)W_1\frac{\overline{\sigma}_2^T}{m_s}\tilde{W}_1$$
$$- \frac{1}{4}\tilde{W}_2^T\overline{D}_1(x)W_1\frac{\overline{\sigma}_2^T}{m_s}W_1 + \frac{1}{4}\tilde{W}_2^T\overline{D}_1(x)W_1\frac{\overline{\sigma}_2}{m_s}\tilde{W}_2 + \tilde{W}_2^TF_2W_1$$
$$- \tilde{W}_2^TF_2\tilde{W}_2 - \tilde{W}_2^TF_1\overline{\sigma}_2^TW_1 + \tilde{W}_2^TF_1\overline{\sigma}_2^T\tilde{W}_1 \tag{A.31}$$

Now it is desired to introduce norm bounds. It can be shown that under the Assumption 7.3

$$||\varepsilon_1(x)|| < b_{\varepsilon_x}b_f||x|| + \frac{1}{2}b_{\varepsilon_x}b_g^2b_{\phi_x}\sigma_{\min}(R)\left(||W_1|| + ||\tilde{W}_2||\right)$$

Also, since $Q(x) > 0$ there exists q such that $x^Tqx < Q(x)$ for $x \in \Omega$. It is shown in Abu-Khalaf and Lewis (2005) that ε_{HJB} converges to zero uniformly as N increases.

Select $\varepsilon > 0$ and $N_0(\varepsilon)$ such that $\sup_{x\in\Omega}||\varepsilon_{HJB}|| < \varepsilon$. Then, assuming $N > N_0$ and writing in terms of $\tilde{Z} = \begin{bmatrix} x \\ \overline{\sigma}_2^T\tilde{W}_1 \\ \tilde{W}_2 \end{bmatrix}$ (A.31) becomes

$$\dot{L} < \frac{1}{4}||W_1||^2||\overline{D}_1(x)|| + \varepsilon + \frac{1}{2}||W_1||b_{\varepsilon_x}b_{\varphi_x}b_g^2\sigma_{\min}(R)$$
$$-\tilde{Z}^T\begin{bmatrix} qI & 0 & 0 \\ 0 & I & \left(-\frac{1}{2}F_1 - \frac{1}{8m_s}\overline{D}_1W_1\right)^T \\ 0 & -\frac{1}{2}F_1 - \left(\frac{1}{8m_s}\overline{D}_1W_1\right) & F_2 - \frac{1}{8}(\overline{D}_1W_1m^T + mW_1^T\overline{D}_1) \end{bmatrix}\tilde{Z} \tag{A.32}$$
$$+\tilde{Z}^T\begin{bmatrix} b_{\varepsilon_x}b_f \\ \dfrac{\varepsilon}{m_s} \\ (\frac{1}{2}\overline{D}_1 + F_2 - F_1\overline{\sigma}_2^T - \frac{1}{4}\overline{D}_1W_1m^T)W_1 + \frac{1}{2}b_{\varepsilon_x}b_g^2b_{\phi_x}\sigma_{\min}(R) \end{bmatrix}$$

$$\text{Define } M = \begin{bmatrix} qI & 0 & 0 \\ 0 & I & \left(-\frac{1}{2}F_1 - \frac{1}{8m_s}\overline{D}_1 W_1\right)^T \\ 0 & -\frac{1}{2}F_1 - \left(\frac{1}{8m_s}\overline{D}_1 W_1\right) & F_2 - \frac{1}{8}(\overline{D}_1 W_1 m^T + m W_1^T \overline{D}_1) \end{bmatrix}$$

$$d = \begin{bmatrix} b_{\varepsilon_x} b_f \\ \frac{\varepsilon}{m_s} \\ \left(\frac{1}{2}\overline{D}_1 + F_2 - F_1 \overline{\sigma}_2^T - \frac{1}{4}\overline{D}_1 W_1 m^T\right) W_1 + \frac{1}{2} b_{\varepsilon_x} b_g^2 b_{\varphi_x} \sigma_{\min}(R) \end{bmatrix}$$

$$c = \frac{1}{4}||W_1||^2 ||\overline{D}_1(x)|| + \varepsilon + \frac{1}{2}||W_1|| b_{\varepsilon_x} b_{\phi_x} b_g{}^2 \sigma_{\min}(R)$$

Let the parameters be chosen such that $M > 0$. Now (A.32) becomes

$$\dot{L} < -||\tilde{Z}||^2 \sigma_{\min}(M) + ||d|| ||\tilde{Z}|| + c + \varepsilon$$

Completing the squares, the Lyapunov derivative is negative if

$$||\tilde{Z}|| > \frac{||d||}{2\sigma_{\min}(M)} + \sqrt{\frac{d^2}{4/\sigma_{\min}^2(M)} + \frac{c+\varepsilon}{\sigma_{\min}(M)}} \equiv B_Z \tag{A.34}$$

It is now straightforward to demonstrate that if L exceeds a certain bound, then, $\dot{L}$ is negative. Therefore, according to the standard Lyapunov extension theorem (Lewis *et al.*, 1999; Khalil, 1996) the analysis above demonstrates that the state and the weights are UUB.

This completes the proof. ■

Proof for Chapter 8

Proof for Theorem 8.1.

Let $\tilde{W}_1 = W_1 - \hat{W}_1$ and $\tilde{W}_2 = W_1 - \hat{W}_2$ denote the errors between the weights. We consider the Lyapunov function candidate

$$L(t) = \int_{t-T}^{t} V(x(\tau))\, d\tau + \frac{1}{2}\tilde{W}_1^T(t) a_1^{-1} \tilde{W}_1(t) + \frac{1}{2}\int_{t-T}^{t} \tilde{W}_2^T(\tau) a_2^{-1} \tilde{W}_2(\tau)\, d\tau \triangleq L_V(x) + L_1(x) + L_2(x) \tag{A.35}$$

The derivative of the Lyapunov function is given by

$$\dot{L}(x) = \int_{t-T}^{t} \dot{V}(x(\tau))\, d\tau + \tilde{W}_1^T(t) a_1^{-1} \dot{\tilde{W}}_1(t) + \int_{t-T}^{t} \tilde{W}_2^T(\tau) a_2^{-1} \dot{\tilde{W}}_2(\tau)\, d\tau \tag{A.36}$$

Next we will evaluate each one of the three terms of $\dot{L}(x)$.

The first term is

$$\dot{L}_V(x) = \int_{t-T}^{t} \dot{V}(x(\tau))\, d\tau = \int_{t-T}^{t} (W_1^T(\tau)\, \nabla\phi_1(x)\dot{x} + \dot{\varepsilon}(x))\, d\tau$$

with the control computed by the actor approximator (8.13) the system dynamics are given by

$$\dot{x} = f(x) - \frac{1}{2} g(x)TR^{-1}g^T(x)\, \nabla\phi_1^T(x)\tilde{W}_2$$

The first term in (A.36) becomes

$$\dot{L}_V(x) = \int_{t-T}^{t} W_1^T\left(\nabla\phi_1(x)f(x) - \frac{1}{2}\overline{D}_1(x)\hat{W}_2\right) d\tau + \varepsilon_1(x)$$

where

$$\varepsilon_1(x(t)) = \int_{t-T}^{t} \nabla\varepsilon^T(x)\left(f(x) - \frac{1}{2}g(x)TR^{-1}g^T(x)\, \nabla\phi_1^T\hat{W}_2\right) d\tau \qquad \text{(A.37)}$$

Now we want to obtain a representation of $\dot{L}_V(x)$ in terms of the parameters of the optimal value function W_1, and the parameter errors $\tilde{W}_1$ and $\tilde{W}_2$. Thus, by adding and subtracting $\frac{1}{2}W_1^T\overline{D}_1(x)W_1$, we obtain

$$\begin{aligned}\dot{L}_V(x) &= \int_{t-T}^{t}\left(W_1^T\nabla\phi_1 f(x) + \frac{1}{2}W_1^T\overline{D}(x)(W_1 - \hat{W}_2) - \frac{1}{2}W_1^T\overline{D}_1(x)W_1\right) d\tau + \varepsilon_1(x)\\ &= \int_{t-T}^{t}\left(W_1^T\nabla\phi_1 f(x) + \frac{1}{2}W_1^T\overline{D}_1(x)\tilde{W}_2 - \frac{1}{2}W_1^T\overline{D}_1(x)W_1\right) d\tau + \varepsilon_1(x)\end{aligned}$$

and using the notation $\sigma_1(x) = \nabla\phi_1 f(x) - \frac{1}{2}\overline{D}_1(x)W_1$

$$\dot{L}_V(x) = \int_{t-T}^{t}\left(W_1^T\sigma_1 + \frac{1}{2}W_1^T\overline{D}_1(x)\tilde{W}_2\right) d\tau + \varepsilon_1(x)$$

From the HJB equation (8.14) one has

$$\int_{t-T}^{t} W_1^T\sigma_1\, d\tau = \int_{t-T}^{t}\left(-Q(x) - \frac{1}{4}W_1^T\overline{D}_1(x)W_1 + \varepsilon_{HJB}(x)\right) d\tau \qquad \text{(A.38)}$$

Then

$$\dot{L}_V(x) = \int_{t-T}^{t} \left(-Q(x) - \frac{1}{4} W_1^T \overline{D}_1(x) W_1 + \varepsilon_{HJB}(x) \right) d\tau \\ + \int_{t-T}^{t} \left(\frac{1}{2} W_1^T \overline{D}_1(x) \tilde{W}_2 \right) d\tau + \varepsilon_1(x) \tag{A.39}$$

Using the tuning law for the critic, the second term in (A.35) becomes

$$\dot{L}_1 = \tilde{W}_1^T(t) a_1^{-1} \dot{\tilde{W}}_1(t) \\ = \tilde{W}_1^T \frac{\Delta\phi_2(t)}{\left(\Delta\phi_2(t)^T \Delta\phi_2(t) + 1\right)^2} \Delta\phi_2(t)^T \hat{W}_1 + \int_{t-T}^{t} \left(Q(x) + \frac{1}{4} \hat{W}_2^T \bar{D}_1 \hat{W}_2 \right) d\tau)$$

Adding (A.37) to the integral in the right hand side, using the notation $m_s = \Delta\phi_2(t)^T \Delta\phi_2(t) + 1$ we obtain

$$\dot{L}_1 = \tilde{W}_1^T \frac{\Delta\phi_2(t)}{m_s^2} \left(\int_{t-T}^{t} (\sigma_2^T(x) \tilde{W}_1 - \sigma_1^T(x) W_1 \, d\tau \right. \\ \left. + \int_{t-T}^{t} \left(\frac{1}{4} \hat{W}_2^T \overline{D}_1(x) \hat{W}_2 - \frac{1}{4} W_1^T \overline{D}_1(x) W_1 + \varepsilon_{HJB}(x) \right) d\tau \right) \tag{A.40}$$

The first integral in (A.40) can be written as

$$\int_{t-T}^{t} (\sigma_2^T(x) \hat{W}_1 - \sigma_1^T(x) W_1) \, d\tau = \int_{t-T}^{t} (-\tilde{W}_1 \nabla\phi_1(x) f(x) - \frac{1}{2} \hat{W}_2^T \overline{D}_1(x) \hat{W}_1 \\ + \frac{1}{2} W_1^T \overline{D}_1(x) W_1) \, d\tau$$

Then one has

$$\dot{L}_1 = \tilde{W}_1^T \frac{\Delta\phi_2(x(t), T)}{m_s^2} \int_{t-T}^{t} \left(-\frac{1}{2} \hat{W}_2^T \overline{D}_1(x) \hat{W}_1 + \frac{1}{4} W_1^T \overline{D}_1(x) W_1 \right) d\tau \\ + \int_{t-T}^{t} \left(\frac{1}{4} \hat{W}_2^T \overline{D}_1(x) \hat{W}_2 - \tilde{W}_1^T \nabla\phi_1(x) f(x) + \varepsilon_{HJB}(x) \right) d\tau$$

Using the definition for the parameter error $W_1 = \hat{W}_2 + \tilde{W}_2$, the first three terms under the integral can be written as

$$\frac{1}{4}W_1^T\overline{D}_1(x)W_1 - \frac{1}{2}\hat{W}_2^T\overline{D}_1(x)\hat{W}_1 + \frac{1}{4}\hat{W}_2^T\overline{D}_1(x)\hat{W}_2$$
$$= \frac{1}{4}(\hat{W}_2 + \hat{W}_2)^T\overline{D}_1(x)(\hat{W}_2 + \tilde{W}_2) - \frac{1}{2}\hat{W}_2^T\overline{D}_1(x)\hat{W}_1 + \frac{1}{4}\hat{W}_2^T\overline{D}_1(x)\hat{W}_2$$

Developing the parenthesis, and making use of the definition $W_1 = \hat{W}_1 + \tilde{W}_1$ we obtain

$$\frac{1}{4}W_1^T\overline{D}_1(x)W_1 - \frac{1}{2}\hat{W}_2^T\overline{D}_1(x)\hat{W}_1 + \frac{1}{4}\hat{W}_2^T\overline{D}_1(x)\hat{W}_2$$
$$= \frac{1}{2}\hat{W}_2^T\overline{D}_1(x)\tilde{W}_1 + \frac{1}{4}\tilde{W}_2^T\overline{D}_1(x)\tilde{W}_2$$

Using this last relation, we obtain

$$\dot{L}_1 = \tilde{W}_1^T\frac{\Delta\phi_2(x(t-T),T,\hat{u}_2)}{m_s^2}\int_{t-T}^{t}\left(\frac{1}{4}\tilde{W}_2^T\overline{D}_1(x)\tilde{W}_2 - \sigma_2^T\tilde{W}_1 + \varepsilon_{HJB}(x)\right)d\tau \tag{A.41}$$

Inserting the results (A.39) and (A.41), and using the notation $(\Delta\phi_2(x(t-T),T,\hat{u}_2))/m_s = \Delta\overline{\phi}_2(x(t-T),T,\hat{u}_2)$, (A.36) becomes

$$\begin{aligned}\dot{L}(x) = &\int_{t-T}^{t}\left(-Q(x) - \frac{1}{4}W_1(\tau)^T\overline{D}_1(x(\tau))W_1(\tau) + \varepsilon_{HJB}(x)\right)d\tau + \varepsilon_1(x)\\
&- \tilde{W}_1(t)^T\,\Delta\overline{\phi}_2(x(t-T),T,\hat{u}_2)\left(\Delta\overline{\phi}_2(x(t-T),T,\hat{u}_2)\right)^T\tilde{W}_1(t)\\
&+ \tilde{W}_1(t)^T\,\Delta\overline{\phi}_2(x(t-T),T,\hat{u}_2)\int_{t-T}^{t}\varepsilon_{HJB}(x(\tau))\,d\tau\\
&+ \int_{t-T}^{t}\left(\frac{1}{2}W_1(\tau)^T\overline{D}_1(x)\tilde{W}_2(\tau)\right)d\tau + \int_{t-T}^{t}\tilde{W}_2^T(\tau)\alpha_2^{-1}\dot{\tilde{W}}_2(\tau)\,d\tau\\
&+ \tilde{W}_1(t)^T\frac{\Delta\phi_2(x(t-T),T,\hat{u}_2)}{m_s}\int_{t-T}^{t}\left(\frac{1}{4}\tilde{W}_2(\tau)^T\overline{D}_1(x(\tau))\tilde{W}_2(\tau)\right)d\tau\end{aligned}$$

Using the relation

$$\frac{\Delta\overline{\phi}_2(x(t-T),T,\hat{u}_2)}{m_s} = \int_{t-T}^{t}\frac{1}{m_s}\overline{\sigma}_2(x(\tau))\,d\tau \equiv \overline{\Phi}_2(t)$$

and making use of weight error definitions this can be written as

$$\dot{L}(x) = \int_{t-T}^{t} \left(-Q(x(\tau)) - \frac{1}{4} W_1(\tau)^T \overline{D}_1(x(\tau)) W_1(\tau) + \varepsilon_{HJB}(x(\tau)) \right) d\tau$$

$$+ \varepsilon_1(x(t)) - \tilde{W}_1(t)^T \, \Delta\overline{\phi}_2(t) \, \Delta\overline{\phi}_2(t)^T \tilde{W}_1(t)$$

$$+ \tilde{W}_1(t)^T \overline{\Phi}_2(t) \int_{t-T}^{t} \varepsilon_{HJB}(x(\tau)) \, d\tau + \int_{t-T}^{t} \left(\frac{1}{2} W_1(\tau)^T \overline{D}_1(x) \tilde{W}_2(\tau) \right) d\tau$$

$$+ \int_{t-T}^{t} \left[\dot{\tilde{W}}_2(\tau)^T a_2^{-1} + \frac{1}{4} \hat{W}_1(\tau)^T \overline{\Phi}_2(\tau)^T \hat{W}_2(\tau)^T \overline{D}_1(x) \right] \tilde{W}_2(\tau) \, d\tau$$

$$- W_1(t)^T \overline{\Phi}_2(t) \int_{t-T}^{t} \left(\frac{1}{4} \tilde{W}_1(\tau)^T \overline{D}_1(x) \tilde{W}_2(\tau) \right) d\tau$$

$$+ \tilde{W}_1(t)^T \overline{\Phi}_2(t) \int_{t-T}^{t} \left(\frac{1}{4} W_1(\tau)^T \overline{D}_1(x) \tilde{W}_2(\tau) \right) d\tau$$

$$+ W_1(t)^T \overline{\Phi}_2(t) \int_{t-T}^{t} \left(\frac{1}{4} \tilde{W}_2(\tau)^T \overline{D}_1(x) \tilde{W}_2(\tau) \right) d\tau$$

$$+ \tilde{W}_1(t)^T \overline{\Phi}_2(t)^T \frac{1}{4} \int_{t-T}^{t} \hat{W}_2(\tau)^T \overline{D}_1(x(\tau)) \tilde{W}_2(\tau) \, d\tau$$

$$- \frac{1}{4} \int_{t-T}^{t} \hat{W}_1(\tau)^T \overline{\Phi}_2(\tau)^T \hat{W}_2(\tau)^T \overline{D}_1(x(\tau)) \tilde{W}_2(\tau) \, d\tau$$

Using the dynamics of the actor parameters

$$\dot{\hat{W}}_2 = -a_2 \left\{ (F_2 \hat{W}_2 - F_1 T \, \Delta\overline{\phi}_2^T \hat{W}_1) - \tfrac{1}{4m_s} \overline{D}_1(x) \hat{W}_2 \, \Delta\overline{\phi}_2^T \hat{W}_1 \right\}$$

the weight error definitions and rearranging the terms, this becomes

$$\dot{L}(x) = \int_{t-T}^{t} \left(-Q(x(\tau)) - \frac{1}{4} W_1(\tau)^T \overline{D}_1(x(\tau)) W_1(\tau) + \varepsilon_{HJB}(x(\tau)) \right) d\tau$$

$$+ \varepsilon_1(x(t)) - \tilde{W}_1(t)^T \, \Delta\overline{\phi}_2(t) \, \Delta\overline{\phi}_2(t)^T \tilde{W}_1(t)$$

$$+ \int_{t-T}^{t} (\tilde{W}_2(\tau)^T F_2 W_1(\tau) - T \tilde{W}_2(\tau)^T F_1 \, \Delta\overline{\phi}_2(\tau)^T W_1(\tau)$$

$$- \tilde{W}_2(\tau)^T F_2 \tilde{W}_2(\tau) + T \tilde{W}_2(\tau)^T F_1 \, \Delta\overline{\phi}_2(\tau)^T \tilde{W}_1(\tau)) \, d\tau$$

$$
\begin{aligned}
&+\tilde{W}_1(t)^T\overline{\Phi}_2(t)\int_{t-T}^{t}\varepsilon_{HJB}(x(\tau))\,d\tau+\int_{t-T}^{t}\left(\frac{1}{2}W_1(\tau)^T\overline{D}_1(x)\tilde{W}_2(\tau)\right)d\tau\\
&-W_1(t)^T\overline{\Phi}_2(t)\int_{t-T}^{t}\left(\frac{1}{4}W_1(\tau)^T\overline{D}_1(x)\tilde{W}_2(\tau)\right)d\tau\\
&+\tilde{W}_1(t)^T\overline{\Phi}_2(t)\int_{t-T}^{t}\left(\frac{1}{4}W_1(\tau)^T\overline{D}_1(x)\tilde{W}_2(\tau)\right)d\tau\\
&+W_1(t)^T\overline{\Phi}_2(t)\int_{t-T}^{t}\left(\frac{1}{4}\tilde{W}_2(\tau)^T\overline{D}_1(x)\tilde{W}_2(\tau)\right)d\tau\\
&+\tilde{W}_1(t)^T\overline{\Phi}_2(t)^T\frac{1}{4}\int_{t-T}^{t}\hat{W}_2(\tau)^T\overline{D}_1(x)\tilde{W}_2(\tau)\,d\tau\\
&-\frac{1}{4}\int_{t-T}^{t}\hat{W}_1(\tau)^T\overline{\Phi}_2(\tau)^T\tilde{W}_2(\tau)^T\overline{D}_1(x(\tau))\tilde{W}_2(\tau)\,d\tau
\end{aligned}
$$

According to Assumption 7.3, we can write (A.37) as

$$
||\varepsilon_1(x)||<\int_{t-T}^{t}\left(b_{\varepsilon_x}b_f||x||+\frac{1}{2}Tb_{\varepsilon_x}b_g^2b_{\phi_x}\sigma_{\min}(R)\big(||W_1||+||\tilde{W}_2||\big)\right)d\tau
$$

Also since $Q(x)>0$ there exist q such that $x^Tqx<Q(x)$ and for $x\in\Omega$. Now (A.36) becomes

$$
\begin{aligned}
\dot{L}(x)\le{}&\frac{1}{4}\int_{t-T}^{t}||W_1||^2||\overline{D}_1(x)||\,d\tau+\int_{t-T}^{t}(x(\tau)^Tqx(\tau)+\varepsilon(\tau))\,d\tau\\
&+\int_{t-T}^{t}\left(b_{\varepsilon_x}b_f||x||+\frac{1}{2}b_{\varepsilon_x}b_g^2b_{\phi_x}T\sigma_{\min}(R)(||W_1||+||\hat{W}_2||)\right)d\tau\\
&-\tilde{W}_1(t)^T\,\Delta\overline{\phi}_2(t)\,\Delta\overline{\phi}_2(t)^T\tilde{W}_1(t)+\int_{t-T}^{t}(\tilde{W}_2(\tau)^TF_2W_1(\tau)\\
&-T\tilde{W}_2(\tau)^TF_1\,\Delta\overline{\phi}_2(\tau)^TW_1(\tau)-\tilde{W}_2(\tau)^TF_2\tilde{W}_2(\tau))\,d\tau\\
&+\tilde{W}_1(t)^T\overline{\Phi}_2(t)\int_{t-T}^{t}\varepsilon_{HJB}(x(\tau))\,d\tau+\int_{t-T}^{t}\left(\frac{1}{2}W_1(\tau)^T\overline{D}_1(x)\tilde{W}_2(\tau)\right)d\tau\\
&-W_1(t)^T\overline{\Phi}_2(t)\int_{t-T}^{t}\left(\frac{1}{4}W_1(\tau)^T\overline{D}_1(x)\tilde{W}_2(\tau)\right)d\tau\\
&+\tilde{W}_1(t)^T\overline{\Phi}_2(t)\int_{t-T}^{t}\left(\frac{1}{4}W_1(\tau)^T\overline{D}_1(x)\tilde{W}_2(\tau)\right)d\tau\\
&+W_1(t)^T\overline{\Phi}_2(t)\int_{t-T}^{t}\left(\frac{1}{4}\tilde{W}_2(\tau)^T\overline{D}_1(x)\tilde{W}_2(\tau)\right)d\tau\\
&+\tilde{W}_1(t)^T\overline{\Phi}_2(t)^T\frac{1}{4}\int_{t-T}^{t}\hat{W}_2(\tau)^T\overline{D}_1(x(\tau))\tilde{W}_2(\tau)\,d\tau\\
&-\frac{1}{4}\int_{t-T}^{t}\hat{W}_1(\tau)^T\overline{\Phi}_2(\tau)^T\hat{W}_2(\tau)^T\overline{D}_1(x(\tau))\tilde{W}_2(\tau)\,d\tau\\
&+\int_{t-T}^{t}(\tilde{W}_2(\tau)^TTF_1(\Delta\overline{\phi}_2(\tau)^T\tilde{W}_1(\tau)-\Delta\overline{\phi}_2(t)^T\tilde{W}_1(t)))\,d\tau\\
&+\int_{t-T}^{t}T\tilde{W}_2(\tau)^TF_1\,\Delta\overline{\phi}_2(t)^T\tilde{W}_1(t)\,d\tau
\end{aligned}
\tag{A.42}
$$

Select $\varepsilon > 0$ and $N_0(\varepsilon)$ such that $\sup_{x\in\Omega} ||\varepsilon_{HJB}|| < \varepsilon$. Then, assuming $N > N_0$ we define

$$\tilde{Z}(t,\tau) = \begin{bmatrix} x(\tau) \\ \Delta\overline{\varphi}_2(x(t-T),T,\hat{u}_2)^T\tilde{W}_1(t) \\ \tilde{W}_2(\tau) \end{bmatrix}, z(t,\tau)$$

$$= \begin{bmatrix} \Delta\overline{\varphi}_2(x(t-T),T,\hat{u}_2)^T\tilde{W}_1(t) \\ \Delta\overline{\varphi}_2(x(\tau-T),T,\hat{u}_2)^T\tilde{W}_1(\tau) \\ \tilde{W}_2(\tau) \\ \tilde{W}_2(\tau)^T\overline{D}_1(x(\tau))\tilde{W}_2(\tau) \end{bmatrix}$$

Then (A.42) becomes

$$\dot{L} \le -\int_{t-T}^{t} \tilde{Z}^T M\tilde{Z}\, d\tau + \int_{t-T}^{t} d\tilde{Z}\, d\tau + \int_{t-T}^{t} (c+\varepsilon)\, d\tau + T\int_{t-T}^{t} z^T Wz\, d\tau$$

where

$$c = \frac{1}{4}||W_{1\max}||^2||\overline{D}_1(x)|| + \varepsilon + \frac{1}{2}||W_{1\min}||Tb_{\varepsilon_x}b_{\varphi_x}b_g{}^2\sigma_{\min}(R), W$$

$$= \begin{bmatrix} 0 & 0 & W_{13} & W_{14} \\ 0 & 0 & W_{23} & W_{24} \\ W_{31} & W_{32} & 0 & 0 \\ W_{41} & W_{42} & 0 & 0 \end{bmatrix}$$

with

$$W_{31} = W_{13}^T = -W_{32} = -W_{23}^T\left(\frac{1}{8m_s}W_{1\max}D_1(x(\tau)) + \frac{F_1}{2}\right),$$

$$W_{41} = W_{14}^T = -W_{42} = -W_{24}^T\left(-\frac{1}{8m_s}D_1(x(\tau))\right), \text{ and}$$

$$D_1(x) = \nabla\phi(x)g(x)R^{-1}g^T(x)\ \nabla\phi^T(x)$$

Also $M = \begin{bmatrix} M_{11} & 0 & 0 \\ 0 & M_{22} & M_{23} \\ 0 & M_{22} & M_{33} \end{bmatrix}$ with $M_{11} = qI, M_{22} = I$,

$$M_{32} = M_{23}^T = -\frac{1}{2}TF_1 - \left(\frac{1}{8m_s(t)}\overline{D}_1(\tau)W_1(\tau)\right),$$

$$M_{33} = F_2 - \frac{1}{4}(W_1(\tau)^T\overline{\Phi}_2(\tau)^T - W_1(t)^T)\overline{\Phi}_2(t)^T\overline{D}_1(x(\tau))$$
$$-\frac{1}{8}\left(\overline{D}_1(\tau)W_1(t)(m)^T(t) + m(t)W_1(t)^T\overline{D}_1(\tau)\right)$$

and $d = \begin{bmatrix} d_1 \\ d_2 \\ d_3 \end{bmatrix}$

with $d_1 = b_{\varepsilon_x} b_f, d_2 = \dfrac{\varepsilon_{HJB}(x(\tau))}{m_s(t)}$,

$$d_3 = \frac{1}{4}(W_1(t)^T\overline{\Phi}_2(t) - W_1(\tau)^T\overline{\Phi}_2(\tau))W_1(\tau)^T\overline{D}_1(x(\tau))$$
$$+ \left(\frac{1}{2}\overline{D}_1(x(\tau)) + F_2 - TF_1\ \Delta\overline{\phi}_2(\tau)^T - \frac{1}{4}\overline{D}_1(x(t))W_1(t)m(t)^T\right)W_1(\tau)$$
$$+ \frac{1}{2}b_{\varepsilon_x} b_{g^2} b_{\phi_x} \sigma_{\min}(R)$$

and $m(t) \equiv \dfrac{\Delta\phi_2(t)}{(\Delta\phi_2(t)^T\ \Delta\phi_2(t) + 1)^2}$

After taking norms and using the relations

$$||z||^2 = \|\Delta\overline{\phi}_2(x(t-T), T, \hat{u}_2)^T\tilde{W}_1(t)\|2 + \|\Delta\overline{\phi}_2(x(\tau-T), T, \hat{u}_2)^T\tilde{W}_1(\tau)\|^2$$
$$+ \|\tilde{W}_2(\tau)\| + \|\tilde{W}_2(\tau)^T\overline{D}_1(x(\tau))\tilde{W}_2(\tau)\|^2$$

and for appropriate selection of ρ_1, ρ_2 one has

$$||z(t,\tau)|| \le \rho_1||\tilde{Z}(t,\tau)|| + \rho_2 T||\tilde{Z}(t,\tau)||^2$$

So (A.42) becomes

$$\dot{L} \le -\int_{t-T}^{t} \|\tilde{Z}\|^2\sigma_{\min}(M)\,d\tau + \int_{t-T}^{t} \|d\|\|\tilde{Z}\|\,d\tau + \int_{t-T}^{t}(c+\varepsilon)\,d\tau$$
$$+ \int_{t-T}^{t}\left(\rho_1\|\tilde{Z}(t,\tau)\| + \rho_2 T\|\tilde{Z}(t,\tau)\|^2\right)^2 T\sigma_{\max}(W)\,d\tau$$
$$\dot{L} \le -\int_{t-T}^{t} \|\tilde{Z}\|^2\sigma_{\min}(M)\,d\tau + \int_{t-T}^{t} \|d\|\|\tilde{Z}\|\,d\tau + \int_{t-T}^{t}(c+\varepsilon)\,d\tau$$
$$+ \int_{t-T}^{t}\left(\rho_1{}^2 T\|\tilde{Z}(t,\tau)\|^2 + \rho_1{}^2T^2\|\tilde{Z}(t,\tau)\|^4 + 2\rho_1\rho_2 T\|\tilde{Z}(t,\tau)\|^3\right)\sigma_{\max}(w)\,d\tau$$
$$\dot{L} \le -\int_{t-T}^{t}(\sigma_{\min}(M) - \rho_1^2 T\sigma_{\max}(W))\|\tilde{Z}\|^2\,d\tau + \int_{t-T}^{t}\|d\|\|\tilde{Z}\|\,d\tau$$
$$+ \int_{t-T}^{t}(c+\varepsilon)\,d\tau + \int_{t-T}^{t}\left(\rho_2{}^2T^2\|\tilde{Z}(t,\tau)\|^4 + 2\rho_1\rho_2 T\|\tilde{Z}(t,\tau)\|^3\right)\sigma_{\max}(W)\,d\tau$$
$$\dot{L} \le \int_{t-T}^{t}\left(\rho_2{}^2T^2\|\tilde{Z}(t,\tau)\|^4\sigma_{\max}(W) + 2\rho_1\rho_2 T\|\tilde{Z}(t,\tau)\|^3\sigma_{\max}(W)\right)d\tau$$
$$+ \int_{t-T}^{t}\left((-\sigma_{\min}(M) + \rho_1{}^2T\sigma_{\max}(W))\|\tilde{Z}\|^2 + \|d\|\|\tilde{Z}\| + (c+\varepsilon)\right)d\tau$$
$$\equiv \left(T\int_{t-T}^{t} f(t,\tau) + \int_{t-T}^{t} g(t,\tau)\right)d\tau$$

where

$$f(t,\tau)=\rho_2^2 T\|\tilde{Z}(t,\tau)\|^4\sigma_{\max}(W)+2\rho_1\rho_2\|\tilde{Z}(t,\tau)\|^3\sigma_{\max}(W)+\rho_1^2\sigma_{\max}(W)\|\tilde{Z}\|^2$$
$$g(t,\tau)=-\sigma_{\min}(M)\|\tilde{Z}\|^2+\|d\|\|\tilde{Z}\|+(c+\varepsilon)$$

It is now desired to show that for small enough T, $\|\tilde{Z}\|$ is UUB. Select $p_B > 0$ large as detailed subsequently. Select T such that $Tf(t,\tau) < \varepsilon_f$ for some fixed $\varepsilon_f > 0 \forall \|\tilde{Z}\| < p_B$. Complete the squares to see that

$$\dot{L} \le \int_{t-T}^{t} \left(-\sigma_{\min}(M)\|\tilde{Z}\|^2 + \|d\|\|\tilde{Z}\| + (c+\varepsilon+\varepsilon_f)\right) d\tau$$

$$\dot{L} \le -\int_{t-T}^{t} \left(\|\tilde{Z}\| - \frac{\|d\|}{2\sigma_{\min}(M)}\right)^2 d\tau + \int_{t-T}^{t} \left(\frac{\|d\|}{2\sigma_{\min}(M)}\right)^2 d\tau$$
$$+ \int_{t-T}^{t} \left(\frac{c+\varepsilon+\varepsilon_f}{\sigma_{\min}(M)}\right) d\tau \equiv -\int_{t-T}^{t} W_3(\|\tilde{Z}\|)\, d\tau$$

Then

$$\|\tilde{Z}\| > \frac{\|d\|}{2\sigma_{\min}(M)} + \sqrt{\frac{d^2}{4\sigma^2{}_{\min}(M)} + \frac{c+\varepsilon+\varepsilon_f}{\sigma_{\min}(M)}} \equiv p_1 \tag{A.43}$$

Implies $W_3(\|\tilde{Z}\|) > 0$ and $\dot{L} < 0$.

It is now desired to find upper and lower bounds on the Lyapunov function L. Define $w(t) = \begin{bmatrix} x(t) \\ \tilde{W}_1(t) \\ \tilde{W}_2(t) \end{bmatrix}$ according to

$$L(t) = \int_{t-T}^{t} V(x(\tau))\, d\tau + \frac{1}{2}\tilde{W}_1^T(t)a_1^{-1}\tilde{W}_1(t) + \frac{1}{2}\int_{t-T}^{t} \tilde{W}_2^T(\tau)a_2^{-1}\tilde{W}_2(\tau)\, d\tau$$

We can find class K (see Definition 8.1) functions k_j and write

$$k_3(\|x(t)\|) = \int_{t-T}^{t} k_1(\|x(\tau)\|)\, d\tau \le \int_{t-T}^{t} V(x(\tau))\, d\tau \le \int_{t-T}^{t} k_2(\|x(\tau)\|)\, d\tau$$
$$= k_4(\|x(t)\|)k_5(\|\tilde{W}_2\|) \le \frac{1}{2}\int_{t-T}^{t} \tilde{W}_2^T(\tau)a_2^{-1}\tilde{W}_2(\tau)\, d\tau \le k_6(\|\tilde{W}_2\|)$$

We now need to find a relationship between $\|w(t)\|$ and $\|\tilde{Z}(t,\tau)\|$ to apply the results of Theorem 4.18 in Khalil (1996). One has

$$\|\tilde{Z}(t,\tau)\| \le \|w(t)\|\|\Delta\overline{\phi}_2(t)\| \le \|w(t)\|.$$

According to Technical Lemma 8.1,

$$\|\tilde{W}_1(t)\| \le \sqrt{\frac{\beta_2 T_{PE}}{\beta_1}}\left\{\left[(1+\delta\beta_2 a_1)\|\Delta\overline{\phi}_2(t)^T\tilde{W}_1\| + \delta\beta_2 a_1 \varepsilon_B\right]\right\}$$

$$\equiv \frac{\sqrt{\beta_2 T_{PE}}}{\beta_1}(1+\delta\beta_2 a_1)\|\Delta\overline{\phi}_2(t)^T\tilde{W}_1\| + \varepsilon_3$$

Now assume that we have enough hidden layer units $N > N_0$ then $\varepsilon_B \to 0$ according to Remark 8.1. So we can write $k_g\|w(t)\| \equiv \left(\sqrt{\frac{\beta_2 T_{PE}}{\beta_1}}(1+\delta\beta_2 a_1)\right)^{-1}$ $\|w(t)\| \le \|\tilde{Z}(t,\tau)\| \le \|w(t)\|$.

Finally, we can bound the Lyapunov function as

$$w^T \underline{S} w \le L \le w^T \overline{S} w$$

Therefore,

$$\|\tilde{Z}\|^2 \underline{\sigma}(\underline{S}) \le \|w\|^2 \underline{\sigma}(\underline{S}) \le L \le \|w\|^2 \overline{\sigma}(\overline{S}) \le \|\tilde{Z}\|^2 \overline{\sigma}(\overline{S})$$

and $\underline{S}\begin{bmatrix} k_3 & 0 & 0 \\ 0 & 1 & 0 \\ 0 & 0 & k_5 \end{bmatrix}$ and $\overline{S} = \begin{bmatrix} k_4 & 0 & 0 \\ 4 & k_8 & 0 \\ 0 & 0 & k_6 \end{bmatrix}$

Take $p_1 > 0$ as defined in (A.43). Select $p_B > \sqrt{\frac{\overline{\sigma}(\overline{S})}{\underline{\sigma}(\underline{S})}}\, p_1$. Then according to Theorem 4.18 in Khalil (1996), $\forall \tilde{Z}(0), \|\tilde{Z}\| \le p_B, \forall 0 \le t$, and $\|\tilde{Z}\| \le \sqrt{\frac{\overline{\sigma}(\overline{S})}{\underline{\sigma}(\underline{S})}}$ $p_1, \forall t \le T_B$.

Now the Technical Lemma 8.1 and the persistence of excitation condition of $\Delta\overline{\phi}_2$ show UUB of $\|\tilde{W}_1\|$. ■

Proofs for Chapter 9

Proof for Theorem 9.2.

The convergence proof is based on Lyapunov analysis. We consider the Lyapunov function

$$L(t) = V(x) + \frac{1}{2}tr(\tilde{W}_1^T a_1^{-1}\tilde{W}_1) + \frac{1}{2}tr(\tilde{W}_2^T a_2^{-1}\tilde{W}_2) + \frac{1}{2}tr(\tilde{W}_3^T a_3^{-1}\tilde{W}_3) \quad \text{(A.44)}$$

The derivative of the Lyapunov function is given by

$$\dot{L}(x) = \dot{V}(x) + \tilde{W}_1^T \alpha_1^{-1}\dot{\tilde{W}}_1 + \tilde{W}_2^T \alpha_2^{-1}\dot{\tilde{W}}_2 + \tilde{W}_3^T \alpha_3^{-1}\dot{\tilde{W}}_3 \quad \text{(A.45)}$$

First term is

$$\dot{L}_V(x) = \dot{V}(x) = W_1^T\left(\nabla\phi_1 f(x) - \frac{1}{2}\overline{D}_1(x)\hat{W}_2 + \frac{1}{2\gamma^2}\overline{E}(x)\hat{W}_3\right)$$
$$+ \nabla\varepsilon^T(x)\left(f(x) - \frac{1}{2}g(x)R^{-1}g^T(x)\ \nabla\phi_1^T\hat{W}_2 + \frac{1}{2\gamma^2}kk^T\ \nabla\phi_1^T\hat{W}_3\right)$$

Then

$$\begin{aligned}
\dot{L}_V(x)\dot{V}(x) &= W_1^T\left(\nabla\phi_1 f(x) - \frac{1}{2}\overline{D}_1(x)\hat{W}_2 + \frac{1}{2\gamma^2}\overline{E}(x)\hat{W}_3\right) + \varepsilon_1(x) \\
&= W_1^T\ \nabla\phi_1 f(x) + \frac{1}{2}W_1^T\overline{D}_1(x)(W_1 - \hat{W}_2) - \frac{1}{2}W_1^T\overline{D}_1(x)W_1 \\
&\quad - \frac{1}{2\gamma^2}W_1^T\overline{E}_1(x)(W_1 - \hat{W}_3) + \frac{1}{2\gamma^2}W_1^T\overline{E}_1(x)W_1 + \varepsilon_1(x) \\
&= W_1^T\ \nabla\phi_1 f(x) + \frac{1}{2}W_1^T\overline{D}_1(x)\tilde{W}_2 - \frac{1}{2}W_1^T\overline{D}_1(x)W_1 \\
&\quad - \frac{1}{2\gamma^2}W_1^T\overline{E}_1(x)\tilde{W}_3 + \frac{1}{2\gamma^2}W_1^T\overline{E}_1(x)W_1 + \varepsilon_1(x) \\
&= W_1^T\sigma_1 + \frac{1}{2}W_1^T\overline{D}_1(x)\tilde{W}_2 - \frac{1}{2\gamma^2}W_1^T\overline{E}_1(x)\tilde{W}_3 + \varepsilon_1(x)
\end{aligned}$$

where

$$\varepsilon_1(x) \equiv \dot{\varepsilon}(x) = \nabla\varepsilon^T(x)(f(x) - \frac{1}{2}g(x)R^{-1}g^T(x)\ \nabla\phi_1^T\hat{W}_2 + \frac{1}{2\gamma^2}kk^T\ \nabla\phi_1^T\hat{W}_3)$$

From the HJI equation (9.10) one has

$$W_1^T\sigma_1 = -Q(x) - \frac{1}{4}W_1^T\overline{D}_1(x)W_1 + \frac{1}{4\gamma^2}W_1^T\overline{E}_1(x)W_1 + \varepsilon_{HJI} \tag{A.46}$$

Then

$$\begin{aligned}
\dot{L}_V(x) &= -Q(x) - \frac{1}{4}W_1^T\overline{D}_1(x)W_1 + \frac{1}{4\gamma^2}W_1^T\overline{D}_1(x)W_1 + \frac{1}{2}W_1^T\overline{E}_1(x)\tilde{W}_2 \\
&\quad - \frac{1}{2\gamma^2}W_1^T\overline{E}_1(x)\tilde{W}_3 + \varepsilon_{HJI}(x) + \varepsilon_1(x) \\
&\equiv \dot{\overline{L}}_V(x) + \frac{1}{2}W_1^T\overline{D}_1(x)\tilde{W}_2 - \frac{1}{2\gamma^2}W_1^T\overline{E}_1(x)\tilde{W}_3 + \varepsilon_1(x)
\end{aligned} \tag{A.47}$$

where

$$\dot{\overline{L}}_V(x) = -Q(x) - \frac{1}{4}W_1^T\overline{D}_1(x)W_1 + \frac{1}{4\gamma^2}W_1^T\overline{D}_1(x)W_1 + \varepsilon_{HJI}(x) + \varepsilon_1(x)$$

Using the tuning law (9.58), the definitions (9.55), (9.57) and (9.65) for the parameter errors, and adding zero from the (A.46) the second term after grouping terms together becomes

$$\begin{aligned}\dot{L}_1 &= \tilde{W}_1^T \frac{\sigma_2}{(\sigma_2^T\sigma_2+1)^2}\left(-\sigma_2^T\tilde{W}_1 + \frac{1}{4}\tilde{W}_2^T\overline{D}_1(x)\tilde{W}_2 - \frac{1}{4\gamma^2}\tilde{W}_3^T\overline{E}_1\tilde{W}_3 + \varepsilon_{HJI}(x)\right)\\ &= \dot{\bar{L}}_1 + \frac{1}{4}\tilde{W}_1^T\frac{\sigma_2}{(\sigma_2^T\sigma_2+1)^2}\left(\tilde{W}_2^T\overline{D}_1(x)\tilde{W}_2 - \frac{1}{\gamma^2}\tilde{W}_3^T\overline{E}_1\tilde{W}_3\right)\end{aligned} \tag{A.48}$$

where

$$\dot{\bar{L}}_1 = \tilde{W}_1^T\frac{\sigma_2}{(\sigma_2^T\sigma_2+1)^2}\left(-\sigma_2^T\tilde{W}_1 + \varepsilon_{HJI}(x)\right) = \tilde{W}_1^T\overline{\sigma}_2\left(-\sigma_2^T\tilde{W}_1 + \frac{\varepsilon_{HJI}(x)}{m_s}\right)$$

By adding the terms of (A.47) and (A.48) and grouping terms together we have

$$\begin{aligned}\dot{L}(x) &= \dot{\bar{L}}_1 + \dot{\bar{L}}_V(x) + \varepsilon_1(x) + \frac{1}{4}\tilde{W}_1^T\frac{\sigma_2}{(\sigma_2^T\sigma_2+1)^2}\left(\tilde{W}_2^T\bar{D}_1(x)\tilde{W}_2 - \frac{1}{\gamma^2}\tilde{W}_3^T\mathrm{E}_1\tilde{W}_3\right)\\ &+\frac{1}{2}W_1^T\bar{D}_1(x)\tilde{W}_2 - \frac{1}{2\gamma^2}W_1^T\bar{E}_1(x)\tilde{W}_3 + \tilde{W}_2^T\alpha_2^{-1}\dot{\tilde{W}}_2 + \tilde{W}_3^T\alpha_3^{-1}\dot{\tilde{W}}_3\end{aligned}$$

After developing the parenthesis and using the definitions for the parameter errors one has

$$\begin{aligned}\dot{L}(x) &= \dot{\bar{L}}_1 + \dot{\bar{L}}_V(x) + \varepsilon_1(x) - \frac{1}{4}\tilde{W}_2^T\overline{D}_1(x)W_1\frac{\overline{\sigma}_2^T}{m_s}W_1 + \frac{1}{4}\tilde{W}_2^T\overline{D}_1(x)W_1\frac{\overline{\sigma}_2^T}{m_s}\tilde{W}_1\\ &+\frac{1}{4}\tilde{W}_2^T\overline{D}_1(x)\tilde{W}_2\frac{\overline{\sigma}_2^T}{m_s}W_1 + \frac{1}{4}\tilde{W}_2^T\overline{D}_1(x)\hat{W}_2\frac{\overline{\sigma}_2^T}{m_s}\hat{W}_1 - \frac{1}{4\gamma^2}\tilde{W}_3^T\overline{E}_1(x)W_1\frac{\overline{\sigma}_2^T}{m_s}\tilde{W}_1\\ &+\frac{1}{4\gamma^2}\tilde{W}_3^T\overline{E}_1(x)W_1\frac{\overline{\sigma}_2^T}{m_s}W_1 - \frac{1}{4\gamma^2}\tilde{W}_3^T\overline{E}_1(x)\tilde{W}_3\frac{\overline{\sigma}_2^T}{m_s}W_1\\ &+\frac{1}{4\gamma^2}\tilde{W}_3^T\overline{E}_1(x)\tilde{W}_3\frac{\overline{\sigma}_2^T}{m_s}\hat{W}_1 + \frac{1}{2}\tilde{W}_2^T\overline{D}_1(x)W_1 - \frac{1}{2\gamma^2}\tilde{W}_3^T\overline{E}_1(x)W_1\\ &-\tilde{W}_2^T\alpha_2^{-1}\dot{\hat{W}}_2 - \tilde{W}_3^T\alpha_3^{-1}\dot{\hat{W}}_3\end{aligned} \tag{A.49}$$

where

$$\overline{\sigma}_2 = \frac{\sigma_2}{\sigma_2^T\sigma_2+1} \quad \text{and} \quad m_s = \sigma_2^T\sigma_2 + 1$$

To select the update law for the action neural networks, write (A.49) as

$$\dot{L}(x) = \dot{L}_V + \dot{\bar{L}}_1 + \varepsilon_1(x) - \tilde{W}_2^T\left[\alpha_2^{-1}\dot{\hat{W}}_2 - \frac{1}{4}\overline{D}_1(x)\hat{W}_2\frac{\overline{\sigma}_2^T}{m_s}\hat{W}_1\right]$$

$$- \tilde{W}_3^T\left[\alpha_3^{-1}\dot{\hat{W}}_3 + \frac{1}{4\gamma^2}\overline{E}_1(x)\hat{W}_3\frac{\overline{\sigma}_2^T}{m_s}\hat{W}_1\right] + \frac{1}{2}\tilde{W}_2^T\overline{D}_1(x)W_1$$

$$+\frac{1}{4}\tilde{W}_2^T\overline{D}_1(x)W_1\frac{\overline{\sigma}_2^T}{m_s}\tilde{W}_1 - \frac{1}{4}\tilde{W}_2^T\overline{D}_1(x)W_1\frac{\overline{\sigma}_2^T}{m_s}W_1 + \frac{1}{4}\tilde{W}_2^T\overline{D}_1(x)W_1\frac{\overline{\sigma}_2^T}{m_s}\tilde{W}_2$$

$$-\frac{1}{2\gamma^2}\tilde{W}_3^T\overline{E}_1(x)W_1 - \frac{1}{4\gamma^2}\tilde{W}_3^T\overline{E}_1(x)W_1\frac{\overline{\sigma}_2^T}{m_s}\tilde{W}_1 + \frac{1}{4\gamma^2}\tilde{W}_3^T\overline{E}_1(x)W_1\frac{\overline{\sigma}_2^T}{m_s}W_1$$

$$-\frac{1}{4\gamma^2}\tilde{W}_3^T\overline{E}_1(x)W_1\frac{\overline{\sigma}_2^T}{m_s}\tilde{W}_3$$

Now define the actor tuning law as

$$\dot{\hat{W}}_2 = -\alpha_2\left\{\left(F_2\hat{W}_2 - F_1\overline{\sigma}_2^T\hat{W}_1\right) - \frac{1}{4}\overline{D}_1(x)\hat{W}_2 m^T\hat{W}_1\right\} \tag{A.50}$$

and the disturbance tuning law as

$$\dot{\hat{W}}_3 = -\alpha_3\left\{\left(F_4\hat{W}_3 - F_3\overline{\sigma}_2^T\hat{W}_1\right) - \frac{1}{4\gamma^2}\overline{E}_1(x)\hat{W}_3 m^T\hat{W}_1\right\} \tag{A.51}$$

This adds to $\dot{L}$ the terms

$$\begin{aligned}
&\tilde{W}_2^T F_2\hat{W}_2 - \tilde{W}_2^T F_1\overline{\sigma}_2^T\hat{W}_1 + \tilde{W}_3^T F_4\hat{W}_3 - \tilde{W}_3^T F_3\overline{\sigma}_2^T\hat{W}_1 \\
&\quad = \tilde{W}_2^T F_2(W_1 - \tilde{W}_2) - \tilde{W}_2^T F_1\overline{\sigma}_2^T(W_1 - \tilde{W}_1) + \tilde{W}_3^T F_4(W_1 - \tilde{W}_2) \\
&\qquad - \tilde{W}_3^T F_3\overline{\sigma}_2^T(W_1 - \tilde{W}_1) \\
&\quad = \tilde{W}_2^T F_2\tilde{W}_1 - \tilde{W}_2^T F_2\tilde{W}_2 - \tilde{W}_2^T F_1\overline{\sigma}_2^T\tilde{W}_1 + \tilde{W}_2^T F_1\overline{\sigma}_2^T\tilde{W}_1 \\
&\qquad + \tilde{W}_3^T F_4 W_1 - \tilde{W}_3^T F_4\tilde{W}_3 - \tilde{W}_3^T F_3\overline{\sigma}_2^T W_1 + \tilde{W}_3^T F_3\overline{\sigma}_2^T\tilde{W}_1
\end{aligned}$$

Overall

$$\dot{L}(x) = -Q(x) - \frac{1}{4}W_1^T\overline{D}_1(x)W_1 + \frac{1}{4\gamma^2}W_1^T\overline{E}_1(x)W_1$$

$$+ \varepsilon_{HJI}(x) + \tilde{W}_1^T\overline{\sigma}_2\left(-\overline{\sigma}_2^T\tilde{W}_1 + \frac{\varepsilon_{HJI}(x)}{m_s}\right) + \varepsilon_1(x)$$

$$+\frac{1}{2}\tilde{W}_2^T\overline{D}_1(x)W_1 + \frac{1}{4}\tilde{W}_2^T\overline{D}_1(x)W_1\frac{\overline{\sigma}_2^T}{m_s}\tilde{W}_1 - \frac{1}{4}\tilde{W}_2^T\overline{D}_1(x)W_1\frac{\overline{\sigma}_2^T}{m_s}W_1$$

$$+\frac{1}{4}\tilde{W}_2^T\overline{D}_1(x)W_1\frac{\overline{\sigma}_2}{m_s}\tilde{W}_2 - \frac{1}{2\gamma^2}\tilde{W}_3^T\overline{E}_1(x)W_1$$

$$-\frac{1}{4\gamma^2}\tilde{W}_2^T\overline{E}_1(x)W_1\frac{\overline{\sigma}_2^T}{m_s}\tilde{W}_1 + \frac{1}{4\gamma^2}\tilde{W}_3^T\overline{E}_1(x)W_1\frac{\overline{\sigma}_2^T}{m_s}W_1 - \frac{1}{4\gamma^2}\tilde{W}_3^T\overline{E}_1(x)W_1\frac{\overline{\sigma}_2^T}{m_s}\tilde{W}_3$$

$$+ \tilde{W}_2^TF_2W_1 - \tilde{W}_2^TF_2\tilde{W}_2 - \tilde{W}_2^TF_1\overline{\sigma}_2^TW_1 + \tilde{W}_2^TF_1\overline{\sigma}_2^T\tilde{W}_1$$

$$+ \tilde{W}_3^TF_4W_1 - \tilde{W}_3^TF_4\tilde{W}_3 - \tilde{W}_3^TF_3\overline{\sigma}_2^TW_1 + \tilde{W}_3^TF_3\overline{\sigma}_2^T\tilde{W}_1$$

Now it is desired to introduce norm bounds. It can be shown that under Assumption 9.1

$$\|\varepsilon_1(x)\| < b_{\varepsilon_x}b_f\|x\| + \frac{1}{2}b_{\varepsilon_x}b_g^2b_{\phi_x}\sigma_{\min}(R)\left(W_{\max} + \|\tilde{W}_2\|\right)$$
$$+\frac{1}{2\gamma^2}b_{\varepsilon_x}b_k^2b_{\phi_x}\left(W_{\max} + \|\tilde{W}_3\|\right)$$

Also since $Q(x) > 0$ there exists q such that $x^Tqx < Q(x)$ locally. It is shown in Abu-Khalaf *et al.* (2006), Abu-Khalaf and Lewis (2005) and Finlayson (1990) that ε_{HJI} converges to zero uniformly as N increases.

Select $\varepsilon > 0$ and $N_0(\varepsilon)$ such that $\sup\|\varepsilon_{HJI}\| < \varepsilon$. Then assuming $N > N_0$ and $\|W_1\| < W_{\max}$, and writing in terms of $\tilde{Z} = \begin{bmatrix} x \\ \overline{\sigma}_2^T\tilde{W}_1 \\ \tilde{W}_2 \\ \tilde{W}_3 \end{bmatrix}$, (A.52) becomes

$$\dot{L} < \frac{1}{4}W_{\max}^2\|\overline{D}_1(x)\| + \frac{1}{4\gamma^2}W_{\max}^2\|\overline{E}_1(x)\| + \varepsilon + \tfrac{1}{2}W_{\max}b_{\varepsilon_x}b_{\phi_x}{b_g}^2\sigma_{\min}(R)$$

$$+\tfrac{1}{2\gamma^2}W_{\max}b_{\varepsilon_x}b_k^2b_{\phi_x}$$

$$-\tilde{Z}^T\begin{bmatrix} qI & 0 & 0 & 0 \\ 0 & I & \left(\frac{1}{2}F_1-\frac{1}{8m_s}\overline{D}_1W_1\right)^T & \frac{1}{2}F_3+\left(\frac{1}{8\gamma^2 m_s}\overline{E}_1W_1\right) \\ 0 & \frac{1}{2}F_1-\left(\frac{1}{8m_s}\overline{D}_1W_1\right) & F_2-\frac{1}{8}(\overline{D}_1W_1m^T+mW_1^T\overline{D}_1) & 0 \\ 0 & \frac{1}{2}F_3+\left(\frac{1}{8\gamma^2 m_s}\overline{E}_1W_1\right) & 0 & F_4+\frac{1}{8\gamma^2}(\overline{E}_1W_1m^T+mW_1^T\overline{E}_1) \end{bmatrix}\tilde{Z} \tag{A.53}$$

$$+\tilde{Z}^T\begin{bmatrix} b_{\varepsilon_x}b_f \\ \dfrac{\varepsilon}{m_s} \\ \left(\frac{1}{2}\overline{D}_1+F_2-F_1\overline{\sigma}_2^T-\frac{1}{4}\overline{D}_1W_1m^T\right)W_1+\frac{1}{2}b_{\varepsilon_x}b_g^2b_{\phi_x}\sigma_{\min}(R) \\ \left(-\frac{1}{2\gamma^2}\overline{E}_1+F_4-F_3\overline{\sigma}_2^T+\frac{1}{4\gamma^2}\overline{E}_1W_1m^T\right)W_1+\frac{1}{2\gamma^2}b_{\varepsilon_x}b_k^2b_{\phi_x} \end{bmatrix}$$

Define

$$M=\begin{bmatrix} qI & 0 & 0 & 0 \\ 0 & I & \left(\frac{1}{2}F_1-\frac{1}{8m_s}\overline{D}_1W_1\right)^T & \frac{1}{2}F_3+\left(\frac{1}{8\gamma^2 m_s}\overline{E}_1W_1\right) \\ 0 & \frac{1}{2}F_1-\left(\frac{1}{8m_s}\overline{D}_1W_1\right) & F_2-\frac{1}{8}(\overline{D}_1W_1m^T+mW_1^T\overline{D}_1) & 0 \\ 0 & \frac{1}{2}F_3+\left(\frac{1}{8\gamma^2 m_s}\overline{E}_1W_1\right) & 0 & F_4+\frac{1}{8\gamma^2}(\overline{E}_1W_1m^{\mathrm{T}}+mW_1^T\overline{E}_1) \end{bmatrix} \tag{A.54}$$

$$T=\begin{bmatrix} b\varepsilon_xb_f \\ \dfrac{\varepsilon}{m_s} \\ \left(\frac{1}{2}\overline{D}_1+F_2-F_1\overline{\sigma}_2^T-\frac{1}{4}\overline{D}_1W_1m^T\right)W_1+\frac{1}{2}b_{\varepsilon_x}b_g^2b_{\phi_x}\sigma_{\min}(R) \\ \left(-\frac{1}{2\gamma^2}\overline{E}_1+F_4-F_3\overline{\sigma}_2^T+\frac{1}{4\gamma^2}\overline{E}_1W_1m^T\right)W_1+\frac{1}{2\gamma^2}b_{\varepsilon_x}b_k^2b_{\phi_x} \end{bmatrix}$$

$$c = \frac{1}{4} W_{\max}{}^{2} \left\| \overline{D}_1(x) \right\| + \frac{1}{4\gamma^2} W_{\max}{}^{2} \left\| \overline{E}_1(x) \right\| \\ + \frac{1}{2} W_{\max} b_{\varepsilon_x} b_{\phi_x} b_g{}^2 \sigma_{\min}(R) + \frac{1}{2\gamma^2} W_{\max} b_{\varepsilon_x} b_k^2 b_{\phi_x}$$

According to Assumption 9.1, c is bounded by max $c_{\max}$, and T is bounded by max $T_{\max}$ that can be expressed in terms of the bounds given there.

Let the parameters F_1, F_2, F_3 and F_4 be chosen such that $M > 0$. To justify this, the matrix M is written in compact form as

$$M = \begin{bmatrix} q & 0 & 0 \\ 0 & I & M_{23} \\ 0 & M_{32} & M_{33} \end{bmatrix} \tag{A.55}$$

where in order to be positive definite the following properties must hold

a. $q > 0$
b. $I > 0$
c. Schur complement for I is

$$D_{22} = I - M_{23} M_{33}^{-1} M_{32} > 0 \tag{A.56}$$

Matrix D_{22} can be made positive definite by selecting $F_2 \gg F_1, F_4 \gg F_3$, since W_1 is bounded above by $W_{\max}$.

Now (A.53) becomes

$$\dot{L} < -\|\tilde{Z}\|^2 \sigma_{\min}(M) + \|T\| \|\tilde{Z}\| + c_{\max} + \varepsilon$$

Completing the squares, the Lyapunov derivative is negative if

$$\|\tilde{Z}\| > \frac{T_{\max}}{2\sigma_{\min}(M)} + \sqrt{\frac{T_{\max}^2}{4\sigma_{\min}^2(M)} + \frac{c_{\max} + \varepsilon}{\sigma_{\min}(M)}} \equiv B_z \tag{A.57}$$

It is now straightforward to demonstrate that if L exceeds a certain bound, then, $\dot{L}$ is negative. Therefore, according to the standard Lyapunov extension theorem (Lewis *et al.*, 1999) the analysis above demonstrates that the state and the weights are UUB.

Note that condition (A.57) holds if the norm of any component of $\tilde{Z}$ exceeds the bound, that is specifically $x > B_Z$ or $\overline{\sigma}_2{}^{\mathrm{T}} \tilde{W}_1 > B_Z$ or $\tilde{W}_2 > B_Z$ or $\tilde{W}_3 > B_Z$. Therefore these quantities are less than $B_Z + z$ for any $z > 0$ (Khalil, 1996).

Now consider the error dynamics and the output as in Technical Lemmas 9.1, 9.2

$$\begin{aligned} \dot{\tilde{W}}_1 &= -a_1 \overline{\sigma}_2 \overline{\sigma}_2^{\mathrm{T}} \tilde{W}_1 + a_1 \overline{\sigma}_2 \frac{\varepsilon_{HJI}}{m_s} + \frac{a_1}{4m_s^2} \tilde{W}_2^T \overline{D}_1(x) \tilde{W}_2 - \frac{a_1}{4\gamma^2 m_s^2} \tilde{W}_2^T \overline{E}_1 \tilde{W}_3 \\ y &= \overline{\sigma}_2^T \tilde{W}_1 \end{aligned} \tag{A.58}$$

and assume $\overline{\sigma}_2$ is persistently exciting. Then Theorem 9.1 is true with

$$\left\| \frac{{\sigma_2}^T}{m_s}\tilde{W}_1 \right\| > \varepsilon_{\max} > \frac{1}{4}\|\tilde{W}_2\|^2 \left\| \frac{\overline{D}_1}{m_s} \right\| - \frac{1}{4\gamma^2}\|\tilde{W}_3\|^2 \left\| \frac{\overline{E}_1}{m_s} \right\| + \varepsilon \left\| \frac{1}{m_s} \right\|$$

This provides an effectssive practical bound for $\|{\overline{\sigma}_2}^T \tilde{W}_1\|$.

This completes the proof. ■

Proofs for Chapter 10

Proof for Theorem 10.2.

The convergence proof is based on Lyapunov analysis. We consider the Lyapunov function

$$L(t) = V_1(x) + V_2(x) + \frac{1}{2}\tilde{W}_1^T a_1^{-1}\tilde{W}_1 + \frac{1}{2}\tilde{W}_2^T a_2^{-1}\tilde{W}_2 + \frac{1}{2}\tilde{W}_3^T a_3^{-1}\tilde{W}_3 + \frac{1}{2}\tilde{W}_4^T a_4^{-1}\tilde{W}_4 \quad \text{(A.59)}$$

where $V_1(x)$ and $V_2(x)$ are the approximate solutions to (10.10) and are given by (10.25) and (10.26) respectively.

The time derivative of the Lyapunov function is given by

$$\dot{L}(x) = \dot{V}_1(x) + \dot{V}_2(x) + \tilde{W}_1^T a_1^{-1}\dot{\tilde{W}}_1 + \tilde{W}_2^T a_2^{-1}\dot{\tilde{W}}_2 + \tilde{W}_3^T a_3^{-1}\dot{\tilde{W}}_3 + \tilde{W}_4^T a_4^{-1}\dot{\tilde{W}}_4 \quad \text{(A.60)}$$

Next we will evaluate each one of the terms of $\dot{L}(x)$.

First term is, differentiating (10.25), and adding and subtracting $\frac{1}{2}W_1^T\overline{E}_2(x)W_2$ and $\frac{1}{2}W_1^T\overline{D}_1(x)W_1$

$$\begin{aligned}
\dot{V}_1(x) &= W_1^T\left(\nabla\phi_1 f(x) - \frac{1}{2}\overline{D}_1(x)\hat{W}_3 - \frac{1}{2}\overline{E}_2(x)\hat{W}_4\right) \\
&\quad + \nabla\varepsilon_1^T(x)\left(f(x) - \frac{1}{2}g(x){R_{11}}^{-1}g^T(x)\nabla\phi_1^T\hat{W}_3 - \frac{1}{2}k{R_{22}}^{-1}k^T\nabla\phi_2^T\hat{W}_4\right) \\
&= W_1^T\nabla\phi_1 f(x) + \frac{1}{2}W_1^T\overline{D}_1(x)(W_1 - \hat{W}_3) + \frac{1}{2}W_1^T\overline{E}_2(x)(W_2 - \hat{W}_4) \\
&\quad - \frac{1}{2}W_1^T\overline{D}_1(x)W_1 - \frac{1}{2}W_1^T\overline{E}_2(x)W_2 + \dot{\varepsilon}_1(x) \\
&= W_1^T\sigma_1 + \frac{1}{2}W_1^T\overline{D}_1(x)\tilde{W}_3 + \frac{1}{2}W_1^T\overline{E}_2(x)\tilde{W}_4 + \dot{\varepsilon}_1(x)
\end{aligned}$$

where $\overline{E}_2(x) \equiv \nabla\phi_1(x)kR_{22}^{-1}k^T\,\nabla\phi_2^T(x), \sigma_1 = \nabla\phi_1(x)(f + gu_1 + kd_2)$

$\overline{D}_1(x) \equiv \nabla\phi_1(x)g(x)R_{11}^{-1}g^T(x)\ \nabla\phi_{11}^T(x)$ and

$$\dot{\varepsilon}_1(x)=\nabla\varepsilon_1^T(x)\left(f(x)-\frac{1}{2}g(x)R_{11}^{-1}g^T(x)\nabla\phi^T\hat{W}_3-\frac{1}{2}kR_{22}^{-1}k^T\nabla\varphi_2^T\hat{W}_4\right) \quad \text{(A.61)}$$

From (10.23) we have

$$W_1^T\sigma_1 = -Q_1(x) - \frac{1}{4}W_1^T\nabla\varphi_1 g(x)R_{11}^{-1}g^T(x)\nabla\varphi_1^T W_1$$
$$-\frac{1}{4}W_2^T\nabla\varphi_2 k(x)R_{22}^{-T}R_{12}R_{22}^{-1}k^T(x)\nabla\varphi_2^T W_2 + \varepsilon_{HJ_1}.$$

Similarly for the second term

$$\dot{V}_2(x) = W_2^T\sigma_2 + \frac{1}{2}W_2{}^T\overline{E}_1(x)\tilde{W}_3 + \frac{1}{2}W_2{}^T\overline{D}_2(x)\tilde{W}_4 + \dot{\varepsilon}_2(x)$$

where $\overline{E}_1(x) \equiv \nabla\phi_2(x)g(x)R_{11}^{-1}g(x)^T\ \nabla\phi_1^T(x)$, $\overline{D}_2(x) \equiv \nabla\phi_2(x)kR_{22}^{-1}k^T\ \nabla\phi_2^T(x)$, $\sigma_2 = \nabla\phi_2(x)(f + gu_1 + kd_2)$, and

$$\dot{\varepsilon}_2(x)=\nabla\varepsilon_2^T(x)\left(f(x)-\frac{1}{2}g(x)R_{11}^{-1}g^T(x)\nabla\phi_1^T\hat{W}_3-\frac{1}{2}kR_{22}^{-1}k^T\nabla\dot{\phi}_2^T\hat{W}_4\right) \quad \text{(A.62)}$$

From (10.24)

$$W_2^T\sigma_2 = -Q_2(x) - \frac{1}{4}W_1^T\ \nabla\phi_1 g(x)R_{11}^{-T}R_{21}R_{11}^{-1}g^T(x)\ \nabla\phi_1^T W_1$$
$$-\frac{1}{4}W_2^T\ \nabla\phi_2 k(x)R_{22}^{-1}k^T(x)\ \nabla\phi_2^T W_2 + \varepsilon_{HJ_2}$$

Then we add $\dot{V}_1(x)$ and $\dot{V}_2(x)$

$$\begin{aligned}\dot{L}_v \equiv \dot{V}_1(x) + \dot{V}_2(x) = -Q_1(x) &- \frac{1}{4}W_1^T\ \nabla\phi_1 g(x)R_{11}^{-1}g^T(x)\ \nabla\phi_1^T W_1\\
&- \frac{1}{4}W_2^T\nabla\phi_2 k(x)R_{22}^{-T}R_{12}R_{22}^{-1}k^T(x)\nabla\phi_2^T W_2 + \varepsilon_{HJ_1}\\
&+ \frac{1}{2}W_1^T\overline{D}_1(x)\tilde{W}_3 + \frac{1}{2}W_1^T\overline{E}_2(x)\tilde{W}_4 + \dot{\varepsilon}_1(x)\\
&- Q_2(x) - \frac{1}{4}W_1^T\nabla\phi_1 g(x)R_{11}^{-T}R_{21}R_{11}^{-1}g^T(x)\nabla\phi_1^T W_1\\
&- \frac{1}{4}W_2^T\nabla\phi_2 k(x)R_{22}^{-T}(x)\nabla\phi_2^T W_2 + \varepsilon_{HJ_2}\\
&+ \frac{1}{2}W_2^T\overline{E}_1(x)\tilde{W}_3 + \frac{1}{2}W_2^T\overline{D}_2(x)\tilde{W}_4 + \dot{\varepsilon}_2(x)\end{aligned}$$

or

$$\dot{L}_v \equiv \dot{\bar{L}}_v(x) + \frac{1}{2}W_1^T\overline{D}_1(x)\tilde{W}_3 + \frac{1}{2}W_1^T\overline{E}_2(x)\tilde{W}_4 + \frac{1}{2}W_1^T\overline{E}_1(x)\tilde{W}_3$$

$$+\frac{1}{2}W_1^T\overline{D}_2(x)\tilde{W}_4 + \dot{\varepsilon}_1(x) + \dot{\varepsilon}_2(x) \tag{A.63}$$

where

$$\dot{\bar{L}}_v = -Q_1(x) - \frac{1}{4}W_1^T\ \nabla\phi_1 g(x)R_{11}^{-T}g^T(x)\ \nabla\phi_1^T W_1$$

$$-\frac{1}{4}W_2^T\ \nabla\phi_2 k(x)R_{22}^{-T}R_{12}R_{22}^{-1}k^T(x)\ \nabla\phi_2^T W_2 + \varepsilon_{HJ_1} - Q_2(x)$$

$$-\frac{1}{4}W_1^T\ \nabla\phi_1 g(x)R_{11}^{-T}R_{21}R_{11}^{-1}g^T(x)\ \nabla\phi_1^T W_1$$

$$-\frac{1}{4}W_2^T\ \nabla\phi_2 k(x)R_{22}^{-1}k^T(x)\ \nabla\phi_2^T W_2 + \varepsilon_{HJ_2}$$

Using the tuning law (10.42) for the first critic and the definitions for the parameter errors (10.41), the third term in (A.60) becomes

$$\dot{L}_1 = \tilde{W}_1^T\frac{\sigma_3}{(\sigma_3^T\sigma_3+1)^2}\Bigg(-\sigma_3^T\tilde{W}_1 + \frac{1}{2}W_1^T\overline{E}_2(x)\tilde{W}_4 + \frac{1}{4}\tilde{W}_3^T\overline{D}_1(x)\tilde{W}_3$$

$$+\frac{1}{4}\tilde{W}_4^T\ \nabla\phi_2 k(x)R_{22}^{-T}R_{12}R_{22}^{-1}k^T(x)\nabla\phi_2^T\tilde{W}_4$$

$$-\frac{1}{2}W_2^T\ \nabla\phi_2 k(x)R_{22}^{-T}R_{12}R_{22}^{-1}k^T(x)\nabla\phi_2^T\tilde{W}_4 + \varepsilon_{HJ_1}(x)\Bigg)$$

Finally, by rearranging and grouping the terms

$$\dot{L}_1 = \dot{\bar{L}}_1 + \frac{1}{4}\tilde{W}_1^T\frac{\sigma_3}{(\sigma_3^T\sigma_3+1)^2}\Big[\tilde{W}_3^T\overline{D}_1(x)\tilde{W}_3$$

$$+\tilde{W}_4^T\ \nabla\phi_2 k(x)R_{22}^{-T}R_{12}R_{22}^{-1}k^T(x)\nabla\phi_2^T\tilde{W}_4 + 2(W_1^T\overline{E}_2(x)\tilde{W}_4$$

$$-W_2^T\ \nabla\phi_2 k(x)R_{22}^{-T}R_{12}R_{22}^{-1}k^T(x)\ \nabla\phi_2^T\tilde{W}_4\Big] \tag{A.64}$$

where

$$\dot{\bar{L}}_1 = \tilde{W}_1^T\bar{\sigma}_3\left(-\sigma_3^{-T}\tilde{W}_1 + \frac{\varepsilon_{HJ_1}(x)}{m_{s_1}}\right), \bar{\sigma}_3 = \frac{\sigma_3}{\sigma_3^T\sigma_3+1}$$

and

$$m_{s_1} = \sigma_3^T\sigma_3 + 1$$

Similarly, by using the tuning law (10.43) for the second critic and the definitions for the parameter errors (10.41), the fourth term in (A.60) becomes

$$\dot{L}_2 = \dot{\bar{L}}_2 + \frac{1}{4}\tilde{W}_2^T \frac{\sigma_4}{(\sigma_4^T\sigma_4+1)^2}\Big[\tilde{W}_4^T\overline{D}_2(x)\tilde{W}_4$$

$$+ \tilde{W}_3^T\,\nabla\phi_1 g(x)R_{11}^{-T}R_{21}R_{11}^{-1}g^T(x)\,\nabla\phi_1^T\tilde{W}_3$$

$$+ 2(W_3^T\overline{E}_1W_2 - W_3^T\,\nabla\phi_1 g(x)R_{11}^{-T}R_{21}R_{11}^{-1}g^T(x)\,\nabla\phi_1^T\tilde{W}_1\Big] \tag{A.65}$$

where $\dot{\bar{L}}_2 = \tilde{W}_2^T\overline{\sigma}_4(-\sigma_4^{-T}\tilde{W}_2 + \frac{\varepsilon_{HJ_2}(x)}{m_{s_2}}), \overline{\sigma}_4 = \frac{\sigma_4}{\sigma_4^T\sigma_4+1}$ and $m_{s_2} = \sigma_4^T\sigma_4+1$

Finally, we need to add the terms of (A.63), (A.64) and (A.65), but in order to select the update laws for the action NNs, we group together the terms of $\tilde{W}_3$ and $\tilde{W}_4$ that are multiplied with the estimated values (two last terms)

$$\dot{L}(x) = \dot{L}_v(x) + \dot{\bar{L}}_1 + \dot{\bar{L}}_2 + \dot{\varepsilon}_1(x) + \dot{\varepsilon}_2(x) + \frac{1}{2}W_2^T\bar{D}_2(x)\tilde{W}_4$$

$$+\frac{1}{2}W_1^T\bar{D}_1(x)\tilde{W}_3 + \frac{1}{2}W_1^T\bar{E}_2(x)\tilde{W}_4 + \frac{1}{2}W_2^T\bar{E}_1(x)\tilde{W}_3$$

$$+\frac{1}{2}W_1^T\bar{E}_2(x)\tilde{W}_4\frac{\bar{\sigma}_3^T}{m_{s_1}}\tilde{W}_1 + \frac{1}{2}W_2^T\bar{E}_2\tilde{W}_3\frac{\bar{\sigma}_4^T}{m_{s_2}}\tilde{W}_2$$

$$-\frac{1}{2}W_2^T\,\nabla\phi_2 k(x)R_{22}^{-T}R_{12}R_{22}^{-1}k^T(x)\,\nabla\phi_2^T\tilde{W}_4\frac{\bar{\sigma}_3^T}{m_{s_1}}\tilde{W}_1$$

$$-\frac{1}{2}W_1^T\,\nabla\phi_1 g(x)R_{11}^{-T}R_{21}R_{11}^{-1}g^T(x)\,\nabla\phi_1^T\tilde{W}_3\frac{\bar{\sigma}_4^T}{m_{s_2}}\tilde{W}_2$$

$$+\frac{1}{4}\tilde{W}_3^T\bar{D}_1(x)W\frac{\bar{\sigma}_3^T}{m_{s_1}}\tilde{W}_1 - \frac{1}{4}\tilde{W}_3^T\bar{D}_1(x)W_1\frac{\bar{\sigma}_3^T}{m_{s_1}}\tilde{W}_1 + \frac{1}{4}\tilde{W}_3^T\bar{D}_1(x)\tilde{W}_3\frac{\bar{\sigma}_3^T}{m_{s_1}}W_1$$

$$-\frac{1}{4}\tilde{W}_4^T\,\nabla\phi_2 k(x)R_{22}^{-T}R_{12}R_{22}^{-1}k^T(x)\,\nabla\phi_2^T W_2\frac{\bar{\sigma}_3^T}{m_{s_1}}\tilde{W}_1$$

$$+\frac{1}{4}\tilde{W}_4^T\,\nabla\phi_2 k(x)R_{22}^{-T}R_{12}R_{22}^{-1}k^T(x)\,\nabla\phi_2^T W_2\frac{\bar{\sigma}_3^T}{m_{s_1}}\tilde{W}_1$$

$$+\frac{1}{4}\tilde{W}_4^T\,\nabla\phi_2 k(x)R_{22}^{-T}R_{12}R_{22}^{-1}k^T(x)\,\nabla\phi_2^T\tilde{W}_4\frac{\bar{\sigma}_3^T}{m_{s_1}}W_1$$

$$+\frac{1}{4}\tilde{W}_4^T\bar{D}_2(x)W_2\frac{\bar{\sigma}_4^T}{m_{s_2}}\tilde{W}_2-\frac{1}{4}\tilde{W}_4^T\bar{D}_2(x)W_2\frac{\bar{\sigma}_4^T}{m_{s_2}}W_2+\frac{1}{4}\tilde{W}_4^T\bar{D}_2(x)\tilde{W}_3\frac{\bar{\sigma}_4^T}{m_{s_2}}W_2$$

$$+\frac{1}{4}\tilde{W}_3^T\,\nabla\phi_1 g(x)R_{11}^{-T}R_{21}R_{11}^{-1}g^T(x)\,\nabla\phi_1^T W_1\frac{\bar{\sigma}_4^T}{m_{s_2}}\tilde{W}_2$$

$$-\frac{1}{4}\tilde{W}_3^T\,\nabla\phi_1 g(x)R_{11}^{-T}R_{21}R_{11}^{-1}g^T(x)\,\nabla\phi_1^T W_1\frac{\bar{\sigma}_4^T}{m_{s_2}}W_2$$

$$+\frac{1}{4}\tilde{W}_3^T\,\nabla\phi_1 g(x)R_{11}^{-T}R_{21}R_{11}^{-1}g^T(x)\,\nabla\phi_1^T\tilde{W}_3\frac{\bar{\sigma}_4^T}{m_{s_2}}W_2$$

$$-\tilde{W}_3^T\left[a_3^{-1}\dot{\hat{W}}_4-\frac{1}{4}\nabla\phi_1 g(x)R_{11}^{-T}R_{21}R_{11}^{-1}g^T(x)\,\nabla\phi_1^T\hat{W}_3\frac{\bar{\sigma}_4^T}{m_{s_2}}\hat{W}_2\right.$$
$$\left.-\frac{1}{4}\bar{D}_1(x)\hat{W}_3\frac{\bar{\sigma}_3^T}{m_{s_1}}\hat{W}_1\right]$$

$$-\tilde{W}_4^T\left[a_4^{-1}\dot{\hat{W}}_4-\frac{1}{4}\nabla\phi_2 k(x)R_{22}^{-T}R_{12}R_{22}^{-1}k^T(x)\,\nabla\phi_2^T\hat{W}_4\frac{\bar{\sigma}_3^T}{m_{s_1}}\hat{W}_1\right.$$
$$\left.-\frac{1}{4}\bar{D}_2(x)\hat{W}_4\frac{\bar{\sigma}_4^T}{m_{s_2}}\hat{W}_2\right]$$

In order for the last two terms to be zero, we define the actor tuning law for the first player as

$$\dot{\hat{W}}_3=-a_3\left\{(F_2\hat{W}_3-F_1\bar{\sigma}_3^T\hat{W}_1)-\frac{1}{4}\overline{D}_1(x)\hat{W}_3\frac{\overline{\sigma}_3^T}{m_{s_1}}\hat{W}_1\right.$$
$$\left.-\frac{1}{4}\nabla\phi_1 g(x)R_{11}^{-T}R_{21}R_{11}^{-1}g^T(x)\nabla\phi_1^T\hat{W}_3\frac{\overline{\sigma}_4^T}{m_{s_2}}\hat{W}_2\right\}$$

and the second player's actor tuning law is defined as

$$\dot{\hat{W}}_4=-a_4\left\{(F_4\hat{W}_4-F_3\bar{\sigma}_4^T\hat{W}_2)-\frac{1}{4}\overline{D}_2(x)\hat{W}_4\frac{\overline{\sigma}_4^T}{m_{s_2}}\hat{W}_2\right.$$
$$\left.-\frac{1}{4}\nabla\phi_2 k(x)R_{22}^{-T}R_{12}R_{22}^{-1}k^T(x)\nabla\phi_2^T\hat{W}_4\frac{\overline{\sigma}_3^T}{m_{s_1}}\hat{W}_1\right\}$$

But this adds to $\dot{L}$ the following terms

$$\tilde{W}_3^T F_2 W_1-\tilde{W}_3^T F_2\tilde{W}_3-\tilde{W}_3^T F_1\bar{\sigma}_3^T W_1+\tilde{W}_3^T F_1\bar{\sigma}_3^T W_1+\tilde{W}_4^T F_4 W_2-\tilde{W}_4^T F_4\tilde{W}_4$$
$$-\tilde{W}_4^T F_3\bar{\sigma}_4^T W_2+\tilde{W}_4^T F_3\bar{\sigma}_4^T\tilde{W}_2$$

Overall

$$
\begin{aligned}
\dot{L}(x) = & -Q_1(x) - \frac{1}{4}W_1^T \nabla\phi_1 g(x) R_{11}^{-T} g^T(x) \nabla\varphi_1^T W_1 \\
& - \frac{1}{4}W_2^T \nabla\phi_2 k(x) R_{22}^{-T} R_{12} R_{22}^{-1} k^T(x) \nabla\phi_2^T W_2 + \varepsilon_{HJB_1} \\
& - Q_2(x) - \frac{1}{4}W_1^T \nabla\phi_1 g(x) R_{11}^{-T} R_{21} R_{11}^{-1} g^T(x) \nabla\phi_1^T W_1 \\
& - \frac{1}{4}W_2^T \nabla\phi_2 k(x) R_{22}^{-T} k^T(x) \nabla\phi_2^T W_2 + \varepsilon_{HJB_2} \\
& + \tilde{W}_1^T \bar{\sigma}_3 (-\sigma_3^T \tilde{W}_1 + \frac{\varepsilon_{HJ_1}(x)}{m_{s_1}}) + \dot{\varepsilon}_1(x) + \tilde{W}_2^T \bar{\sigma}_4 (-\sigma_4^T \tilde{W}_2 + \frac{\varepsilon_{HJ2}(x)}{m_{s_2}}) + \dot{\varepsilon}_2(x) \\
& + \frac{1}{2}W_1^T \bar{D}_1(x) \tilde{W}_3 + \frac{1}{2}W_1^T \bar{E}_2(x) \tilde{W}_4 + \frac{1}{2}W_2^T \bar{E}_1(x) \tilde{W}_3 + \frac{1}{2}W_2^T \bar{D}_2(x) \tilde{W}_4 \\
& + \frac{1}{2}W_1^T \bar{E}_2(x) \tilde{W}_4 + \frac{\bar{\sigma}_3^T}{m_{s_1}} \tilde{W}_1 + \frac{1}{2}W_2^T \bar{E}_2 \tilde{W}_3 \frac{\bar{\sigma}_4^T}{m_{s_2}} \tilde{W}_2 \\
& - \frac{1}{2}W_2^T \nabla\phi_2 k(x) R_{22}^{-T} R_{12} R_{22}^{-1} k^T(x) \nabla\phi_2^T \tilde{W}_4 \frac{\bar{\sigma}_3^T}{m_{s_1}} \tilde{W}_1 \\
& - \frac{1}{2}W_1^T \nabla\phi_1 g(x) R_{11}^{-T} R_{21} R_{11}^{-1} g^T(x) \nabla\phi_1^T \tilde{W}_3 \frac{\bar{\sigma}_4^T}{m_{s_2}} \tilde{W}_2 \\
& + \frac{1}{4}\tilde{W}_3^T \bar{D}_1(x) W_1 \frac{\bar{\sigma}_3^T}{m_{s_1}} \tilde{W}_1 - \frac{1}{4}\tilde{W}_3^T \bar{D}_1(x) W_1 \frac{\bar{\sigma}_3^T}{m_{s_1}} W_1 + \frac{1}{4}\tilde{W}_3^T \bar{D}_1(x) \tilde{W}_3 \frac{\bar{\sigma}_3^T}{m_{s_1}} W_1 \\
& + \frac{1}{4}\tilde{W}_4^T \nabla\phi_2 k(x) R_{22}^{-T} R_{12} R_{22}^{-1} k^T(x) \nabla\phi_2^T W_2 \frac{\bar{\sigma}_3^T}{m_{s_1}} \tilde{W}_1 \\
& - \frac{1}{4}\tilde{W}_4^T \nabla\phi_2 k(x) R_{22}^{-T} R_{12} R_{22}^{-1} k^T(x) \nabla\phi_2^T W_2 \frac{\bar{\sigma}_3^T}{m_{s_1}} W_1 \\
& + \frac{1}{4}\tilde{W}_4^T \nabla\phi_2 k(x) R_{22}^{-T} R_{12} R_{22}^{-1} k^T(x) \nabla\phi_2^T W_4 \frac{\bar{\sigma}_3^T}{m_{s_1}} W_1 \\
& + \frac{1}{4}\tilde{W}_4^T \bar{D}_2(x) W_2 \frac{\bar{\sigma}_4^T}{m_{s_2}} \tilde{W}_2 - \frac{1}{4}\tilde{W}_4^T \bar{D}_2(x) W_2 \frac{\bar{\sigma}_4^T}{m_{s_2}} W_2 + \frac{1}{4}\tilde{W}_4^T \bar{D}_2(x) \tilde{W}_4 \frac{\bar{\sigma}_4^T}{m_{s_2}} W_2 \\
& + \frac{1}{4}\tilde{W}_3^T \nabla\phi_1 g(x) R_{11}^{-T} R_{21} R_{11}^{-1} g^T(x) \nabla\phi_1^T W_1 \frac{\bar{\sigma}_4^T}{m_{s_2}} \tilde{W}_2 \\
& + \frac{1}{4}\tilde{W}_3^T \nabla\phi_1 g(x) R_{11}^{-T} R_{21} R_{11}^{-1} g^T(x) \nabla\phi_1^T W_1 \frac{\bar{\sigma}_4^T}{m_{s_2}} \tilde{W}_2 \\
& + \frac{1}{4}\tilde{W}_3^T \nabla\phi_1 g(x) R_{11}^{-T} R_{21} R_{11}^{-1} g^T(x) \nabla\phi_1^T \tilde{W}_3 \frac{\bar{\sigma}_4^T}{m_{s_2}} W_2 \\
& + \tilde{W}_3^T F_2 W_1 - \tilde{W}_3^T F_2 \tilde{W}_3 - \tilde{W}_3^T F_1 \bar{\sigma}_3^T W_1 + \tilde{W}_3^T F_1 \bar{\sigma}_3^T W_1 \\
& + \tilde{W}_4^T F_4 W_2 - \tilde{W}_4^T F_4 \tilde{W}_4 - \tilde{W}_4^T F_3 \bar{\sigma}_4^T W_2 + \tilde{W}_4^T F_3 \bar{\sigma}_4^T \tilde{W}_2
\end{aligned}
\tag{A.66}
$$

Now it is desired to introduce norm bounds. It can be shown that under Assumption 10.1 (A.61) and (A.62) imply

$$\|\dot{\varepsilon}_1(x)\| < b_{\varepsilon_{1x}} b_f \|x\| + \frac{1}{2} b_{\varepsilon_{1x}} b_g^2 b_{\phi_{1x}} \sigma_{\min}(R_{11})\left(\|W_1\| + \|\tilde{W}_3\|\right) + \frac{1}{2} b_{\varepsilon_{1x}} b_k^2 b_{\phi_{2x}} \sigma_{\min}(R_{22})\left(\|W_2\| + \|\tilde{W}_4\|\right)$$

$$\|\dot{\varepsilon}_2(x)\| < b_{\varepsilon_{2x}} b_f \|x\| + \frac{1}{2} b_{\varepsilon_{2x}} b_g^2 b_{\phi_{1x}} \sigma_{\min}(R_{11})\left(\|W_1\| + \|\tilde{W}_3\|\right) + \frac{1}{2} b_{\varepsilon_{2x}} b_k^2 b_{\phi_{2x}} \sigma_{\min}(R_{22})\left(\|W_2\| + \|\tilde{W}_4\|\right)$$

Also since $Q_1(x) > 0$ and $Q_2(x) > 0$ there exists q_1 and q_2 such that $x^T q_1 x < Q_1(x)$ and $x^T q_2 x < Q_2(x)$ for $x \in \Omega$

Select $\bar{\varepsilon}_1 > 0, \bar{\varepsilon}_2 > 0$ and $K_0(\bar{\varepsilon}_1), K_0(\bar{\varepsilon}_2)$ such that $\sup_{x\in\Omega} \|\varepsilon_{HJ_1}\| < \bar{\varepsilon}_1$ and $\sup_{x\in\Omega} \|\varepsilon_{HJ_2}\| < \bar{\varepsilon}_2$. Then assuming $K > K_0$ and writing in terms of $\tilde{Z} = \begin{bmatrix} x \\ \bar{\sigma}_3^T \tilde{W}_1 \\ \bar{\sigma}_4^T \tilde{W}_2 \\ \tilde{W}_3 \\ \tilde{W}_4 \end{bmatrix}$

and known bounds, (A.66) becomes

$$\begin{aligned}
\dot{L}(x) = {} & \frac{1}{4} \|W_1\|^2 \|\nabla\phi_1 g(x) R_{11}^{-1} g^T(x) \nabla\phi_1^T\| \\
& + \frac{1}{4} \|W_2\|^2 \|\nabla\phi_2 k(x) R_{22}^{-T} R_{12} R_{22}^{-1} k^T(x) \nabla\phi_2^T\| + \bar{\varepsilon}_1 \\
& + \frac{1}{4} \|W_1\|^2 \|\nabla\phi_1 g(x) R_{11}^{-T} R_{21} R_{11}^{-1} g^T(x) \nabla\phi_1^T\| \\
& + \frac{1}{4} \|W_2\|^2 \|\nabla\phi_2 k(x) R_{22}^{-T} k^T(x) \nabla\phi_2^T\| + \bar{\varepsilon}_2 \\
& + \frac{1}{2} b_{\varepsilon_{x1}} b_g^2 b_{\phi_{1x}} \sigma_{\min}(R_{11}) \|W_1\| + \frac{1}{2} b_{\varepsilon_{x1}} b_k^2 b_{\phi_{2x}} \sigma_{\min}(R_{22}) \|W_2\| \\
& + \frac{1}{2} b_{\varepsilon_{x2}} b_g^2 b_{\phi_{1x}} \sigma_{\min}(R_{11}) \|W_1\| + \frac{1}{2} b_{\varepsilon_{x2}} b_k^2 b_{\phi_{2x}} \sigma_{\min}(R_{22}) \|W_2\| \\
& - \tilde{Z}^T \begin{bmatrix} m_{11} & m_{12} & m_{13} & m_{14} & m_{15} \\ m_{21} & m_{22} & m_{23} & m_{24} & m_{25} \\ m_{31} & m_{32} & m_{33} & m_{34} & m_{35} \\ m_{41} & m_{42} & m_{43} & m_{44} & m_{45} \\ m_{51} & m_{52} & m_{53} & m_{54} & m_{55} \end{bmatrix} \tilde{Z} + \tilde{Z} \begin{bmatrix} d_1 \\ d_2 \\ d_3 \\ d_4 \\ d_5 \end{bmatrix}
\end{aligned} \tag{A.67}$$

where the components of the matrix M are given by

$$
\begin{aligned}
m_{11} &= q_1 + q_2 \\
m_{22} &= m_{33} = I \\
m_{44} &= F_2 - \frac{1}{8}(\overline{D}_1 W_1 m_1^T + m_1 W_1^T \overline{D}_1 + \nabla\phi_1 g(x) R_{11}^{-1} R_{21} R_{11}^{-1} g^T(x) \nabla\phi_1^T W_2 m_2^T) \\
&\quad + m_2 W_2^T \nabla\phi_1 g(x) R_{11}^{-T} R_{21} R_{11}^{-1} g^T(x) \nabla\phi_1^T) \\
m_{55} &= F_4 - \frac{1}{8}(\nabla\phi_2 k(x) R_{22}^{-1} R_{12} R_{22}^{-1} k^T(x) \nabla\phi_2^T W_1 m_1^T \\
&\quad + m_1 W_1^T \nabla\phi_2 k(x) R_{22}^{-T} R_{12} R_{22}^{-1} k^T(x) \nabla\phi_2^T \\
&\quad + m_2 W_2^T \overline{D}_2(x) + \overline{D}_2(x) W_2 m_2^T) \\
m_{42} &= -\frac{1}{2} F_1 - \frac{1}{8 m_{s1}} \overline{D}_1 W_1 = m_{24}^T \\
m_{52} &= -\frac{1}{8 m_{s_1}} \nabla\phi_2 k(x) R_{22}^{-T} R_{12} R_{22}^{-1} k^T(x) \nabla\phi_2^T W_2 - \frac{1}{4 m_{s_1}} \overline{E}_2 W_1 = m_{25}^T \\
m_{43} &= -\frac{1}{4 m_{s_2}} \nabla\phi_1 g(x) R_{11}^{-T} R_{21} R_{11}^{-1} g^T(x) \nabla\phi_1^T W_1 \\
&\quad - \frac{1}{8 m_{s_2}} \nabla\phi_1 g(x) R_{11}^{-T} R_{21} R_{11}^{-1} g^T(x) \nabla\phi_1^T W_1 = m_{34}^T \\
m_{53} &= -\frac{1}{2} F_3 - \frac{1}{8 m_{s_2}} \overline{D}(x) W_2 = m_{53}^T \\
m_{12} &= m_{21} = m_{31} = m_{13} = m_{32} \\
&= m_{23} = m_{41} = m_{14} = m_{51} = m_{15} = m_{54} = m_{45} = 0
\end{aligned}
$$

and the components of vector D are given by

$$
\begin{aligned}
d_1 &= (b_{\varepsilon_{x_1}} + b_{\varepsilon_{x_2}}) b_f \\
d_2 &= \frac{\varepsilon_{HJ_1}}{m_{s_1}} \\
d_3 &= \frac{\varepsilon_{HJ_2}}{m_{s_2}} \\
d_4 &= F_2 W_1 - F_1 \overline{\sigma}_3^T W_1 + \frac{1}{2} \overline{D}_1 W_1 + \frac{1}{2} \overline{E}_1 W_2 + \frac{1}{2} b_{\varepsilon_{x_1}} \\
&\quad + b_g^2 b_{\phi_{1x}} \sigma_{\min} - \frac{1}{4} \overline{D}_1 W_1 m_1^T W_1 \\
&\quad - \frac{1}{4} \nabla\phi_1 g(x) R_{11}^{-T} R_{21} R_{11}^{-1} g^T(x) \nabla\phi_1^T W_1 m_2^T W_2 \\
d_5 &= F_4 W_2 - F_3 \overline{\sigma}_4^T W_2 + \frac{1}{2} \overline{D}_2 W_2 + \frac{1}{2} \overline{E}_2 W_1 + \frac{1}{2} b_{\varepsilon_{x_1}} \\
&\quad + b_g^2 b_{\phi_{2x}} \sigma_{\min}(R_{22}) - \frac{1}{4} \overline{D}_2(x) W_2 m_2^T W_2 \\
&\quad - \frac{1}{4} \nabla\phi_2 k(x) R_{22}^{-T} R_{12} R_{22}^{-1} k^T(x) \nabla\phi_2^T W_2 m_1^T W_1
\end{aligned}
$$

Define

$$
\begin{aligned}
c = &\frac{1}{4}\|W_1\|^2\|\nabla\phi_1 g(x)R_{11}^{-1}g^T(x)\ \nabla\phi_1^T\| \\
&+\frac{1}{4}\|W_2\|^2\|\nabla\phi_2 k(x)R_{22}^{-T}R_{12}R_{22}^{-1}k^T(x)\ \nabla\phi_2^T\| \\
&+\frac{1}{4}\|W_1\|^2\|\nabla\phi_1 g(x)R_{11}^{-T}R_{21}R_{11}^{-1}g^T(x)\ \nabla\phi_1^T\| \\
&+\frac{1}{4}\|W_2\|^2\|\nabla\phi_2 k(x)R_{22}^{-1}k^T(x)\ \nabla\phi_2^T\| \\
&+\frac{1}{2}b_{\varepsilon_{x1}}b_g^2 b_{\phi_{1x}}\sigma_{\min}(R_{11})\|W_1\| + \frac{1}{2}b_{\varepsilon_{x1}}b_k^2 b_{\phi_{2x}}\sigma_{\min}(R_{22})\|W_2\| \\
&+\frac{1}{2}b_{\varepsilon_{x2}}b_g^2 b_{\phi_{1x}}\sigma_{\min}(R_{11})\|W_1\| + \frac{1}{2}b_{\varepsilon_{x2}}b_k^2 b_{\phi_{2x}}\sigma_{\min}(R_{22})\|W_2\|
\end{aligned}
$$

According to Assumption 10.1, c is bounded by $c_{\max}$ and D is bounded by $D_{\max}$ that can be expressed in terms of the bounds given there.

Let the parameters be chosen such that $M > 0$. To justify this, the matrix M is written in compact form as

$$
M\begin{bmatrix} q_1+q_2 & 0 & 0 \\ 0 & I_2 & M_{23} \\ 0 & M_{32} & M_{33} \end{bmatrix} \tag{A.68}
$$

where in order to be positive definite the following properties (Lewis *et al.*, 2012) must hold

a. $q_1 + q_2 > 0$
b. $I_2 > 0$
c. Schur complement for I_2 is $D_{22} = I_2 - M_{23}M_{33}^{-1}M_{32} > 0$ that hold after proper selection of F_1, F_2, F_3 and F_4.
d. Schur complement for M_{33} is $D_{33} = M_{33} - M_{32}I_2^{-1}M_{23} > 0$ that hold after proper selection of F_1, F_2, F_3 and F_4.

Now (A.67) becomes

$$
\dot{L} < -\|\tilde{Z}\|^2\sigma_{\min}(M) + D_{\max}\|\tilde{Z}\| + c_{\max} + \bar{\varepsilon}_1 + \bar{\varepsilon}_2
$$

Completing the squares, the Lyapunov derivative is negative if

$$
\|\tilde{Z}\| > \frac{D_{\max}}{2\sigma_{\min}(M)} + \sqrt{\frac{D^2{}_{\max}}{4\sigma_{\min}^2(M)} + \frac{c_{\max} + \bar{\varepsilon}_1 + \bar{\varepsilon}_2}{\sigma_{\min}(M)}} \equiv B_Z \tag{A.69}
$$

It is now straightforward to demonstrate that if L exceeds a certain bound, then, $\dot{L}$ is negative. Therefore, according to the standard Lyapunov extension theorem the analysis above demonstrates that the state and the weights are UUB (Lewis *et al.*, 1995; Khalil, 1996).

Note that condition (A.69) holds if the norm of any component of $\tilde{Z}$ exceeds the bound, that is specifically $x > B_Z$ or $\overline{\sigma}_3^T \tilde{W}_1 > B_Z$ or $\overline{\sigma}_4^T \tilde{W}_2 > B_Z$ or $\tilde{W}_3 > B_Z$ or $\tilde{W}_4 > B_Z$.

Now consider the error dynamics and the output as in Chapter 7 and assume $\overline{\sigma}_3$ and $\overline{\sigma}_4$ are persistently exciting. Substituting (10.23), (10.24) in (10.31) and (10.32) respectively we obtain the error dynamics

$$\begin{aligned}\dot{\tilde{W}}_1 &= -a_1\bar{\sigma}_3\bar{\sigma}_3^T\tilde{W}_1 + a_1\bar{\sigma}_3\frac{\varepsilon_{HJ_1}}{m_{s_1}} + \frac{a_1}{4m_{s_1}^2}\tilde{W}_3^T\bar{D}(x)\tilde{W}_3 \\ &\quad + \frac{a_1}{4m_s^2}\tilde{W}_4^T\,\nabla\phi_2 R_{22}^{-T}R_{12}R_{22}^{-1}k^T(x)\,\nabla\phi_2^T\tilde{W}_4 \\ &\quad + \frac{a_1}{2m_s^2}(W_1^T\bar{E}_2(x)\tilde{W}_4 - W_2^T\,\nabla\phi_2 R_{22}^{-T}R_{12}R_{22}^{-1}k^T(x)\,\nabla\phi_2^T\tilde{W}_4) \\ y_1 &= \bar{\sigma}_3^T\tilde{W}_1\end{aligned}$$

and

$$\begin{aligned}\dot{\tilde{W}}_2 &= -a_2\bar{\sigma}_4\bar{\sigma}_4^T\tilde{W}_2 + a_2\bar{\sigma}_4\frac{\varepsilon_{HJ_2}}{m_{s_2}} + \frac{a_2}{4m_{s_1}^2}\tilde{W}_4^T\bar{D}_2(x)\tilde{W}_4 \\ &\quad + \frac{a_2}{4m_{s_2}^2}\tilde{W}_3^T\,\nabla\phi_1 g(x)R_{11}^{-T}R_{21}R_{11}^{-1}g^T(x)\,\nabla\phi_1^T\tilde{W}_3 \\ &\quad + \frac{a_2}{2m_s^2}(\tilde{W}_3^T\bar{E}_1 W_2 - \tilde{W}_3^T\,\nabla\phi_1\, g(x)R_{11}^{-T}R_{21}R_{11}^{-1}g^T(x)\,\nabla\phi_1^T\tilde{W}_1) \\ y_2 &= \bar{\sigma}_4^T\tilde{W}_2\end{aligned}$$

Then Theorem 10.1 is true with

$$\begin{aligned}&\left\|\frac{\sigma_3^T}{m_{s_1}}\tilde{W}_1\right\| > \varepsilon_{\max_1} > \frac{1}{4}||\tilde{W}_3||^2\left\|\frac{\overline{D}_1}{m_{s_1}}\right\| + \frac{1}{2}||W_1||\left\|\frac{\overline{E}_2(x)}{m_{s_1}}\right\|||\tilde{W}_4|| \\ &\quad + \frac{1}{4}||\tilde{W}_4||^2\left\|\frac{\nabla\phi_2 k(x)R_{22}^{-T}R_{12}R_{22}^{-1}k^T(x)\nabla\phi_2^T}{m_{s_1}}\right\| + \bar{\varepsilon}_1\left\|\frac{1}{m_{s_1}}\right\| \\ &\quad - \frac{1}{2}||\tilde{W}_4||\left\|\frac{\nabla\phi_2 k(x)R_{22}^{-T}R_{12}R_{22}^{-1}k^T(x)\nabla\phi_2^T}{m_{s_1}}\right\|||\tilde{W}_4||\end{aligned}$$

and

$$\left\| \frac{\sigma_4^T}{m_{s_2}} \tilde{W}_2 \right\| > \varepsilon_{\max_2} > \frac{1}{4} ||\tilde{W}_4||^2 \left\| \frac{\overline{D}_2}{m_{s_2}} \right\| + \frac{1}{2} ||W_2|| \left\| \frac{\overline{E}_1(x)}{m_{s_2}} \right\| ||\tilde{W}_3||$$

$$+ \frac{1}{4} ||\tilde{W}_3||^2 \left\| \frac{\nabla\phi_1 g(x) R_{11}^{-T} R_{21} R_{11}^{-1} g^T(x) \nabla\phi_1^T}{m_{s_2}} \right\| + \overline{\varepsilon}_2 \left\| \frac{1}{m_{s_2}} \right\|$$

$$- \frac{1}{2} ||W_2|| \left\| \frac{\nabla\phi_1 g(x) R_{11}^{-T} R_{21} R_{11}^{-1} g^T(x) \nabla\phi_1^T}{m_{s_2}} \right\| ||\tilde{W}_3||$$

This provides effective bounds for $\left\| \overline{\sigma}_3^T \tilde{W}_1 \right\|$ and $\left\| \overline{\sigma}_4^T \tilde{W}_2 \right\|$.

This completes the proof. ■

References

Abu-Khalaf M., Lewis F. L. 'Nearly optimal control laws for nonlinear systems with saturating actuators using a neural network HJB approach', *Automatica*. 2005;**41**(5):779–791.

Abu-Khalaf M., Lewis F. L. 'Neurodynamic programming and zero-sum games for constrained control systems', *IEEE Transactions on Neural Networks*. 2008;**19**(7):1243–1252.

Abu-Khalaf M., Lewis F. L., Huang J. 'Policy iterations and the Hamilton-Jacobi-Isaacs equation for H-infinity state feedback control with input saturation', *IEEE Transactions on Automatic Control*. 2006;**51**(12):1989–1995.

Adams R., Fournier J. *Sobolev Spaces*. New York: Academic Press; 2003.

Al-Tamimi A., Abu-Khalaf M., Lewis F. L. 'Model-free Q-learning designs for discrete-time zero-sum games with application to H-infinity control', *Automatica*. 2007a;**43**(3):473–482.

Al-Tamimi A., Abu-Khalaf M., Lewis F. L. 'Adaptive critic designs for discrete-time zero-sum games with application to H-infinity control', *IEEE Transactions on Systems, Man and Cybernetics–B*. 2007b;**37**(1).

Al-Tamimi A., Lewis F. L., Abu-Khalaf M. 'Discrete-time nonlinear HJB solution using approximate dynamic programming: convergence proof', *IEEE Transactions on Systems, Man, Cybernetics, Part B*. 2008;**38**(4):943–949, special issue on ADP/RL.

Astrom K. J., Wittenmark B. *Adaptive Control*. Addison-Wesley; 1995.

Baird L. C. III. 'Reinforcement learning in continuous time: Advantage updating', *Proceedings of the ICNN*. 1994, Orlando FL.

Balakrishnan S. N., Ding J., Lewis F. L. 'Issues on stability of ADP feedback controllers for dynamical systems', *IEEE Transactions on Systems, Man, Cybernetics, Part B*. 2008;**38**(4):913–917, special issue on ADP/RL, invited survey paper.

Ball J., Helton W. 'Viscosity solutions of Hamilton-Jacobi equations arising in nonlinear H_x-control', *Journal of Mathematical Systems, Estimation and Control*. 1996;**6**(1):1–22.

Balzer L. A. 'Accelerated convergence of the matrix sign function method of solving Lyapunov, Riccati and other equations', *International Journal of Control*. 1980;**32**(6):1076–1078.

Banks H. T., Ito K. 'A numerical algorithm for optimal feedback gains in high dimensional linear quadratic regulator problems', *SIAM Journal on Control Optimal*. 1991;**29**(3):499–515.

Bardi M., Capuzzo-Dolcetta I. *Optimal Control and Viscosity Solutions of Hamilton-Jacobi-Bellman Equations*. Boston, MA: Birkhâuser; 1997.

Barto A. G., Sutton R. S., Anderson C. 'Neuron-like adaptive elements that can solve difficult learning control problems', *IEEE Transactions on Systems, Man, and Cybernetics*. SMC-1983;**13**:834–846.

Başar T., Bernard P. *Optimal Control and Related Minimax Design Problems*. Boston, MA: Birkhâuser; 1995.

Başar T., Olsder G. J. *Dynamic Noncooperative Game Theory*, 2nd ed. (Classics in Applied Mathematics; 23), SIAM; 1999.

Beard R., Saridis G., Wen J. 'Galerkin approximations of the generalized Hamilton-Jacobi-Bellman equation', *Automatica*. 1997;**33**(12):2159–2177.

Bellman R. E. *Dynamic Programming*. Princeton, NJ: Princeton University Press; 1957.

Bertsekas D. P., Tsitsiklis J. N. *Neuro-Dynamic Programming. MA*: Athena Scientific; 1996.

Bhasin S., Johnson M., Dixon W. E. 'A model free robust policy iteration algorithm for optimal control of nonlinear systems', *Proceedings of the 49th IEEE Conference on Decision and Control*. 2010, 3060–3065.

Bradtke S. J., Ydestie B. E., Barto A. G. 'Adaptive linear quadratic control using policy iteration', *Proceedings of the ACC*. 1994, 3475–3476.

Brewer J. W. 'Kronecker products and matrix calculus in system theory', *IEEE Transactions on Circuit and System*. 1978;**25**(9):772–781.

Busoniu L., Babuska R., De Schutter B., Ernst D. *Reinforcement Learning and Dynamic Programming Using Function Approximators*. Boca Raton, FL: CRC Press; 2009.

Byers R. 'Solving the algebraic Riccati equation with the matrix sign', *Linear Algebra and its Applications*. 1987;**85**:267–279.

Callier F. M., Desoer C. A. *Linear Systems Theory*. New York: Springer-Verlag; 1991.

Cao X. *Stochastic Learning and Optimization*. Berlin: Springer-Verlag; 2007.

Chen B. M., Lin Z., Shamash Y. *Linear Systems Theory: A Structural Decomposition Approach*. Boston, MA: Birkhauser; 2004.

Cherfi L., Abou-Kandil H., Bourles H. 'Iterative method for general algebraic Riccati equation', *Proceedings of the ACSE*'05. 2005.

Damm T. *Rational Matrix Equations in Stochastic Control*. Berlin, Germany: Springer-Verlag; 2004.

Darwin C. *On the Origin of Species by Means of Natural Selection*. London: John Murray; 1859.

Dierks T., Jagannathan S. 'Optimal control of affine nonlinear continuous-time systems using an online Hamilton-Jacobi-Isaacs formulation', *Proceedings of the IEEE Conference on Decision and Control*. 2010, 3048–3053.

Doya K. 'Reinforcement learning in continuous time and space', *Neural Computation*. 2000;**12**(1):219–245.

Doya K., Kimura H., Kawato M. 'Neural mechanisms of learning and control', *IEEE Control Systems Magazine*. 2001;**21**(4):42–54.

Doyle J., Glover K., Khargonekar P., Francis B. 'State-space solutions to standard H2 and H-infinity control problems', *IEEE Transactions on Automatic Control*. 1989;**34**:831–847.

Feng Y., Anderson B. D., Rotkowitz M. 'A game theoretic algorithm to compute local stabilizing solutions to HJBI equations in nonlinear Hot control', *Automatica*. 2009;**45**(4):881–888.

Ferrari S., Stengel R. 'An adaptive critic global controller', *Proceedings of the American Control Conference*. 2002, 2665–2670.

Finlayson B. A. *The Method of Weighted Residuals and Variational Principles*. New York: Academic Press; 1990.

Freeman R. A., Kokotovic P. *Robust Nonlinear Control Design: State-Space and Lyapunov Techniques*. Boston, MA: Birkhauser; 1996.

Gajic Z., Li T. Y. 'Simulation results for two new algorithms for solving coupled algebraic Riccati equations', *Third International Symposium on Differential Games*. 1988., Sophia, Antipolis, France.

Ge S. S., Wang C. 'Adaptive neural control of uncertain MIMO nonlinear systems', *IEEE Transactions Neural Networks*. 2004;**15**(3):674–692.

Guo C. H., Lancaster P. 'Analysis and Modification of Newton's Method for Algebraic Riccati Equations', *Mathematics of Computation*. 1998;**67**(223): 1089–1105.

Hanselmann T., Noakes L., Zaknich A. 'Continuous-time adaptive critics', *IEEE Transactions on Neural Networks*. 2007;**18**(3):631–647.

Hasan M. A., Yang J. S., Hasan A. A. 'A method for solving the algebraic Riccati and Lyapunov equations using higher order matrix sign function algorithms', *Proceedings of the ACC*. 1999, 2345–2349.

Hewer G. 'An iterative technique for the computation of the steady state gains for the discrete optimal regulator', *IEEE Transactions on Automatic Control*. 1971;**16**(4):382–384.

Hornik K., Stinchcombe M., White H. 'Universal approximation of an unknown mapping and its derivatives using multilayer feedforward networks', *Neural Networks*. 1990;**3**:551–560.

Howard R. A. *Dynamic Programming and Markov Processes*. Cambridge, MA: MIT Press; 1960.

Huang J., Lin C. F. 'Numerical approach to computing nonlinear H_∞ control laws', *Journal of Guidance, Control and Dynamics*. 1995;**18**(5):989–994.

Jungers M., De Pieri E., Abou-Kandil H. 'Solving coupled algebraic Riccati equations from closed-loop Nash strategy, by lack of trust approach', *International Journal of Tomography and Statistics*. 2007;**7**(F07):49–54.

Ioannou P., Fidan B. *Adaptive Control Tutorial*, Philadelphia, PA: SIAM, Advances in Design and Control; 2006.

Kailath T. 'Some new algorithms for recursive estimation in constant linear systems', *IEEE Transactions on Information Theory*. 1973;**19**(6):750–760.

Khalil H. K. *Nonlinear Systems*. New Jersey, Prentice-Hall; 1996.

Kirk D. E. *Optimal Control Theory–an introduction*. Mineola, NY: Dover Pub. Inc.; 2004.

Kleinman D. 'On an iterative technique for Riccati equation computations', *IEEE Transactions on Automatic Control*. 1968;**13**(1):114–115.

Kolmogorov A. N., Fomin S. V. *Elements of the Theory of Functions and Functional Analysis*. Mineola, NY: Dover Pub. Inc.; 1999.

Krstic M., Deng H. *Stabilization of Nonlinear Uncertain Systems*. Springer; 1998.

Lancaster P., Rodman L. *Algebraic Riccati Equations*., UK: Oxford University Press; 1995.

Landelius T. *Reinforcement Learning and Distributed Local Model Synthesis, PhD Dissertation*, Linkoping University, Sweden; 1997.

Lanzon A., Feng Y., Anderson B. D. O., Rotkowitz M. 'Computing the positive stabilizing solution to algebraic Riccati equations with an indefinite quadratic term via a recursive method', *IEEE Transactions on Automatic Control*. 2008;**53**(10):2280–2291.

Laub A. J. 'A schur method for solving algebraic Riccati equations', *IEEE Transactions on Automatic Control*. 1979;**24**(6):913–921.

Leake R. J., Liu Ruey-Wen. 'Construction of suboptimal control sequences', *Journal on SIAM Control*. 1967;**5**(1):54–63.

Levine D. S., Brown V. R., Shirey V. T. *Oscillations in Neural Systems*. Mahwah, NJ: Lawrence Erlbaum Associates Publ; 2000.

Lewis F. L., Vrabie D., Syrmos V. L. *Optimal Control*, 3rd ed. New Jersey: John Wiley; 2012.

Lewis F. L., Vrabie D. 'Reinforcement learning and adaptive dynamic programming for feedback control', *IEEE Circuits & Systems Magazine, Third Quarter*. 2009, 32–48.

Lewis F. L., Lendaris G., Liu Derong. 'Special issue on approximate dynamic programming and reinforcement learning for feedback control', *IEEE Transactions on Systems, Man, and Cybernetics, Part B*. 2008;**38**(4).

Lewis F. L., Jagannathan S., Yesildirek A. *Neural Network Control of Robot Manipulators and Nonlinear Systems*. London: Taylor & Francis; 1999.

Lewis F. L., Liu. K., Yesildirek A. 'Neural net controller with guaranteed tracking performance', *IEEE Transactions on Neural Networks*. 1995;**6**(3): 703–715.

Li Z. H., Krstic M. 'Optimal design of adaptive tracking controllers for nonlinear systems', *Proceedings of the ACC*, 1997, 1191–1197.

Limebeer D. J. N., Anderson B. D. O., Hendel H. 'A Nash game approach to mixed control', *IEEE Transactions on Automatic Control*. 1994;**39**(1):69–82.

Liu X., Balakrishnan S. N. 'Convergence analysis of adaptive critic based optimal control', *Proceedings of the American Control Conference*. 2000, 1929–1933.

Ljung L. *System Identification*. New Jersey: Prentice-Hall; 1999.

Luenberger D. G. *Introduction to Dynamic Systems*. New York: John Wiley; 1979.

MacFarlane A. G. J. 'An eigenvector solution of the optimal linear regulator problem', *Journal of Electronics and Control*. 1963;**14**:643–654.

Mehta P., Meyn S. 'Q-learning and Pontryagin's minimum principle', *Proceedings of the IEEE Conference on Decision and Control*. 2009, 3598–3605.

Mendel J. M., MacLaren R.W. 'Reinforcement learning control and pattern recognition systems', in Mendel J. M., Fu K. S. (eds). *Adaptive, Learning, and Pattern Recognition Systems: Theory and Applications*. New York: Academic Press; 1970. pp. 287–318.

Moris K., Navasca C. 'Iterative solution of algebraic Riccati equations for damped systems', *Proceedings of the IEEE Conference on Decision and Control*. 2006, 2436–2440.

Moore K. L. *Iterative Learning Control for Deterministic Systems*. London: Springer-Verlag; 1993.

Murray J. J., Cox C. J., Lendaris G. G., Saeks R. 'Adaptive dynamic programming', *IEEE Transactions on Systems, Man and Cybernetics*. 2002;**32**(2):140–153.

Nevistic V., Primbs J. 'Constrained nonlinear optimal control: A converse HJB approach'. *Technical Report 96-021, California Institute of Technology*. 1996.

Potter J. E. 'Matrix quadratic solutions', *SIAM Journal on Applied Mathematics*. 1966;**14**:496–501.

Powell W. B. *Approximate Dynamic Programming*. New Jersey: John Wiley; 2007.

Prokhorov D., Wunsch D. 'Adaptive critic designs', *IEEE Transactions on Neural Networks*. 1997;**8**(5):997–1007.

Sandberg I. W. 'Notes on uniform approximation of time-varying systems on finite time intervals', *IEEE Transactions on Circuits and Systems—I: Fundamental Theory and Applications*. 1998;**45**(8):863–865.

Schultz W. 'Neural coding of basic reward terms of animal learning theory, game theory, microeconomics and behavioral ecology', *Current Opinion in Neurobiology*. 2004;**14**:139–147.

Schultz W., Dayan P., Read Montague P. 'A neural substrate of prediction and reward', *Science*. 1997;**275**:1593–1599.

Schultz W., Tremblay L., Hollerman J. R. 'Reward processing in primate orbitofrontal cortex and basal ganglia', *Cerebral Cortex*. 2000;**10**:272–283.

Si J., Barto A., Powell W., Wunsch D. *Handbook of Learning and Approximate Dynamic Programming*. New Jersey: John Wiley; 2004.

Sontag E. D., Sussmann H. J. 'Nonsmooth control-Lyapunov functions', *Proceedings of the IEEE Conference on Decision and Control*. 1995, 2799–2805.

Stevens B. L., Lewis F. L. *Aircraft Control and Simulation*, 2nd ed. New Jersey: John Wiley; 2003.

Stoorvogel A. A. *The H-infinity Control Problem: A State Space Approach*. New York: Prentice-Hall; 1992.

Sutton R. 'Learning to predict by the method of temporal differences', *Machine Learning*. 1988;**3**:9–44.

Sutton R. S., Barto A. G. *Reinforcement Learning–An Introduction*. Cambridge, MT: MIT Press; 1998.

Sutton R. S., Barto A. G., Williams R. J. 'Reinforcement learning is direct adaptive optimal control', *IEEE Control Systems Magazine*. 1992, 19–22.

Tao G. *Adaptive Control Design and Analysis. Adaptive and Learning Systems for Signal Processing, Communications and Control Series*. Hoboken, NJ: Wiley-Interscience; 2003.

Tsitsiklis J. N. 'On the convergence of optimistic policy iteration', *Journal of Machine Learning Research*. 2002;**3**:59–72.

Vamvoudakis K. G., Lewis F. L. 'Online solution of nonlinear two-player zero-sum games using synchronous policy iteration', *Proceedings of the IEEE Conference on Decision and Control*. Dec. 2010b, 3040–3047, Atlanta.

Vamvoudakis K. G., Lewis F. L. 'Multi-player non-zero sum games: Online adaptive learning solution of coupled Hamilton-Jacobi Equations', *Automatica*. 2011;**47**:1556–1569.

Vamvoudakis K. G. *Online Learning Algorithms for Differential Dynamic Games and Optimal Control*, Ph.D. Thesis, School of Electrical Engineering, The University of Texas at Arlington, May 2011.

Vamvoudakis K. G., Lewis F. L. 'Online actor-critic algorithm to solve the continuous-time infinite horizon optimal control problem', *Automatica*. 2010a;**46**(5):878–888.

Van Der Schaft A. J. 'L2-gain analysis of nonlinear systems and nonlinear state feedback Hot control', *IEEE Transactions on Automatic Control*. 1992;**37**(6): 770–784.

Vrabie D., Lewis F. L. 'Neural network approach to continuous-time direct adaptive optimal control for partially-unknown nonlinear systems', *Neural Networks*. 2009;**22**(3):237–246.

Vrabie D., Lewis F. L. 'Adaptive dynamic programming algorithm for finding online the equilibrium solution of the two-player zero-sum differential game', *Proceedings of the International Joint Conference on Neural Networks*. July 2010, 1–8, Barcelona.

Vrabie D., Lewis F. L., Abu-Khalaf M. 'Biologically inspired scheme for continuous-time approximate dynamic programming', *Transactions of the Institute of Measurement and Control*. 2008;**30**:207–223.

Vrabie D., Pastravanu O., Abu-Khalaf M., Lewis F. L. 'Adaptive optimal control for continuous-time linear systems based on policy iteration', *Automatica*. 2009;**45**:477–484.

Vrabie D. *Online Adaptive Optimal Control for Continuous-Time Systems*, Ph.D. Thesis, School of Electrical Engineering, The University of Texas at Arlington, Dec. 2009.

Wang F. Y., Zhang H., Liu D. 'Adaptive dynamic programming: An introduction', *IEEE Computational Intelligence Magazine*. 2009, 39–47.

Wang Y., Zhou R., Wen C. 'Robust load-frequency controller design for power systems', *IEE Proceedings C*. 1993;**140**(1):11–16.

Watkins C. J. C. H. (1989). *Learning from Delayed Rewards*, PhD. Thesis, University of Cambridge, England.

Watkins C. J. C. H., Dayan P. 'Q-learning', *Machine Learning*. 1992;**8**:279–292.

Werbos P. J. *Beyond Regression: New Tools for Prediction and Analysis in the Behavior Sciences*, Ph.D. Thesis, 1974.

Werbos P. 'Neural networks for control and system identification', *Proceedings of the IEEE Conference on Decision and Control*. 1989, 260–265.

Werbos P. J. 'A menu of designs for reinforcement learning over time', in Miller W. T., Sutton R. S., Werbos P. J. (eds). *Neural Networks for Control*, Cambridge: MIT Press; 1991. pp. 67–95.

Werbos P. J. 'Approximate dynamic programming for real-time control and neural modeling', in White D. A., Sofge D. A. (eds). *Handbook of Intelligent Control*, New York: Van Nostrand Reinhold; 1992. pp. 493–525.

Werbos P. 'Intelligence in the brain: A theory of how it works and how to build it', *Neural Networks-Special Issue: Goal-Directed Neural Systems*. 2009;**22**(3): 200–212.

Wei Q., Zhang H. 'A new approach to solve a class of continuous-time onlinear quadratic zero-sum game using ADP', *Proceedings of the IEEE International Conference on Networking, Sensing and Control (ICNSC'08)*. 2008;**6**(8): 507–512.

Wheeler R. M., Narendra K. S. 'Decentralized learning in finite Markov Chains', *IEEE Transactions on Automatic Control*. 1986;**31**(6).

White D. A., Sofge D. A. *Handbook of Intelligent Control*. New York: Van Nostrand Reinhold, 1992.

Zames G. 'Feedback and optimal sensitivity: Model reference transformations, multiplicative seminorms and approximate inverses', *IEEE Transactions on Automatic Control*. 1981;**26**:301–320.

Zhang H., Huang J., Lewis F. L. 'Algorithm and stability of ATC receding horizon control', *Proceedings of the IEEE Symposium on ADPRL*. 2009, 28–35, Nashville.

Zhou K., Khargonekar P. P. 'An algebraic Riccati equation approach to H-infinity optimization', *Systems and Control Letters*. 1988;**11**:85–91.

Zhou K., Doyle J. C., Glover K. *Robust and Optimal Control*. Upper Saddle River, NJ: Prentice Hall; 1996.

Index

www.ingramcontent.com/pod-product-compliance
Lightning Source LLC
LaVergne TN
LVHW011917200926
848484LV00002B/646

* 9 7 8 1 8 4 9 1 9 4 8 9 1 *